Innovations in Durability of Sustainable Concrete Materials

Innovations in Durability of Sustainable Concrete Materials

Editors

Woubishet Zewdu Taffese
Sandra Barbosa Nunes

MDPI

Basel • Beijing • Wuhan • Barcelona • Belgrade • Novi Sad • Cluj • Manchester

Editors
Woubishet Zewdu Taffese
Center for Intelligent Infrastructure
Missouri University of Science and Technology
Rolla
United States

Sandra Barbosa Nunes
Department of Engineering Structures
Delft University of Technology
Delft
Netherlands

Editorial Office
MDPI
St. Alban-Anlage 66
4052 Basel, Switzerland

This is a reprint of articles from the Special Issue published online in the open access journal *Sustainability* (ISSN 2071-1050) (available at: www.mdpi.com/journal/sustainability/special_issues/X1W965952N).

For citation purposes, cite each article independently as indicated on the article page online and as indicated below:

Lastname, A.A.; Lastname, B.B. Article Title. *Journal Name* **Year**, *Volume Number*, Page Range.

ISBN 978-3-0365-9105-6 (Hbk)
ISBN 978-3-0365-9104-9 (PDF)
doi.org/10.3390/books978-3-0365-9104-9

Contents

About the Editors

Woubishet Zewdu Taffese

Woubishet Zewdu Taffese holds the position of Postdoctoral Fellow at the Center for Intelligent Infrastructure, situated at the Missouri University of Science and Technology in Missouri, USA. Additionally, he holds the position of Affiliated Researcher at the Arcada University of Applied Sciences in Helsinki, Finland. His research interests are centered at the intersection of applied data science and civil engineering. Before his current position, Dr. Taffese extensively researched concrete durability through the application of data-driven methodologies. At present, his research focuses on analyzing the structural performance of concrete elements through data-driven approaches and leveraging digital twinning to enhance sustainability and resilience within the domain of civil infrastructure.

Sandra Barbosa Nunes

Sandra Conceição Barbosa Nunes is an Assistant Professor in the field of Concrete Structures at the Faculty of Civil Engineering and Geosciences, Delft University of Technology. Her past research focused mainly on concrete technology, following the major advances of the last decades, namely, self-compacting concrete, eco-efficient concretes, and ultra-high performance fibre-reinforced cementitious composites. Currently, Sandra is working on a new research path on data-driven concrete science to extract knowledge and gain new insights into the behaviour of cement-based materials. She is the chair of the RILEM Technical Committee DCS: Data-driven concrete science.

Preface

In the realm of modern construction, the integration of sustainability with durability stands as a pivotal goal. Concrete, a cornerstone of infrastructure, is evolving to meet this imperative. Welcome to the Reprint of "Innovations in Durability of Sustainable Concrete Materials", a collection that embodies the spirit of this transformation. Within these pages lie the results of thorough research and innovation through ten manuscripts that examine various facets of sustainable concrete technology. These explorations range from novel cementitious materials to state-of-the-art sustainable reinforcement techniques, from innovative approaches for durability assessment to encompassing life-cycle analyses and advanced modeling methods for property characterization. This compilation represents a rich repository of knowledge, encapsulating a broad expanse of advancements. We stand at the brink of a paradigm shift, and these manuscripts illuminate the path, showcasing the cutting-edge research propelling the concrete industry toward a sustainable and enduring future.

This compilation, the result of diligent collaboration and research, seeks to provide a comprehensive understanding of the advancements that lie at the intersection of durability and sustainability. It is our hope that this curated collection not only informs but also inspires further exploration and sparks the innovative spirit within all those dedicated to building a sustainable tomorrow. As the Editors of this collection, we extend our gratitude to the authors for their invaluable contributions and commitment to advancing the knowledge in this field. We also extend our appreciation to the readers, researchers, and practitioners whose relentless pursuit of sustainable solutions is shaping the future of construction. May this compilation inspire and foster a deeper understanding of the innovations that lie at the nexus of durability and sustainability in concrete materials.

Woubishet Zewdu Taffese and Sandra Barbosa Nunes
Editors

sustainability

MDPI

Article

Durability of Engineered Cementitious Composites Incorporating High-Volume Fly Ash and Limestone Powder

Kazim Turk [1], Ceren Kina [2] and Moncef L. Nehdi [3,*]

1 Department of Civil Engineering, Engineering Faculty, Inonu University, 44280 Malatya, Türkiye
2 Department of Civil Engineering, Faculty of Engineering and Natural Sciences, Malatya Turgut Ozal University, 44210 Malatya, Türkiye
3 Department of Civil Engineering, McMaster University, Hamilton, ON L8S 4L8, Canada
* Correspondence: nehdim@mcmaster.ca

Abstract: This study investigates the effects of using limestone powder (LSP) and high-volume fly ash (FA) as partial replacement for silica sand (SS) and portland cement (PC), respectively, on the durability properties of sustainable engineered cementitious composites (ECC). The mixture design of ECC included FA/PC ratio of 1.2, 2.2 and 3.2, while LSP was used at 0%, 50% and 100% of SS by mass for each FA/PC ratio. Freeze-thaw and rapid chloride ions penetrability (RCPT) tests were performed to assess the durability properties of ECC, while the compressive and flexural strength tests were carried out to appraise the mechanical properties. Moreover, mercury intrusion porosimetry (MIP) tests were performed to characterize the pore structure of ECC and to associate porosity with the relative dynamic modulus of elasticity, RCPT and mechanical strengths. It was found that using FA/PC ratio of more than 1.2 worsened both the mechanical and durability properties of ECC. Replacement of LSP for SS enhanced both mechanical strengths and durability characteristics of ECC, owing to refined pore size distribution caused by the microfiller effect. It can be further inferred from MIP test results that the total porosity had a vital effect on the resistance to freezing–thawing cycles and chloride ions penetration in sustainable ECC.

Keywords: engineered cementitious composites; limestone powder; fly ash; mercury intrusion porosimetry; chloride ions; freeze–thaw; dynamic modulus of elasticity

check for updates

Citation: Turk, K.; Kina, C.; Nehdi, M.L. Durability of Engineered Cementitious Composites Incorporating High-Volume Fly Ash and Limestone Powder. *Sustainability* **2022**, *14*, 10388. https://doi.org/10.3390/su141610388

Academic Editors: Woubishet Zewdu Taffese and Sandra Barbosa Nunes

Received: 17 July 2022
Accepted: 17 August 2022
Published: 20 August 2022

1. Introduction

Engineered cementitious composites (ECC) are a special class of fiber-reinforced high-performance cementitious composite with properties tailored as per fundamental principles of micromechanics and fracture mechanics [1]. ECCs exhibit excellent strain hardening behavior and ultimate tensile strain capacity that can attain several hundred times that of traditional concrete [2]. Most ECC mixtures reported in the open literature incorporate short randomly distributed polyvinyl alcohol (PVA) fibers, which ensure superior tensile strain hardening behavior under increasing tensile loading through formation of multiple cracks. The interfacial interactions between the cementitious matrix and PVA fibers have an important role in this tensile strain hardening behavior [3]. Generally, the crack width is controlled below 60 μm and the strain capacity can reach up to 8% at the peak strength [4,5]. Considering such increased ductility, narrow crack widths, as well as excellent tensile properties, ECC can overcome many durability problems in structures exposed to harsh environments [6–8].

Durability is a crucial parameter, which affects the service life performance of concrete structures. It is directly linked to the crack width that governs the permeability of concrete. To enhance the water-tightness and durability, the cementitious matrix needs to be densified and the crack width needs to be minimized. ECC is further endowed with self-healing ability of its micro-cracks, thus further mitigating transport mechanisms.

Moreover, the crack bridging benefit of fibers in ECC [9] provides enhanced durability performance, including reduced permeability, and better resistance to freezing–thawing cycles, carbonization, chloride ions penetrability, sulfate attack and other degradation mechanisms [10–12] compared to that of ordinary concrete.

The production of ECC mixtures is facilitated via the incorporation of supplementary cementitious materials, such as fly ash (FA), silica fume and metakaolin as partial replacements for cement. Usually, the use of these mineral admixtures in concrete decreases and refines the permeable voids through the microfiller effect, pozzolanic reaction and densification of the cement paste-aggregate interfacial zone. This results in enhanced resistance to damage mechanisms such as chloride-ions or carbonation induced corrosion, damage by freezing–thawing cycles, sulfate attack, alkali-aggregate reactions, leaching etc., thus yielding overall superior durability properties of concrete [13–16]. FA has been the most widely used mineral admixture in typical ECC mixtures owing to its favorable properties including improved workability, mechanical strength and durability [17–21]. It was reported in the study of Yang et al. [22] that the reduction in the crack width of ECC incorporating FA was due to restricting the slippage of fibers through the effect of high interfacial frictional bond.

The inclusion of limestone powder (LSP) in cementitious systems is desirable considering its relatively low cost and reduction of CO^2 emissions from cement production while maintaining adequate mechanical and durability properties. Turk and Demirhan [23] investigated the effect of using LSP instead of quartz sand on the permeability of ECC mixtures. They found that the carbonation resistance was improved via LSP addition. Likewise, Lecomte et al. [24] concluded that LSP fillers increased the sorptivity while porosity characteristics changed according to the grinding quality of LSP. Some researchers [25–27] found that an increase in the replacement rate of LSP for cement increased the carbonation depth in concrete.

In the production of ECC, special gradation silica sand (SS), which is relatively costly and difficult to source compared to other conventional sands, is used as aggregate in ECC in general. A novel aspect in this study is the use of LSP as aggregate owing to its low cost and advantageous properties. While there is significant work on the use of LSP in conventional concrete, there is currently a dearth of studies in the available literature on the effects of LSP on the mechanical and durability properties of ECC. The limited existing experimental works explored the inclusion of mineral admixtures, including FA, SF or slag in ECC. Within this scope, the goal of this study is to use LSP as aggregate with FA as binder for the partial replacement of SS and PC, respectively, to develop more sustainable ECC, which can help widespread usage of this promising material using local materials. Hence, in the present study, dedicated experimental research was designed to investigate the effects of using both LSP and high-volume of FA on the durability properties of ECC, as can be observed in Figure 1. LSP was used as aggregate at 0%, 50% and 100% mass replacement for silica sand to study the effectiveness of LSP as replacement for SS, while FA was used as a binder partial replacement for cement at FA/PC ratio of 1.2, 2.2 and 3.2. Additionally, in this study, to reveal the effects of using high-volume FA content, FA/PC ratios of 2.2 and 3.2 were selected in the design of ECC mixtures, while the FA/PC ratio used is 1.2 for standard ECC. To assess the durability properties of the sustainable ECC mixtures thus developed, mercury intrusion porosimetry (MIP), resistance to freezing–thawing (F-T) cycles, and rapid chloride ions penetrability (RCPT) tests were performed. Moreover, compressive and flexural strength tests were carried out to assess mechanical properties. The consistency between the mechanical and durability test results of ECC specimens was discussed. Given the importance of the pore size distribution and porosity for durability and mechanical properties, the correlation between MIP and mechanical/durability properties was examined.

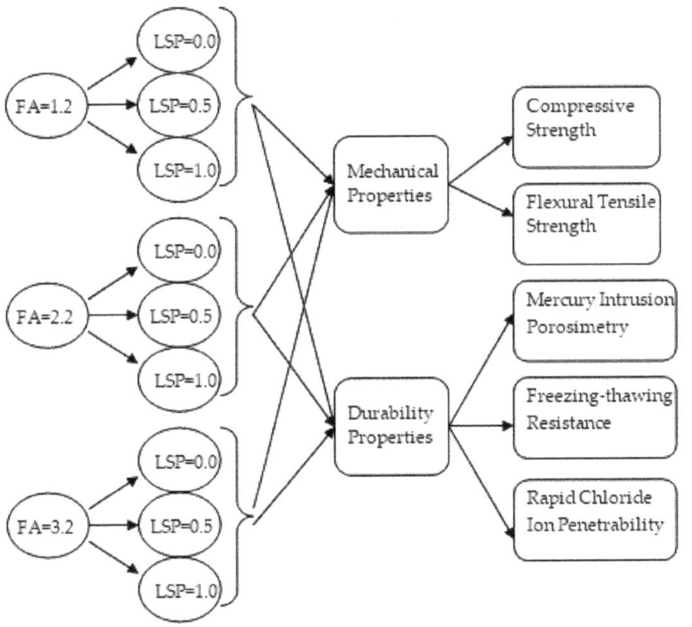

Figure 1. Schematic flow diagram of this study.

2. Experimental Program

2.1. Materials

The binders used in the manufacturing of ECC mixtures included Type I ordinary portland cement (PC) and Class-C fly ash (FA) conforming to guidelines of ASTM C150 [28] and ASTM C618 [29], respectively. Table 1 shows the physical and chemical properties of the binders. Silica sand (SS) with a maximum particle size of 1.18 mm and limestone powder (LSP) with an average particle size of 425 μm were used as aggregate. The specific gravity of LSP and SS were 2.65 and 2.70, respectively, and their particle size distribution is depicted in Figure 2. To meet the strain-hardening performance requirements for ECC, PVA fibers with the properties presented in Table 2 were used. In order to improve the interfacial bond between the cementitious matrix and PVA fibers, a proprietary hydrophobic oiling matter was used to coat the surface of PVA fibers during the manufacturing process [30]. A high-range-water-reducing admixture (HRWR) with a specific gravity of 1.06 was added into ECC mixtures to enhance the fresh properties.

Table 1. Chemical characteristics and physical properties of PC and FA.

	Chemical Characteristics (Oxides and Phase)										
	CaO	SiO_2	Al_2O_3	Fe_2O_3	MgO	SO_3	$K_2O + 0.66\ Na_2O$	C_3S	C_2S	C_3A	C_4AF
PC (%)	61.5	19.6	4.8	3.3	3	3.5	0.7	55	15	7	10
FA (%)	14.92	41.76	22.91	9.23	2.95	1.6	2.05	-	-	-	-

	Physical Properties						
	Loss of Ignition	Insoluble Residue	$SiO_2 + Al_2O_3 + Fe_2O_3$	Autoclave Expansion, %	Specific gravity	Surface area (m^2/kg)	Amount retained on 45 micron, %
PC (%)	1.9	0.44	27.7	0.09	3.15	371	3
FA (%)	0.8	-	73.9	0	2.43	115	19

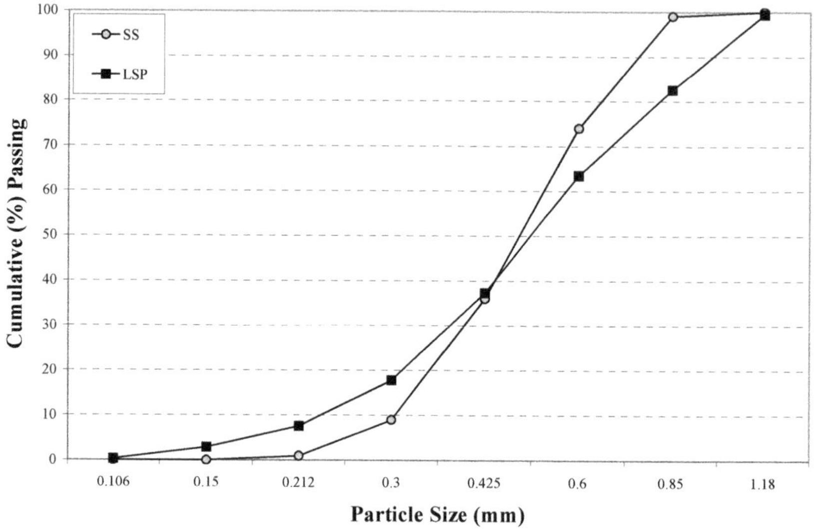

Figure 2. Particle size distribution of LSP and SS.

Table 2. Properties of PVA fiber.

Fiber	Length (mm)	Diameter (μm)	Tensile Strength (MPa)	Modulus of Elasticity (GPa)	Density (kg/m^3)	Max. Elongation (%)	Melting Temperature (°C)
PVA	8	36	1600	40	1300	6.5	225

2.2. Mixture Design and Sample Preparation

In this study, three reference ECC mixtures incorporating SS and with a FA/PC ratio of 1.2, 2.2 and 3.2 were made as given in Table 3. In total, nine ECC mixtures were designed considering 0%, 50% and 100% LSP mass replacement for SS, whilst the FA/PC ratio was varied at 1.2, 2.2 and 3.2. All the mixtures had a water-to-cementitious materials ratio of 0.26, identical total mass of cementitious materials and identical PVA fiber dosage (2% by volume). To mitigate workability as a source of variability, the slump flow of sustainable fresh ECC mixtures was adjusted within the close range of 27–28 cm using a high-range water-reducing admixture (HRWR). All ingredients were kept constant, except the FA/PC ratio (1.2, 2.2, 3.2) and the content of SS and LSP.

Table 3. Mixture proportions of ECC and mortar.

Mixture ID	Unit Weight, (kg/m^3)							Slump-Flow (cm)
	PC	FA	Water	Sand		HRWR	PVA	
				SS	LSP			
FA1.2_LSP0.0	570	684	332	432	0	9.0	26	27.1
FA1.2_LSP0.5	570	684	332	218	218	10.1	26	27.6
FA1.2_LSP1.0	570	684	332	0	440	10.3	26	27.5
FA2.2_LSP0.0	392	862	331	402	0	6.5	26	27.0
FA2.2_LSP0.5	392	862	331	202	202	7.2	26	27.2
FA2.2_LSP1.0	392	862	331	0	408	7.3	26	27.1
FA3.2_LSP0.0	299	955	329	386	0	6.5	26	27.2
FA3.2_LSP0.5	299	955	329	194	194	6.6	26	27.9
FA3.2_LSP1.0	299	955	329	0	392	6.7	26	27.7

A high-shear mixer with a twenty-liter capacity was used in the preparation of the ECC mixtures. The cementitious materials and sand were first dry mixed at 100 rpm for a minute. Then, water and HRWR were added, and mixed for additional 3 min at 300 rpm. Finally, PVA fibers were added, and mixing resumed for 3 min at 150 rpm. The specimens were cast via pouring the fresh mixtures into molds. Subsequently, the specimens were compacted using a vibrating table for 2 min. At 24 h after casting, all specimens were demolded and cured for 7 days in sealed plastic bags at 25 ± 2 °C to prevent moisture loss. Then, the specimens were stored in laboratory conditions at temperature of 25 ± 2 °C and relative humidity of 50 ± 5% until the testing age of 28 and 90 days. Similar curing process for ECC mixtures was adopted in previous studies [31–33]. Three replicate specimens were prepared for each designed mixtures and testing age.

2.3. *Test Methods*

2.3.1. Mechanical Properties

The mechanical properties of ECC mixtures were investigated in terms of compressive and flexural tensile strengths. For each test and curing age (28 and 90 days), three replicate specimens were tested for all designed mixtures and the average values were recorded as the experimental results. The compressive strength test was carried out using 50 × 50 × 50 mm^3 cube specimens according to ASTM C39 [34]. Four-point bending test was performed to evaluate the flexural strength of ECC specimens with a cross section of 100 × 100 mm^2 and a length of 380 mm as per the provisions of ASTM C78 [35]. The loading rate of the MTS testing machine with tri-axial cell testing capability was 0.003 mm/s. The mid-span deflection of ECC prisms was recorded via an LVDT placed at the middle of the span and the applied force was measured using a computerized data recording system.

2.3.2. Durability Properties

Mercury intrusion porosimetry (MIP) test was performed to characterize the pore size distribution of the sustainable ECC specimens. A Micrometrics AutoPore IV 9500 Series porosimeter capable of producing pressures up to 414 MPa was used for pore size distribution analysis based on the mercury intrusion method. The contact angle and assumed surface tension of mercury were 130° and 0.484 N/m (ASTM D4404 [36]), respectively. Cubic fragments with the size of 10 × 10 × 10 mm^3 from all ECC specimens were first dried to constant weight at 50 °C and stored in a vacuum chamber prior to testing. The average of test results of three specimens was recorded as the MIP test result for each mixture.

The freezing and thawing (F-T) test was carried out and the pulse velocity test method was used to monitor the deterioration of specimens during F-T cycling; thus, the dynamic elastic modulus of ECC specimens was calculated at each interval of nominally 30 cycles of freezing and thawing. The resistance to F-T cycles of ECC specimens was assessed in accordance with ASTM C666 "Standard Test Method for Resistance of Concrete to Rapid Freezing and Thawing procedure (b)-rapid freezing in air and thawing in water" [37] using a freeze–thaw cabinet. The test was continued until 330 cycles. After each 30 cycles, the relative dynamic modulus of elasticity of ECC specimens was estimated as follows:

$$P_i = \frac{f_i^2}{f_0^2} \times 100 \tag{1}$$

where f_i and f_0 are the ultrasonic pulse velocity at ith cycle and before the start of freeze–thaw cycling, respectively.

The electrical conductivity of concrete was measured through the RCPT test, which is an indirect measure of chloride ions penetrability [38]. The resistance of concrete to chloride ions penetration is paramount for durability in chloride-laden environments and greatly depends on the pore structure characteristics. Cylindrical specimens with a diameter of 100 mm and thickness of 50 mm were tested as per the guidelines of ASTM C1202 [39] at the age of 28 curing days. The test specimens were put between parallel stainless steel plates, which provide electrical connection between the concrete and steel plates by conductive saturated sponges. The resistance was evaluated via the application of a known alternating current of about 250 μA across the specimen and acquiring the resulting voltage. The measured resistance was divided by the depth of the specimen and multiplied by its cross-sectional area to obtain the resistivity. The average values of three identical specimens were reported as test results for each mixture.

3. Results and Discussions

3.1. Compressive Strength and Flexural Tensile Strength

The compressive and flexural strengths of the sustainable ECC specimens incorporating FA as partial replacement for PC at ratios of 1.2, 2.2 and 3.2, and 0%, 50% and 100% LSP as aggregate instead of SS are shown in Figure 3. ECC specimens with FA/PC = 1.2 and 0%, 50% and 100% mass replacement of SS with LSP attained higher compressive and flexural strengths, while strength values were lowest when the FA/PC ratio was 3.2 (see Figure 3). It can be deduced that both the compressive and flexural strengths of ECC specimens decreased with the increase in FA/PC ratio, regardless of the LSP content. The lowest compressive strength was measured as 38.13 MPa for the ECC specimen FA3.2_LSP0.0 with a decrease of about 50.98% with regard to the corresponding value for specimen of FA1.2_LSP0.0. The ECC specimen of FA3.2_LSP0.5 had the lowest flexural strength of 6.75 MPa with a decrease of about 15.95% with regard to the corresponding value for specimen FA1.2_LSP0.0. This may be attributed to the reduction of the cement content resulting in less formation of hydration products and decreased availability of calcium hydroxide for the pozzolanic reaction of fly ash. Several studies (e.g., [40,41]) found that FA replacement for cement slowed down the hardening rate of specimens at the early hydration stage, but at later ages its contribution to mechanical strength may become more pronounced owing to the advancement of the pozzolanic reaction [42].

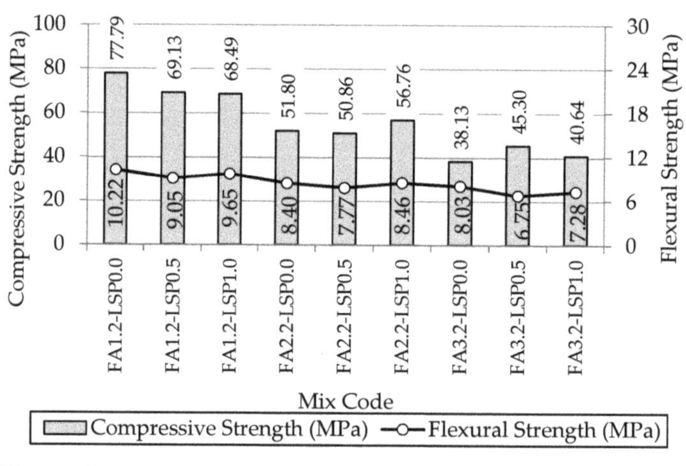

Figure 3. Compressive strength and flexural strength test results at 28 days for ECC specimens.

Using LSP substitution for SS as aggregate improved the compressive strength, except for ECC specimens made with FA/PC = 1.2. This may be ascribed to the filler effect of LSP considering its finer grain size distribution compared to that of SS. The enhancement in strength could also be related to the reduction in the total porosity caused by LSP acting as a microfiller, the formation of carbo-aluminates [43], and LSP providing preferential specific surface area for the nucleation and growth of hydration products [44]. The inclusion of 100% LSP instead of SS into ECC mixtures caused a slight increase in the flexural strength of ECC specimens with FA/PC = 2.2. Conversely, the content of LSP had no significant influence on the flexural strength of ECC specimens for all other FA/PC ratios likely because all ECC specimens included sufficient fine materials resulting in uniform distribution of PVA fibers.

3.2. Mercury Intrusion Porosimetry

The strength and durability of cementitious materials are generally related to microstructure features [45]. The evolution of the pore structure of ECC specimens was investigated via MIP testing in this study. Figure 4 shows the total porosity of ECC specimens with 1.2, 2.2 and 3.2 ratio of FA/PC and variable LSP mass replacement for SS at 0%, 50% and 100%. It can be observed that, regardless of the LSP content, an increase in the FA/PC ratio appears to be associated with increased porosity. For instance, the porosity of ECC specimens made with 3.2 ratio of FA/PC was 33.77% on average, while for the ECC specimens with 1.2 ratio of FA/PC, this value was 24.61% on average. This can be explained by the fact that the substitution of a higher amount of PC with FA diluted the cement, resulting in less formation of hydration products, which affected the compressive and tensile strength of the ECC specimens at 28 and 90 days (See Figure 3). Long et al. [41] also found that the increase in the content of FA/PC from 1.2 to 1.5 increased the porosity and negatively affected the microstructure. On the other hand, for ECC specimens with 1.2 and 3.2 ratio of FA/PC and 50% replacement of SS with LSP caused a reduction in porosity at a ratio of 0.51% and 4.27%, respectively, while 100% replacement of SS with LSP induced a 0.68% decrease in the porosity of ECC specimens made with FA/PC of 2.2%. Moreover, for FA/PC = 2.2, the replacement of LSP by SS in proportion of 100% by mass had an almost similar effect on the porosity compared to corresponding values for ECC specimens, including no LSP. The experimental results show that the replacement of SS with LSP can densify the internal pore structure and reduce the porosity of ECC, thus resulting in an enhancement in compressive strength, which conforms to the compressive strength test results performed in this study (see Figure 3).

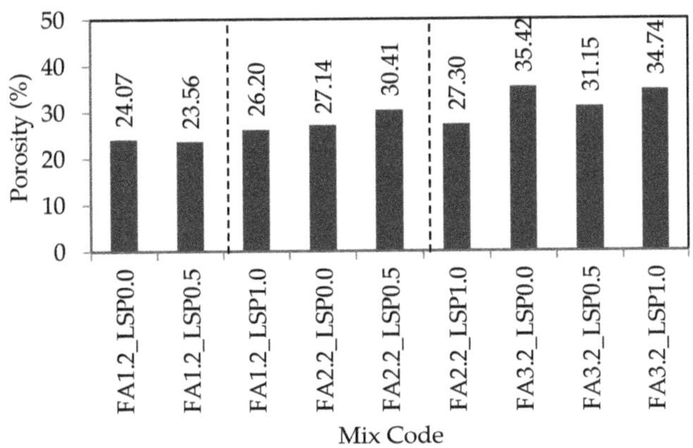

Figure 4. Porosity of ECC specimens.

Figure 5a,b illustrate the pore size distribution and pore volume fraction of ECC matrices, respectively. The pore structure can be grouped into three sizes: micropores (<2 nm), mesopores (2–50 nm) and the macropores (>50 nm) that cause a great effect on both strength and durability of concrete as per International Union of Pure and Applied Chemistry (IUPAC) [45]. The limitations of the MIP test allow the capture of a minimum pore size of 5.6 nm. Thus, in this study, the pore structure includes only mesopores and macropores. It can be observed in Figure 5a that the cumulative intrusion curves exhibited similar behavior at pore sizes below 0.8 μm. However, as the pore diameter decreased, the cumulative mercury intrusion volume for ECC specimens varied versus the FA/PC ratio. The cumulative mercury intrusion volume for the specimens with 3.2 ratio of FA/PC was highest. That is, FA2.2 and FA1.2 specimens had lower pore volume than that of

FA3.2 specimens, regardless of the LSP content, and this is notable for the pore sizes below 0.8 μm. Moreover, the volume fraction of macro-pores of ECC specimens including 3.2 ratio of FA/PC was approximately 48.5%, on average, while it was 42.5% and 31.0% for ECC specimens, which have a FA/PC ratio of 2.2 and 1.2, respectively. It can be explained by the fact that the replacement of a higher amount of PC with FA led to coarsening the micro-pores of ECC matrix, which could be due to the formation of less hydration products due to insufficient dosage of PC. This detrimental effect also caused a reduction in the mechanical strength of specimens, which is consistent with the compressive strength test results at 28 and 90 days.

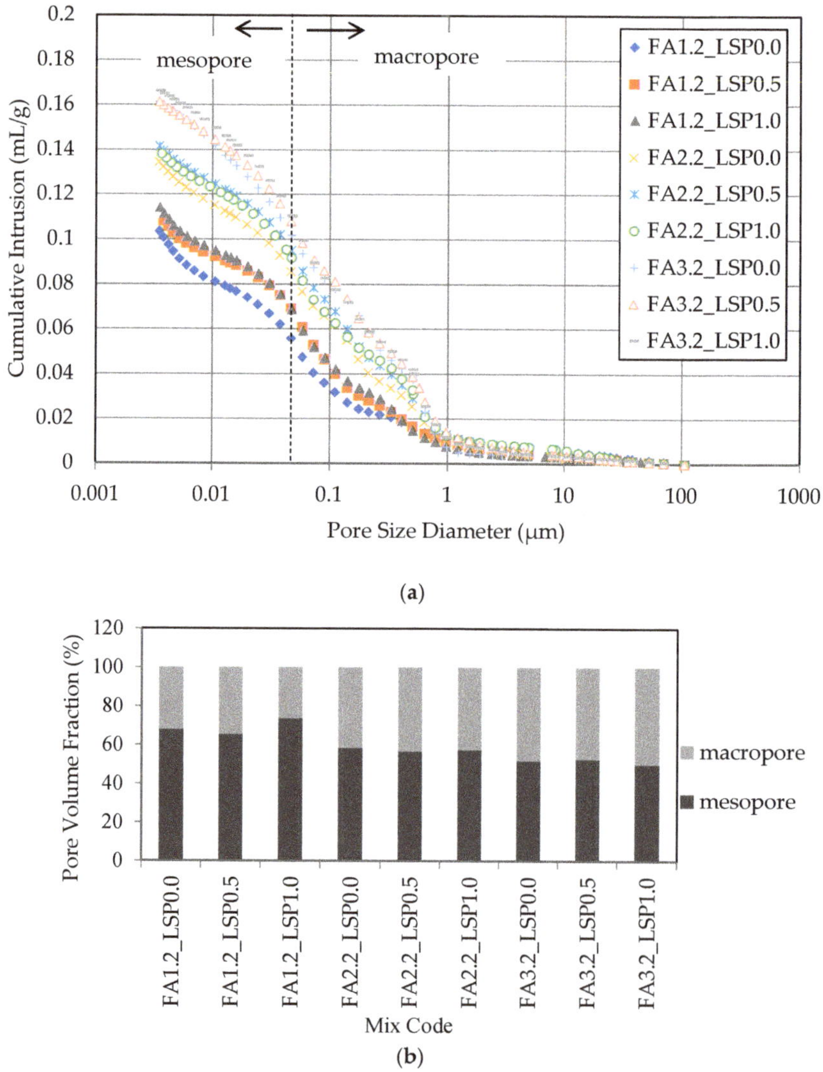

Figure 5. Pore structures in ECC matrix: (**a**) pore size distribution and (**b**) pore volume fraction.

As illustrated in Figure 5b, for ECC specimens with a 1.2 ratio of FA/PC, 100% replacement of SS with LSP induced a positive effect on the formation of mesopores. It could be inferred that the substitution of SS with LSP caused a refinement of pore distribution of the ECC matrix owing to the filler effect and fineness of LSP, especially for FA/PC = 1.2. However, it was found that there was not a certain trend of the effect of LSP inclusion on the volume fraction and pore size of ECC specimens including FA/PC = 2.2 and 3.2. Even for FA/PC = 2.2 and 3.2, ECC specimens with only SS had similar pore structure and size compared to that of specimens with 50% and 100% LSP content. According to [46,47], the use of a proper amount of fine LSP did not only reduce the amount of large pores, but also led to the formation of a higher amount of small pores with a diameter smaller than 100 nm. Valcuende et al. [48] also found that the inclusion of LSP instead of normal aggregate to the cementitious matrix lowered the threshold pore diameter as well as the volume of large capillary pores.

3.3. Freezing–Thawing Resistance

The resistance of concrete to freezing–thawing (F-T) cycles is highly dependent on the internal pore distribution. Increasing the porosity generally compromises the resistance of concrete to F-T cycles [3]. As is known, the change in the value of relative dynamic modulus of elasticity (RDME) of concrete over the duration of F-T cycles gives prior knowledge about the strength. Hence, it can be informative in the assessment of the engineering properties of concrete [49]. Figure 6 depicts the RDME of ECC specimens with a 1.2, 2.2 and 3.2 ratio of FA/PC, as well as the replacement of SS with LSP in proportions of 0%, 50% and 100% by mass. It can be observed that the increase in FA/PC ratio caused a significant reduction in the F-T resistance of ECC, regardless of LSP content. That is, ECC specimens with a 1.2 ratio of FA/PC had the highest average RDEM values at 95.41%, followed by 88.88% and 82.38% for ECC specimens made with FA/PC of 2.2 and 3.2, respectively, after 330 F-T cycles. This can be explained by the fact that as FA content increased beyond a certain threshold, the matrix became more porous and mechanical strength decreased as discussed earlier (see Figure 3).

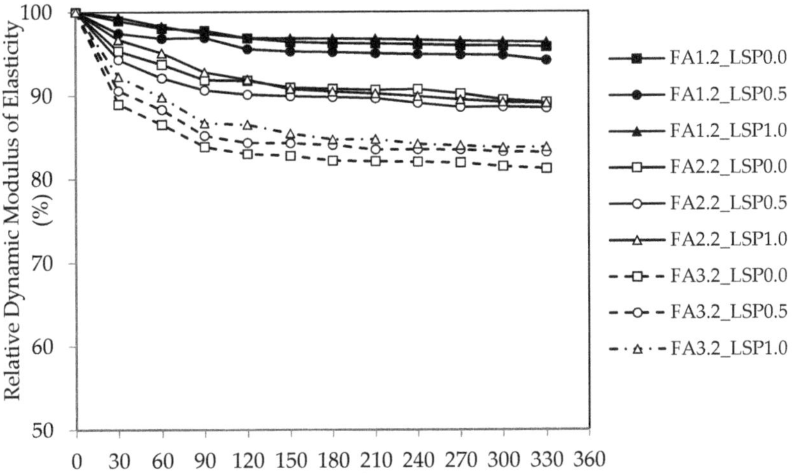

Figure 6. Relative dynamic modulus of elasticity of ECC specimens after different freeze–thaw cycles.

Referring to Figure 6, the RDEM of ECC specimens made with FA/PC = 1.2 remained almost the same during the F-T cycles. Since the RDEM is dependent on the internal compactness of matrix, the near linear trend of RDEM values of ECC specimens with FA/PC = 1.2 indicates that the addition of FA can densify the ECC matrix owing to its pozzolanic reaction with a sufficient amount of CH and resulting in maintaining the integrity after F-T cycles. Şahmaran et al. [50] also found that fly ash provided denser and more compact microstructure. On the other hand, the ECC specimens with a 2.2 ratio of FA/PC exhibited a slightly continuous downward trend, while those with FA/PC = 3.2 had a more apparent decreasing trend compared to that of ECC specimens with FA/PC = 1.2 ratio. Two reasons may be responsible for the reduction in RDEM when FA increased. First, the substitution of higher amount of PC with FA could result in fewer amount of calcium hydroxide to react with silica in FA. Second, from the aspect of pore size volume fraction, as it was found in the MIP test results (see Figure 6), the volume fraction of macro-pores of ECC specimens made with FA/PC = 3.2 was higher than that of the other ECC specimens, which could be detrimental in terms of F-T resistance, and this is consistent with previous study [51]. It was also concluded from this work that the F-T resistance of sample decreased when the pore size diameter increased from 40 nm to 2000 nm. On the other hand, it can be remarked that the RDEM values for ECC specimens with 2.2 and 3.2 ratio of FA/PC were high (88.88% and 82.38%, respectively) after 330 F-T cycles though they exhibited a downtrend. This may be attributed to the insufficient cement dosage and, thus, the filler effect of the unreacted FA particles.

As illustrated in Figure 6, for ECC specimens with a 3.2 ratio of FA/PC, the inclusion of LSP instead of SS into ECC mixtures increased the RDEM values. For instance, the RDEM of ECC specimens, which had a replacement of SS with LSP in proportions of 50% and 100% by mass were 83.17% and 83.79%, respectively, while it was 81.17% for the specimens with no LSP after 330 cycles. This can be explained by the fact that finer particles of LSP compared to SS and the associated filler effect could be more pronounced in the use of higher FA content in terms of inhibiting the interaction between the formed pores. On the other hand, for the ECC specimens having 1.2 and 2.2 ratio of FA/PC, there was no clear trend for the use of LSP, that is, RDEM of ECC specimens were comparable after 330 cycles, especially that in the microstructure of ECC specimens with 1.2 ratio of FA/PC, the pozzolanic reaction was dominant. However, the replacement of LSP with SS in the proportion of 100% by mass had positive influence on RDEM values of the specimens with 2.2 ratio of FA/PC until 120 F-T cycles.

3.4. *Rapid Chloride Ion Penetrability*

The RCPT test depends on the pore solution chemistry and pore structure characteristics of concrete. The presence of pores and cracks cause an aisle and, thus, the charge passed increases [38]. It was emphasized [52] that the chloride penetrability of specimens measured from RCPT was an important parameter to assess the quality of concrete. The migration of more chloride ions into the ECC specimens results in the passing more current through the specimens. An increase in the total charge passed shows that the ECC specimen is more penetrable. Figure 7 indicates the charge passed for ECC specimens with 1.2, 2.2 and 3.2 ratio of FA/PC, as well as the replacement of SS with LSP in proportions of 0%, 50% and 100% by mass. Based on ASTM C1202 [39], RCPT is categorized into five groups: high (>4000 coulombs), moderate (2000–4000 coulombs), low (1000–2000 coulombs), very low (100–1000 coulombs) and negligible (<100 coulombs). Referring to Figure 7, the chloride ion penetration of FA/PC = 1.2 specimens was the lowest, but the increase in FA/PC led to an increase in the charge passed, regardless of the LSP content. For instance, the ECC specimens with FA/PC = 1.2 achieved 'low' ion permeability rating, while it was 'low' to 'moderate' for the specimens with FA/PC = 2.2. However, the use of FA/PC = 3.2 caused higher value of total charge passed, yet the specimens remained below the 'moderate' permeability classification range except for ECC specimens with only SS. This may be explained by the fact that the inclusion of a higher amount of FA made ECC more porous, leading to the formation of more pore networks for charge to pass through. Therefore, the increased RCPT of ECC specimens with FA/PC = 3.2 could also be explained by having higher amounts of inert fly ash particles because of the replacement of PC with high ratio of FA and lack of CH for pozzolanic reaction. The highest RCPT test results obtained from the ECC specimens having 3.2 ratio of FA/PC may be attributed to slow initiation of the pozzolanic reactions because of the low CH content.

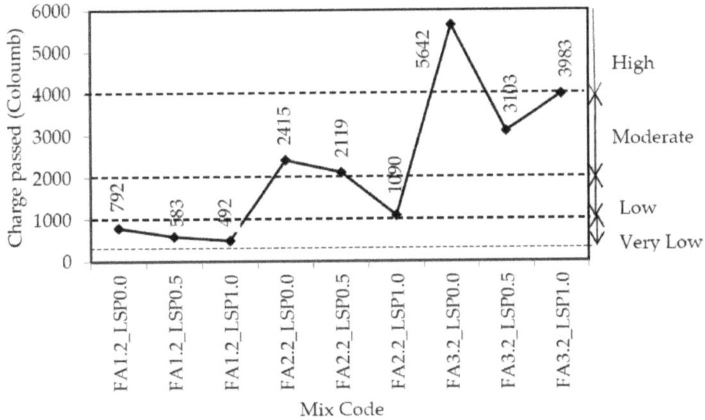

Figure 7. RCPT results of ECC specimens.

The curves in Figure 7 show that the replacement of LSP with SS caused a reduction in the chloride permeability of ECC specimens for all FA/PC ratios. The decrease in total charge passed was the highest for FA2.2_LSP1.0 with 54.86% followed by FA1.2_LSP1.0 with 37.88% compared to the ECC specimens with the same FA/PC content and no LSP. With the inclusion of LSP instead of SS into the mixtures, the ECC specimens with FA/PC = 2.2 and 3.2 were classified as 'low' and 'moderate' according to ASTM C1202, respectively. The lower charge passed through ECC specimens with LSP can be attributed to its filler effect and fineness compared to SS. Moreover, the calcium mono-carbo-aluminates formed by the addition of LSP can fill additional pores [53] and, thus, provide a denser matrix and past aggregate interfacial zone [54].

3.5. The Relationship between MIP and Mechanical/Durability Properties

As already mentioned, MIP is performed to determine the pore size distribution, porosity and average pore diameter of cementitious materials, which play an important role in their performance including strength and durability. Therefore, this study focused on developing a mathematical relation to calculate the compressive strength, flexural strength, chloride permeability and freeze–thaw resistance of ECC via the total porosity of ECC.

Observing Figure 8a, the exponential relationship between the compressive strength (f_c) and porosity (p) yielded the equation: $f_c = 212.87e^{-0.049p}$. The coefficient of determination value for this equation was estimated as 0.8824. On the other hand, the fitted second order equation for the flexural strength (f_f) and porosity (p) was $f_f = 0.0155p^2 - 1.048p + 25.34$ with an R^2 value of 0.6515, which is a low value in terms of validity. Mohr et al. [55] and Armaghani et al. [56] also described the relationship between permeability and compressive strength of concrete by a power function.

As for the correlation between the total porosity and the durability properties of ECC (Figure 8b), there was a significant linear relationship between the RDEM and porosity of ECC in the form $RDEM = -1.1151p + 120.37$ and the R-squared value was 0.9116. It is worth noting that the effect of porosity on freeze–thaw resistance of ECC was prominent in this study. According to the correlation between RCPT and porosity of ECC, the fitted second order equation was found as $RCPT = 23.636p^2 - 983.84p + 10740$ with the R-squared value of 0.9747, which shows that the resistance of ECC to RCPT was strongly affected by the total porosity in its microstructure. Therefore, these empirical formulas showed that the total porosity was a parameter that had the closest relationship to RCPT. Finally, it can be emphasized that the correlation between the porosity and durability properties rendered better fitting results compared to those of the mechanical properties. Zhang and Li et al. [57] found a linear correlation between chloride permeability and pore structure of concretes with coefficient of determination of 0.93.

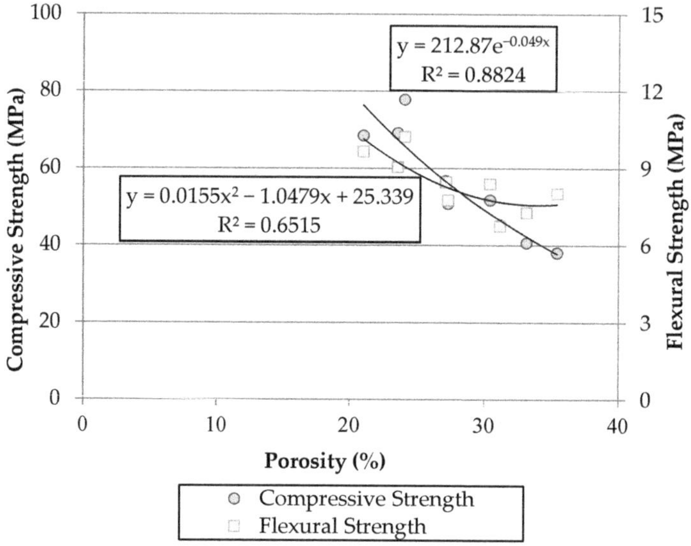

(a)

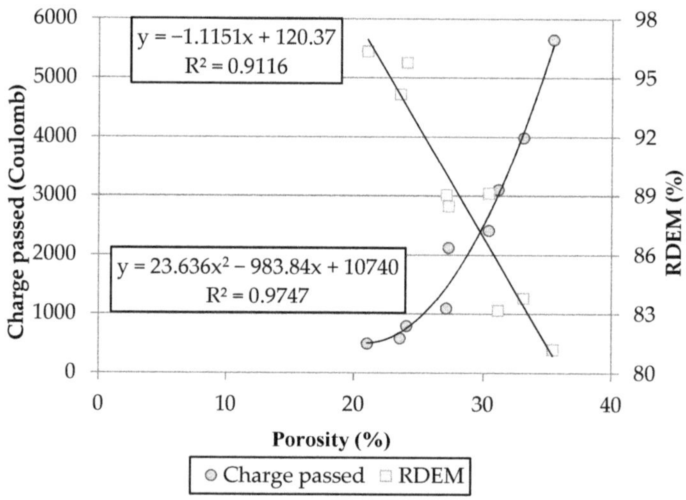

(b)

Figure 8. Correlation between the total porosity and (**a**) mechanical, (**b**) durability properties of ECC.

4. Conclusions

This work investigated the influences of high-volume FA along with LSP partial or total replacement for SS by weight on the durability properties of ECC. Within this scope, mercury intrusion porosimetry, rapid chloride ion penetrability and freezing–thawing tests were conducted. Moreover, the compressive and flexure strengths of ECC were assessed to appraise the hardened properties of ECC. The following conclusions can be listed according to the experimental results:

1. The compressive and flexural strengths of ECC specimens decreased with increase in FA/PC ratio, regardless of LSP content, while the addition of LSP as aggregate enhanced the compressive strength in general. As for flexural strength of ECC spec-

imens, the use of LSP instead of SS caused slight improvement for ECC specimens with 2.2 ratio of FA/PC, whilst the full replacement of LSP for SS had no clear effect for all FA/PC ratios.

2. The MIP results showed that using higher ratio of FA/PC increased the total porosity of ECC specimens, resulting in higher volume fraction of macropores. On the other hand, in general, the partial replacement of LSP with SS caused a reduction in porosity and a refinement of pore distribution of ECC matrix. In particular, 100% replacement of LSP with SS for ECC specimens with FA/PC of 1.2 implied a positive influence on the formation of mesopores.
3. The freezing–thawing resistance of ECC specimens decreased when FA/PC ratio increased, whilst for ECC specimens with a 3.2 ratio of FA/PC the full replacement of LSP with SS caused an increase in RDEM of ECC. On the other hand, for 1.2 and 2.2 ratio of FA/PC, the ECC specimens with LSP exhibited almost the same resistance with the ECC specimens with 100% SS.
4. The charge passed increased as the ratio of FA/PC increased for ECC specimens, regardless of LSP content. However, the use of LSP induced a decrease in the chloride permeability of ECC specimens for all FA/PC ratios.
5. The correlation between the porosity and the test results was higher for freezing–thawing and chloride permeability tests with the R-square value of 0.9116 and 0.9747, respectively, compared to those of the compressive and flexure strengths.
6. According to the test results, the replacement of LSP with silica sand can be advised for the development of sustainable ECC due to causing an improvement in pore size distribution and cumulative porosity of matrices resulting in an increase in the freezing–thawing resistance and chloride permeability as well as strength.

Author Contributions: K.T.: conceptualization; formal analysis; investigation; writing—original draft. C.K.: data curation; formal analysis; methodology; writing—original draft. M.L.N.: project administration; resources; supervision; writing—review and editing. All authors have read and agreed to the published version of the manuscript.

Funding: This research was funded by the Scientific and Technological Research Council of Turkey, Grant Number: TUBITAK-BIDEB-2219 International Postdoctoral Research Scholarship Program (2012-2).

Institutional Review Board Statement: Not applicable.

Informed Consent Statement: Not applicable.

Data Availability Statement: The data that support the findings of this study are available from the corresponding author upon request.

Conflicts of Interest: The authors declare no conflict of interest.

References

1. Li, V.C.; Kanda, T. Engineered Cementitious Composites for Structural Applications. *J. Mater. Civ. Eng.* **1998**, *10*, 66–69. [CrossRef]
2. Zhou, Y.; Xi, B.; Yu, K.; Sui, L.; Xing, F. Mechanical properties of hybrid ultra-high performance engineered cementitous composites incorporating steel and polyethylene fibers. *Materials* **2018**, *11*, 1448. [CrossRef] [PubMed]
3. Li, V.C.; Leung, C.K.Y. Steady-State and Multiple Cracking of Short Random Fiber Composites. *J. Eng. Mech.* **1992**, *118*, 2246–2264. [CrossRef]
4. Liu, H.; Zhang, Q.; Gu, C.; Su, H.; Li, V.C. Influence of micro-cracking on the permeability of engineered cementitious composites. *Cem. Concr. Compos.* **2016**, *72*, 104–113. [CrossRef]
5. Weimann, M.B.; Li, V.C. Hygral Behavior of Engineered Cementitious Composites (ECC). *Int. J. Restor. Build. Monum.* **2003**, *9*, 513–534.
6. Şahmaran, M.; Li, V.C. Durability of mechanically loaded engineered cementitious composites under highly alkaline environments. *Cem. Concr. Compos.* **2008**, *30*, 72–81. [CrossRef]
7. Ma, H.; Yi, C.; Wu, C. Review and outlook on durability of engineered cementitious composite (ECC). *Constr. Build. Mater.* **2021**, *287*, 122719. [CrossRef]
8. Turk, K.; Nehdi, M.L. Coupled effects of limestone powder and high-volume fly ash on mechanical properties of ECC. *Constr. Build. Mater.* **2018**, *164*, 185–192. [CrossRef]

9. Ranjith, S.; Venkatasubramani, R.; Sreevidya, V. Comparative Study on Durability Properties of Engineered Cementitious Composites with Polypropylene Fiber and Glass Fiber. *Arch. Civ. Eng.* **2017**, *63*, 83–101. [CrossRef]
10. Wang, Y.; Wang, G.; Guan, Z.; Ashour, A.; Ge, W.; Zhang, P.; Lu, W.; Cao, D. The effect of freeze–thaw cycles on flexural behaviour of FRP-reinforced ECC beams. *Arch. Civ. Mech. Eng.* **2021**, *21*, 101. [CrossRef]
11. Zhang, W.; Yin, C.; Ma, F.; Huang, Z. Mechanical properties and carbonation durability of engineered cementitious composites reinforced by polypropylene and hydrophilic polyvinyl alcohol fibers. *Materials* **2018**, *11*, 1147. [CrossRef] [PubMed]
12. Hu, X.; Xia, K. Experimental study on the permeability of basalt fibre engineered cementitious composite (ECC). *Mater. Struct.* **2022**, *55*, 85. [CrossRef]
13. Alnahhal, M.F.; Alengaram, U.J.; Jumaat, M.Z.; Alsubari, B.; Alqedra, M.A.; Mo, K.H. Effect of aggressive chemicals on durability and microstructure properties of concrete containing crushed new concrete aggregate and non-traditional supplementary cementitious materials. *Constr. Build. Mater.* **2018**, *163*, 482–495. [CrossRef]
14. Hossain, M.M.; Karim, M.R.; Hasan, M.; Hossain, M.K.; Zain, M.F.M. Durability of mortar and concrete made up of pozzolans as a partial replacement of cement: A review. *Constr. Build. Mater.* **2016**, *116*, 128–140. [CrossRef]
15. Lepech, M.D.; Li, V.C. Water permeability of engineered cementitious composites. *Cem. Concr. Compos.* **2009**, *31*, 744–753. [CrossRef]
16. Liu, H.; Zhang, Q.; Li, V.; Su, H.; Gu, C. Durability study on engineered cementitious composites (ECC) under sulfate and chloride environment. *Constr. Build. Mater.* **2017**, *133*, 171–181. [CrossRef]
17. Yücel, H.E.; Öz, H.Ö.; Güneş, M.; Kaya, Y. Rheological properties, strength characteristics and flexural performances of engineered cementitious composites incorporating synthetic wollastonite microfibers with two different high aspect ratios. *Constr. Build. Mater.* **2021**, *306*, 124921. [CrossRef]
18. Quan, X.; Wang, S.; Liu, K.; Zhao, N.; Xu, J.; Xu, F.; Zhou, J. The corrosion resistance of engineered cementitious composite (ECC) containing high-volume fly ash and low-volume bentonite against the combined action of sulfate attack and dry-wet cycles. *Constr. Build. Mater.* **2021**, *303*, 124599. [CrossRef]
19. Mansoori, A.; Behfarnia, K. Evaluation of mechanical and durability properties of engineered cementitious composites exposed to sulfate attack and freeze–thaw cycle. *Asian J. Civ. Eng.* **2021**, *22*, 417–429. [CrossRef]
20. Turk, K.; Kina, C. Freeze-thaw resistance and sorptivity of self-compacting mortar with ternary blends. *Comput. Concr.* **2018**, *21*, 149–156. [CrossRef]
21. Benli, A.; Turk, K.; Kina, C. Influence of Silica Fume and Class F Fly Ash on Mechanical and Rheological Properties and Freeze-Thaw Durability of Self-Compacting Mortars. *J. Cold Reg. Eng.* **2018**, *32*, 04018009. [CrossRef]
22. Yang, E.H.; Yang, Y.; Li, V.C. Use of high volumes of fly ash to improve ECC mechanical properties and material greenness. *ACI Mater. J.* **2007**, *104*, 620–628. [CrossRef]
23. Turk, K.; Demirhan, S. Effect of limestone powder on the rheological, mechanical and durability properties of ECC. *Eur. J. Environ. Civ. Eng.* **2017**, *21*, 1151–1170. [CrossRef]
24. Marzouki, A.; Lecomte, A.; Beddey, A.; Diliberto, C.; Ben Ouezdou, M. The effects of grinding on the properties of Portland-limestone cement. *Constr. Build. Mater.* **2013**, *48*, 1145–1155. [CrossRef]
25. Balayssac, J.P.; Détriché, C.H.; Grandet, J. Effects of curing upon carbonation of concrete. *Constr. Build. Mater.* **1995**, *9*, 91–95. [CrossRef]
26. Parrott, L.J. Some effects of cement and curing upon carbonation and reinforcement corrosion in concrete. *Mater. Struct./Mater. Constr.* **1996**, *29*, 164–173. [CrossRef]
27. Demirhan, S.; Turk, K.; Ulugerger, K. Fresh and hardened properties of self consolidating Portland limestone cement mortars: Effect of high volume limestone powder replaced by cement. *Constr. Build. Mater.* **2019**, *196*, 115–125. [CrossRef]
28. *ASTM C 150/C150M-12*; Standard Specification for Portland Cement. ASTM Standards: West Conshohocken, PA, USA, 2019; pp. 1–9.
29. *ASTM C618-19*; Standard Specification for Coal Fly Ash and Raw or Calcined Natural Pozzolan for Use in Concrete. ASTM Standards: West Conshohocken, PA, USA, 2019.
30. Li, V.C.; Wu, C.; Wang, S.; Ogawa, A.; Saito, T. Interface tailoring for strain-hardening polyvinyl alcohol-engineered cementitious composite (PVA-ECC). *ACI Mater. J.* **2002**, *99*, 463–472. [CrossRef]
31. Şahmaran, M.; Lachemi, M.; Hossain, K.M.A.; Li, V.C. Internal curing of engineered cementitious composites for prevention of early age autogenous shrinkage cracking. *Cem. Concr. Res.* **2009**, *39*, 893–901. [CrossRef]
32. Yang, E.H.; Sahmaran, M.; Yang, Y.; Li, V.C. Rheological control in production of engineered cementitious composites. *ACI Mater. J.* **2009**, *106*, 357–366. [CrossRef]
33. Ahmaran, M.; Li, V.C. Influence of microcracking on water absorption and sorptivity of ECC. *Mater. Struct./Mater. Constr.* **2009**, *42*, 593–603. [CrossRef]
34. *ASTM C39/C39M-20*; Standard Test Method for Compressive Strength of Cylindrical Concrete Specimens. ASTM Standards: West Conshohocken, PA, USA, 2020.
35. *ASTM C78/C78M-18*; Standard Test Method for Flexural Strength of Concrete (Using Simple Beam with Third-Point Loading). ASTM Standards: West Conshohocken, PA, USA, 2018.
36. *ASTM D 4404*; Standard Test Method for Determination of Pore Volume and Pore Volume Distribution of Soil and Rock by Mercury Intrusion Porosimetry. ASTM Standards: West Conshohocken, PA, USA, 2018.

37. *ASTM C666/C666M-03*; Standard Test Method for Resistance of Concrete to Rapid Freezing and Thawing. ASTM Standards: West Conshohocken, PA, USA, 2008.
38. Shi, C. Effect of mixing proportions of concrete on its electrical conductivity and the rapid chloride permeability test (ASTM C1202 or ASSHTO T277) results. *Cem. Concr. Res.* **2004**, *34*, 537–545. [CrossRef]
39. *ASTM C1202*; Standard Test Method for Electrical Indication of Concrete's Ability to Resist Chloride Ion Penetration. ASTM Standards: West Conshohocken, PA, USA, 2012.
40. Zhang, Z.; Qian, S.; Ma, H. Investigating mechanical properties and self-healing behavior of micro-cracked ECC with different volume of fly ash. *Constr. Build. Mater.* **2014**, *52*, 17–23. [CrossRef]
41. Long, W.J.; Li, H.D.; Mei, L.; Li, W.; Xing, F.; Khayat, K.H. Damping characteristics of PVA fiber-reinforced cementitious composite containing high-volume fly ash under frequency-temperature coupling effects. *Cem. Concr. Compos.* **2021**, *118*, 103911. [CrossRef]
42. Şahmaran, M.; Yaman, I.Ö.; Tokyay, M. Transport and mechanical properties of self consolidating concrete with high volume fly ash. *Cem. Concr. Compos.* **2009**, *31*, 99–106. [CrossRef]
43. Zhang, L.V.; Suleiman, A.R.; Nehdi, M.L. Self-healing in fiber-reinforced alkali-activated slag composites incorporating different additives. *Constr. Build. Mater.* **2020**, *262*, 120059. [CrossRef]
44. Yuan, B.; Yu, Q.L.; Brouwers, H.J.H. Assessing the chemical involvement of limestone powder in sodium carbonate activated slag. *Mater. Struct./Mater. Constr.* **2017**, *50*, 136. [CrossRef]
45. Aligizaki, K.K. *Pore Structure of Cement-Based Materials: Testing Interpretation and Requirements (Modern Concrete Technology)*; Taylor & Francis: London, UK, 2005; 432p.
46. Wu, K.; Shi, H.; Xu, L.; Gao, Y.; Ye, G. Effect of Mineral Admixture on Mechanical Properties of Concrete by Adjusting Interfacial Transition Zone Microstructure. *Kuei Suan Jen Hsueh Pao/J. Chin. Ceram. Soc.* **2017**, *45*, 623–630. [CrossRef]
47. Liu, S.H.; Wang, J. Influence of limestone powder on pore structure of mortar. *Jianzhu Cailiao Xuebao/J. Build. Mater.* **2011**, *14*, 532–535. [CrossRef]
48. Valcuende, M.; Parra, C.; Marco, E.; Garrido, A.; Martínez, E.; Cánoves, J. Influence of limestone filler and viscosity-modifying admixture on the porous structure of self-compacting concrete. *Constr. Build. Mater.* **2012**, *28*, 122–128. [CrossRef]
49. Öznur Öz, H.; Erhan Yücel, H.; Güneş, M. Freeze-Thaw Resistance of Self Compacting Concrete Incorporating Basic Pumice. *Int. J. Theor. Appl. Mech.* **2016**, *1*, 285–291.
50. Şahmaran, M.; Özbay, E.; Yücel, H.E.; Lachemi, M.; Li, V.C. Frost resistance and microstructure of Engineered Cementitious Composites: Influence of fly ash and micro poly-vinyl-alcohol fiber. *Cem. Concr. Compos.* **2012**, *34*, 156–165. [CrossRef]
51. Kamada, E. Frost damage and pore structure of concrete. *Proc. Jpn. Concr. Inst.* **1988**, *10*, 51–60.
52. Misra, S.; Yamamoto, A.; Tsutsumi, T.; Motohashi, K. Application of rapid chloride permeability test to quality control of concrete. *Am. Concr. Inst. ACI Spec. Publ.* **1994**, *SP-145*, 487–502.
53. Heikal, M.; El-Didamony, H.; Morsy, M.S. Limestone-filled pozzolanic cement. *Cem. Concr. Res.* **2000**, *30*, 1827–1834. [CrossRef]
54. Liu, S.; Yan, P. Effect of limestone powder on microstructure of concrete. *J. Wuhan Univ. Technol. Mater. Sci. Ed.* **2010**, *25*, 328–331. [CrossRef]
55. Mohr, P.; Hansen, W.; Jensen, E.; Pane, I. Transport properties of concrete pavements with excellent long-term in-service performance. *Cem. Concr. Res.* **2000**, *30*, 1903–1910. [CrossRef]
56. Armaghani, J.M.; Larsen, T.J.; Romano, D.C. Aspects of Concrete Strength and Durability. *Transp. Res. Rec.* **1992**, 63–69.
57. Zhang, M.H.; Li, H. Pore structure and chloride permeability of concrete containing nano-particles for pavement. *Constr. Build. Mater.* **2011**, *25*, 608–616. [CrossRef]

sustainability

MDPI

Article

Development and Characterization of Eco-Efficient Ultra-High Durability Concrete

Keila Robalo [1], Hugo Costa [2], Ricardo Carmo [2,*] and Eduardo Júlio [3]

[1] Civil Engineering Research and Innovation for Sustainability, University of Cape Verde, Praia 7943-010, Cape Verde
[2] Civil Engineering Research and Innovation for Sustainability, SUScita, ISEC—Polytechnic Institute of Coimbra, 3030-199 Coimbra, Portugal
[3] Civil Engineering Research and Innovation for Sustainability, Instituto Superior Técnico, University of Lisbon, 1049-001 Lisbon, Portugal
* Correspondence: carmo@isec.pt

Abstract: Ultra-High-Performance Concrete (UHPC) is characterized by having an ultra-compact matrix resulting in ultra-high mechanical properties, low permeability to water and gases, and improved ductility provided by the addition of fibers. The production of structures with this type of concrete is advantageous in some situations, especially in aggressive environments since it significantly increases durability. However, high dosages of Portland cement and silica fume are commonly adopted, increasing not only the cost but also the environmental impact, jeopardizing its use, mainly in the present context where the sustainability of the construction sector is a global priority. In this sense, improving the eco-efficiency of this type of concrete is mandatory. The objective of this work is to develop eco-ultra-high-durability concrete (eco-UHDC). The UHDC matrix was optimized, focusing mainly on durability and looking for the lowest environmental impact, where several parameters were varied: cement replacement ratio, additions in binder matrix and its relative proportions, water/binder ratio, type of fibers, and its proportion. The developed eco-UHDC was characterized both in fresh and hardened states, in terms of mechanical properties, time-dependent properties, and durability. This last topic includes the characterization of durability parameters under laboratory conditions and in a real environment, namely, in the tidal zone of the coast of Cape Verde. The results of resistance to carbonation and chloride penetration were used to predict the service life of structures produced with these eco-UHDC. The optimization of the UHDC matrix allowed the development of mixtures with only 60% of cement in relation to the total amount of powder of the matrix, maintaining good workability and the desired mechanical characteristics (compressive strength higher than 100 MPa and flexural strength higher than 12 MPa). The results also showed that considering only the requirements related to durability, the cover of structures produced with these optimized mixtures can be lower than the values recommended by Eurocode 2, with differences that can reach 55%, mainly when pozzolan of Cape Verde is used as partial replacement of Portland cement.

Keywords: ultra-high performance fiber reinforced concrete; durability; sustainability; carbonation resistance; creep and shrinkage

check for updates

Citation: Robalo, K.; Costa, H.; Carmo, R.; Júlio, E. Development and Characterization of Eco-Efficient Ultra-High Durability Concrete. *Sustainability* **2023**, *15*, 2381. https://doi.org/10.3390/su15032381

Academic Editors: Woubishet Zewdu Taffese and Sandra Barbosa Nunes

Received: 13 December 2022
Revised: 21 January 2023
Accepted: 23 January 2023
Published: 28 January 2023

1. Introduction

Recently, the environmental impact of the construction industry has been addressed by many researchers, and the concern to reduce it is growing because this industry consumes enormous quantities of non-renewable raw materials and is also one of main contributors of greenhouse gas release. Ultra-High-Performance Concrete (UHPC) can have ultra-high compressive strength, over 150 MPa, combined with ultra-high durability [1]. For example, the chloride diffusion coefficient of a UHPC can be on the order of 2×10^{-12} m^2/s, significantly lower than that of high-performance concrete, around 4 to 6×10^{-12} m^2/s [2–7].

However, the huge dosages of Portland cement and silica fume required in its composition increases the economic and environmental costs [3,8–10]. In addition, that level of performance requires special curing conditions, only possible in an industrial environment. Mehta (2009) recommended, according to [3,11], three approaches to improve the sustainability of the concrete industry: (1) consume less concrete, developing innovative architectural and structural projects for both the construction of new structures and rehabilitation of existing structures; (2) reduce the cement Portland dosage in concrete mixes using a smart proportioning approach; and (3) reducing the cement Portland dosage in concrete mixes using a larger volume of one or more additions. The last two approaches are related to the formulation of concrete. Several scientific works have been developed [11–19] with the goal of optimizing the proportion of UHPC constituents, which falls under the second approach. Regarding the third approach, the use of supplementary cementitious materials of micrometric and nanometric size has also been widely studied to partially replace the cement Portland in the production of UHPC, citing as examples the studies carried out by Chan and Chu [20], Droll [21], Tuan et al. [22], Ghafari et al. [4], Zhao and Sun [23], Yu et al. [10], Ferdosian et al. [24], etc.

A typical UHPC mix contains cement Portland, silica fume, quartz powder silica sand, steel microfibers, superplasticizer, and water. Typically, a high proportion of cement is used, between 800 and 1100 kg/m^3, approximately three times more than the dosage of current concrete [3]. A wide variety of dosages of microparticles of cementitious materials, mostly pozzolanic or latent hydraulic reaction, has been used to partially replace cement, such as, for example, silica fume, fly ash, ground granulated blast furnace slag, rice husk ash, and metakaolin.

The efficiency of UHPC is particularly dependent on its density, and this can be achieved by optimizing the packing of particles, which consists of an almost perfect granular distribution incorporating a homogeneous gradient of ultrafine, fine, and medium particles in the mixture. In this context, the use of nanometric cementitious materials is highly effective, such as the addition of nano-silica, pozzolanic nanoparticles, nano-metakaolin, and nano-clay [3,25–29]. The studies carried out, considering both the various guidelines on material properties, the UHPC formulation, and the concept of sustainability gave rise to a new family of UHPC, the eco-efficient UHPC. It should be noted that this type of concrete exhibits exceptional mechanical properties and durability but has a significantly lower environmental impact compared to the traditional UHPC formulation.

The present work constitutes an advance in relation to the mentioned studies by improving, even more, the eco-efficiency of the UHPC, which is today accepted to have a compressive strength between 100 and 120 MPa at 28 days [30]. The main objective is to develop eco-ultra-high durability concrete (eco-UHDC), mostly designed with current constituents and with normal curing conditions, where the main property to enhance is not the compressive strength. Therefore, it is not essential to reach a compressive strength of 150 MPa but instead is essential to assure high durability and eco-efficiency, which is why it was named eco-UHDC or eco-UHDFRC (eco-ultra-high durability fiber-reinforced concrete). To enhance the durability of structures and, simultaneously, to reduce the carbon footprint related to cement consumption, the eco-UHDC may be applied only in the cover layer of the structural members where it is most needed to protect the structure, being the core produced with current concrete or even with concrete with significantly lower cement content.

To achieve the mentioned goal, the strategy was to improve the eco-efficiency of the material, minimizing the cement dosage and using more ecological additions to replace it (reduction of about 50% of the cement dosage comparatively to the traditional UHPC formulations), optimizing the dosage of superplasticizer to increase compactness, consequently resulting in a concrete with low porosity. For this purpose, several preliminary mixtures were defined, and, based on the results obtained, the UHDC compositions with the best durability/eco-efficiency ratio and good time-dependent behavior were chosen. The developed eco-UHDC mixtures were characterized in the fresh and hardened state, in terms of mechanical properties (tensile and compressive strength, Young's modulus, and bending fracture energy), time-dependent properties (shrinkage and creep), and durability (resistance to carbonation, resistance to chloride penetration, chloride content on concrete surface, water absorption by capillarity, and electrical resistivity). Durability parameters were determined using specimens under laboratory conditions and specimens exposed to the maritime environment of Cape Verde, in the tidal zone. The service life of structures related to the risk of reinforcement steel corrosion were also estimated, considering the durability performance of the several eco-UHDC developed, namely, the resistance to carbonation and penetration of chloride ions.

2. Experimental Program

2.1. Materials

In order to promote a wide range of solutions and parameters that allow defining the formulation strategy, the following constituents were selected for the experimental program: (a) cement type CEM I 52.5R, several current additions (fly ash, limestone filler, silica fume, and quartz flour), and pozzolan from Cape Verde as alternative to fly ash; (b) fine siliceous sand of 0/1 mm size-fraction (granulometry), selected from a preliminary experimental analysis based on three fine sands; (c) three types of microfibers (basalt fibers, glass fibers, and steel fibers, the latter being the traditionally used in UHPC) and one polymeric type of macrofibers; water and superplasticizer (ether–carboxylate based), named M526.

Table 1 and Figure 1 present the main characteristics of the fibers adopted in the study: steel straight microfibers (OL13/0.16); hybrid polymeric macrofibers (S25), constituted by polypropylene and polyethylene; alkali-resistant glass microfibers; and basalt microfibers.

Table 1. Characteristics of fibers according to the data sheet of companies.

	Steel Fibers	Polymeric Fibers	Glass Fibers	Basalt Fibers
Tensile strength	3000 MPa	650 MPa	1700 MPa	3500 MPa
Young's modulus	200 GPa	5 GPa	72 GPa	95 GPa
Density	7.85 g/cm^3	0.92 g/cm^3	2.68 g/cm^3	2.67 g/cm^3
Diameter	0.16 mm	0.3 to 0.5 mm	14 μm	18 μm
Length	13 mm	25 mm	12 mm	13 mm

Figure 1. Fibers used in the formulation of eco-UHDC: (**a**) polymeric fibers, (**b**) steel fibers, (**c**) glass fibers, and (**d**) basalt fibers.

2.2. Formulation and Optimization of Eco-UHDC

Once the properties of the constituent materials of the various UHDC mixtures were characterized, several preliminary mixtures were developed to calibrate some parameters of the formulation, aiming to optimize the binder paste by combining the intended performance level with the reduction of the ecological and economical costs. It should be noted that, at that first stage, the goal was to evaluate the performance of UHDC matrices without reinforcing fibers. It is usual for the powder dosage in the matrix to be about 1200 to 1400 kg/m^3, and in this study, the dosage of the powder in the binder paste was fixed at 1100 kg/m^3, allowing the addition of high fiber content, maintaining good workability. The target air content was set at 1.0% to 1.5%. Table 2 shows the different formulated mixtures. The mixture proportioning of materials was performed by weight, considering the design parameters used in the literature regarding W/C ratio and W/B ratio, among others, and optimizing the packing density.

Table 2. Dosages of constituents of the developed eco-UHDC.

		C = 75%; FA = 15%; Other Additions = 10%		**C = 60%; LF = 25%; Other Additions = 15%**					**C = 60%; LF = 15%; FA =25%**
Mixtures		**SK1**	**SK2**	**SK3**	**SK4**	**SK5**	**SK6**	**SK7**	**SK8**
Binder powder dosage	kg/m^3	1100							
Water/binder powder (W/B)		0.195		0.18	0.21				
Cement (C)	(kg/m^3)	825	825	660	660	660	660	660	660
Limestone filler (LF)	(kg/m^3)	110		275	275	275	275	275	165
Fly ash (FA)	(kg/m^3)	165	165	165	165				275
Silica fume	(kg/m^3)						165		
Pozzolan—Cape Verde	(kg/m^3)							165	
Quartz flour	(kg/m^3)		110			165			
Water	(kg/m^3)	215	215	198	231	231	231	231	231
Admixture (Spl)	(kg/m^3)	29	35	30	13	15	25	18	17
Sand 0/1 mm	(kg/m^3)	983	969	1006	961	974	919	937	936

Several additions were used in the formulation, as mentioned. The strategy was defined by considering two levels of cement content (825 and 660 kg/m^3, corresponding to 75% and 60% of the powder content), combined with micro-filler powders (limestone or quartz flour) to assure filler effect and with other pozzolanic materials (fly ash, silica fume, or natural ground pozzolan). Preliminary tests proved that lower cement dosages result in higher risk of segregation to assure proper flowability, where the mechanical performance is also affected. Regarding the water/(binder powder) ratio (W/B), which was defined as 0.21 for the mixtures with 60% of cement as powder content, except SK3 where 0.18 ratio was used; however, this parameter requires, together with the mixtures with SK1 and SK2 (cement corresponding to 75% of the powder content and W/B of 0.195), higher superplasticizer (Spl) dosage (between 29 and 35 kg/m^3). As known, silica fume also requires high proportions of superplasticizer compared to the fly ash and pozzolan additions. The superplasticizer proportion was adjusted for each mixture to assure similar flowability and air content between mixtures. Mixture SK8 was formulated with similar parameters to those of SK4, changing the corresponding proportions of fly ash and limestone filler.

Based on the results of tests carried out both in the fresh and in the hardened states presented in the following section, and taking into account mainly the eco-efficiency (which in this case is simply considered the minimum cement dosage), SK4 was selected for the later phase of the study where different fiber types were added, resulting in various fiber-reinforced eco-UHDC mixtures, named eco-UHDFRC: mixtures with 1 and 2% of steel fibers; mixtures with 1 and 2% of basalt fibers; mixtures with 1 and 2% of glass fibers; mixtures with 2 and 4% of polymeric fibers; mixtures with 1% of glass fibers and 2% of polymeric fibers (Table 3). The dosage of fibers in the mixture was defined based on the literature to avoid the workability loss of the concrete, considering the difficulties in mixing and in the lack of flowability caused by the agglomeration of fibers when a certain proportion of fiber content is exceeded. Depending on the variation of the type or quantity of fibers, small adjustments were made, always necessary, mainly in the dosage of superplasticizer and in the air content to approximate the measured values to target values initially defined, assuring objective and coherent further analysis. Small adjustments of sand proportions resulted mainly in those variations, depending also on the constituent's density.

Table 3. Dosages of constituents of the developed eco-UHDFRC.

Mixtures	C	LF	FA	Water (kg/m^3)	Spl	Sand 0/1 mm	Fibers Types and % by Volume	(kg/m^3)
SK4s1	660	275	165	231	13	934	Steel—1%	79
SK4s2					13	907	Steel—2%	157
SK4b1					18	920	Basalt—1%	27
SK4b2					21	888	Basalt—2%	53
SK4p1					13	907	Polymeric—2%	18
SK4p2					14	853	Polymeric—4%	37
SK4g1					18	920	Glass—1%	27
SK4g2					21	874	Glass—2%	53
SK4gp					18	868	Glass—1% + Polymeric—2%	27 + 18

2.3. *Fresh State and Mechanical Properties*

After the formulation, the production and characterization of eco-UHDC mixtures were performed. In the fresh state, the eco-UHDC mixtures were characterized regarding flowability by the mini-slump flow test according to EN 1015-3 [31] and the density and the air content by the pressure gauge method according to EN 1015-6 [32] and EN 1015-7 [33], respectively.

The eco-UHDC mixtures were also characterized in a hardened state. Tests were performed to access flexural and compressive strength on mortars performed on prismatic specimens (40 mm × 40 mm × 160 mm) according to EN 1015-11 [34], and the results were obtained by averaging three specimens for flexural and six half specimens for compressive at 7, 28, and 56 days of age. After calibrating the mixtures, those presenting higher strengths, higher eco-efficiency (lower cement content), and good workability were selected. Those selected mixtures were characterized in terms of the remaining relevant mechanical properties, such as Young's modulus and fracture energy in bending. The average value of the Young's modulus of eco-UHDC was determined according to the specification LNEC E-397 [35], adjusted to prismatic specimens 50 mm × 50 mm × 300 mm. The tensile fracture energy of the UHDFRC was determined according to EN 14651 [36] to evaluate the contribution of the fibers in tension. For this purpose, two prismatic specimens 150 mm ×150 mm × 600 mm were used, with a notch at mid-span 25 mm high and 5 mm wide on the side parallel to the casting direction (Figure 2). The three-point bending test was performed at a speed rate of 3 mm/s; the distance between supports was 500 mm and four LVDTs (Linear Variable Differential Transformer) were used, two placed at mid-span on opposite faces, and the other two aligned to the supports, as shown in Figure 2.

Figure 2. Test set-up and instrumentation.

2.4. Time-Dependent Tests

The time-dependent properties of selected eco-UHDC were characterized. The shrinkage of the selected eco-UHDC was performed experimentally according to the procedure described in EN 12808-4 [37] using prismatic specimens 40 mm × 40 mm × 160 mm. During the casting, metallic inserts were placed on the ends of the specimens to facilitate the shrinkage measurements. The specimens were demolded 24 h after casting and their initial lengths were recorded. During the test, the specimens were kept in the thermo–hygrometric chamber set at a temperature T = 20 °C (±2 °C) and relative humidity RH = 50 % (±5%). The shrinkage test was performed only on the eco-UHDC (without fibers) since SK4 has the same binding matrix as the fiber-containing concrete, and it was intended to evaluate the influence of the paste parameters on the shrinkage of the matrix (without fibers). In addition, it is known that steel fibers tend to significantly reduce the shrinkage of the concrete because the shrinkage deformation is restricted by the fiber-matrix bonds and fibers also give rise to a more cohesive and bonded matrix, mitigating the deformation [38–40].

The creep was also characterized experimentally through the compression creep test according to the specification LNEC E-399 [41], using prismatic specimens 100 mm × 100 mm × 400 mm. The specimens were loaded at 28 days and subjected, throughout the test period, to curing in a thermo–hygrometric chamber set at temperature T = 20 °C (±2 °C) and relative humidity RH = 50% (±5%). The creep is a time-dependent property of concrete, characterized by increasing strains over time at a constant stress loading level. This increase is a result of the viscoelastic behavior of the material itself. This property is essentially affected by the hydrated binder matrix and by the aggregates, which mainly have a restriction role [42,43]. Creep is greatly affected by the Young's modulus of the aggregates used in concrete production. Aggregates with higher Young's modulus provide greater opposition to creep deformation of the binder matrix [44]. The creep coefficient of UHPC is usually lower than that of ordinary concrete. Cement type, its dosage, and water/binder ratio significantly affect concrete creep [45], but this relation is not linear, as creep depends on the amount of hydrated binder in the matrix. Therefore, the evolution of the creep coefficient over time, $\varphi(t)$, was evaluated only for SK7 (mixture with pozzolan addition of Cape Verde and W/B = 0.21) and SK4 (mixture with fly ash addition and W/B = 0.21) to determine which of the two additions is the best option to replace the cement, considering the creep.

2.5. Durability Tests

The selected eco-UHDC mixtures were submitted to laboratory tests, aiming to evaluate the performance relative to durability: resistance to carbonation, resistance to chloride penetration, chloride content on the concrete surface, water absorption by capillarity, and resistivity. These tests were performed as described in the scheme shown in Figure 3.

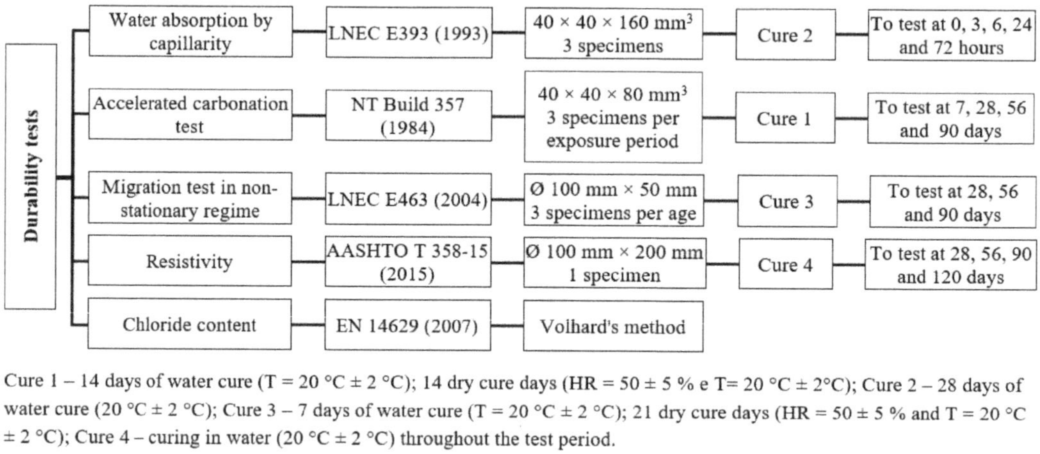

Figure 3. The durability tests performed.

The determination of water absorption by capillarity was determined only in SK4 (eco-UHDC with fly ash) and SK7 (eco-UHDC with pozzolan from Cape Verde) to analyze the influence of the two additions on the water absorption mechanism by capillarity. The eco-UHDC reinforced with steel fibers and polymeric fibers were only characterized regarding resistance to carbonation, resistance to chloride penetration, and resistivity, aiming to evaluate the influence of the incorporation of the fibers in these parameters. It was not considered relevant to study the absorption of water by capillarity because the matrix of these concretes is the same as that of SK4 and, according to Neves and Gonçalves [46], the fibers have a negligible influence on the absorption of water by capillarity.

The chloride penetration resistance and the resistivity test were also carried out, both in eco-UHDC (SK4) with and without fibers, as well as in SK3, SK8, and SK7 mixtures, aiming to evaluate the influence of increasing the dosage of fly ash, decreasing of the water/binder ratio, and the replacement of fly ash by pozzolan from Cape Verde in these performance indicators.

The chloride ion content on the concrete surface (Cs) was determined to assess the risk of corrosion of structures produced with eco-UHDC, more specifically with SK7 and SK4 mixtures. This was performed using the second halves of the non-steady state chloride migration test specimens that were not sprayed with the silver nitrate solution. A portion parallel to the edge of the surface that was exposed to the chloride solution, approximately 5 mm deep or less, was removed from these concrete specimens; this value depended on the results of the depth of penetration of chloride ions.

In addition to the durability tests in a laboratory environment, durability tests were also carried out in real environmental conditions, namely in the tidal zone of Cape Verde coast between July 2018 and January 2021. For this purpose, the following eco-UHDC series SK4 were used: SK4 without fibers, SK4s2 with steel fibers, and SK4p2 with polymeric fibers. For each concrete, two cubic specimens with 100 mm edges were produced. These specimens were subjected to wet curing for 28 days and were subsequently placed in the tidal zone. Chloride ion penetration was measured in all directions since none of the faces were waterproofed. The specimens, after exposure, were divided in half, and each half was used to perform a specific test: one to determine the penetration depth of chlorides and the other to determine the chloride content.

3. Results and Discussion

3.1. Preliminary Mixtures

3.1.1. Properties in Fresh State

The results of the mini-slump flow test and air content in the eco-UHDC mixtures are shown in Table 4. The eco-UHDC had a spread between 300 and 350 mm, presenting good workability and high flowability. The mixtures with higher cement dosage have greater workability and fluidity. Quartz flour and fly ash were the additions that provided better flowability due to the spherical shape of their particles. On the other hand, it was found that to meet the workability requirements in the mixtures with the remaining additions, it was necessary to increase the dosage of superplasticizer, the mixture with silica fume being the one that required the greatest amount of superplasticizer (approximately 25 kg/m^3, which is about 92% higher than the mixture with fly ash).

Table 4. Flow table and air content test results of eco-UHDC and eco-UHDFRC.

		Air Content (%)	Slump Flow Spread (cm)
Eco-UHDC	SK1	1	34
	SK2	0.9	35
	SK3	1.1	31
	SK4	1.5	33
	SK5	0.6	34
	SK6	1.7	30
	SK7	1.5	32
	SK8	1.2	35
Eco-UHDFRC	SK4s1	1	33
	SK4s2	1.2	32
	SK4b1	1.5	30
	SK4b2	1.9	24
	SK4p1	2.8	29
	SK4p2	3.1	25
	SK4g1	1.4	28
	SK4g2	2	26
	SK4gp	1.5	27

Basalt and glass fibers are ultra-fine and have a high density relative to the number of fibers in the matrix, which impairs workability and the release of air in the matrix. Those fibers provide higher stability to the mixtures but decrease the flowability and, consequently, the air content can increase up to circa 3%. The polymeric fibers have larger dimensions and allow an addition of greater fiber volume without significant loss of workability.

3.1.2. Compressive and Tensile Strength of Eco-UHDC

The average results of the compressive (f_{cm}) and tensile strength (f_{ctm}) of eco-UHDC obtained through the test of three identical specimens for each age are shown in Figures 4 and 5.

Analyzing the graphs in Figures 4 and 5, it is possible to verify that all eco-UHDC matrices show an evolution, with age, of the compressive and flexural strengths. The eco-UHDC mixtures with 660 kg/m^3 of cement present average values of compressive strength at 56 days between 92 and 116 MPa, and the tensile strength is between 12 and 16 MPa. The compressive and flexural strengths of SK1 and SK2 were not determined. The presented results demonstrate the potential for optimizing the production of eco-UHDC with a lower dosage of cement because the mixtures with only 60% cement present similar performances to the mixtures with 75% of cement. The literature focuses this point, as the UHPC must have a reduced W/B ratio (between 0.15 and 0.25), but with the reduction of cement dosage, compared to traditional formulations, the W/C ratio can have a higher level without impairing the performance of the concrete. The mixtures that obtained the highest strengths and simultaneously had the lowest amount of cement are SK3 (mixtures with 60% cement, 25% limestone filler and 15% fly ash and W/B = 0.18), SK4 and SK6 (mixtures with W/B = 0.21, 60% cement, 25% limestone filler, and the remaining 15% fly ash, and silica fume, respectively in SK4 and SK6), and SK8 (mixtures with 60% cement, 15% limestone filler, 25% fly ash and W/B = 0.21). It is quite evident the strength evolves with age due to the pozzolanic effect in several mixtures.

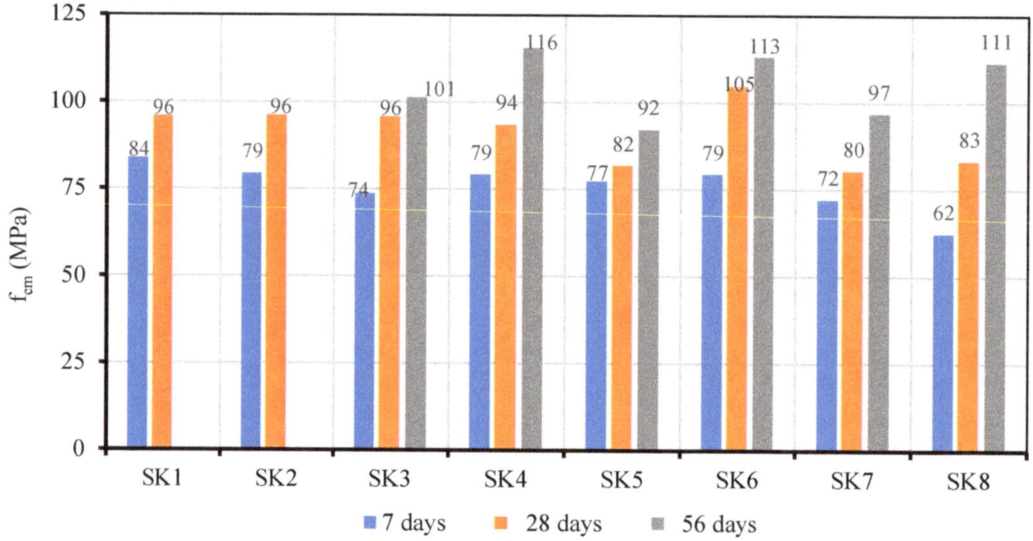

Figure 4. Average compressive strengths of eco-UHDC at 7, 28, and 56 days.

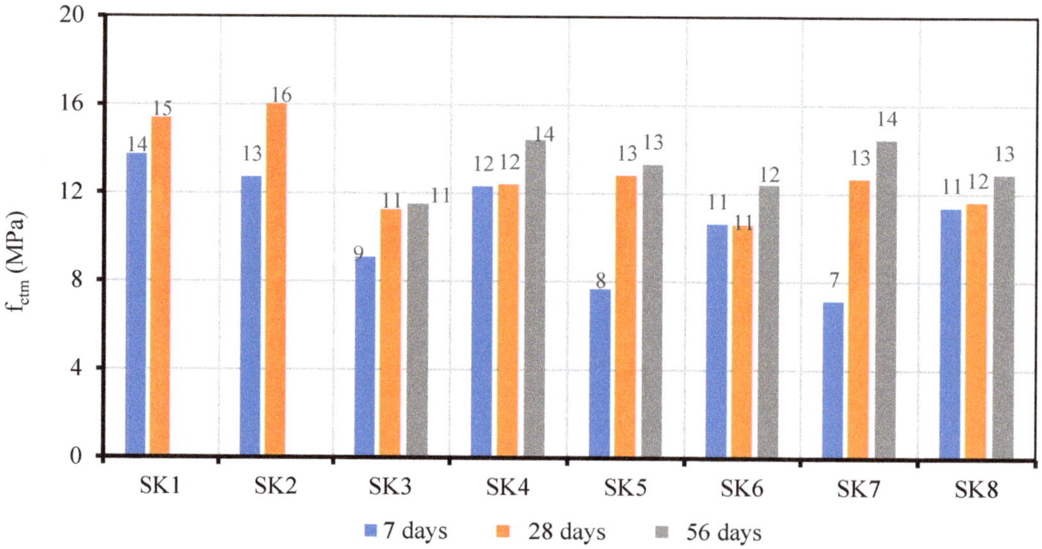

Figure 5. Average flexural strengths of eco-UHDC at 7, 28, and 56 days.

3.1.3. Compressive and Tensile Strength of Eco-UHDFRC

Figures 6 and 7 present the average compressive (f_{cm}) and tensile strength (f_{ctm}) of eco-UHDFRC for each of the ages.

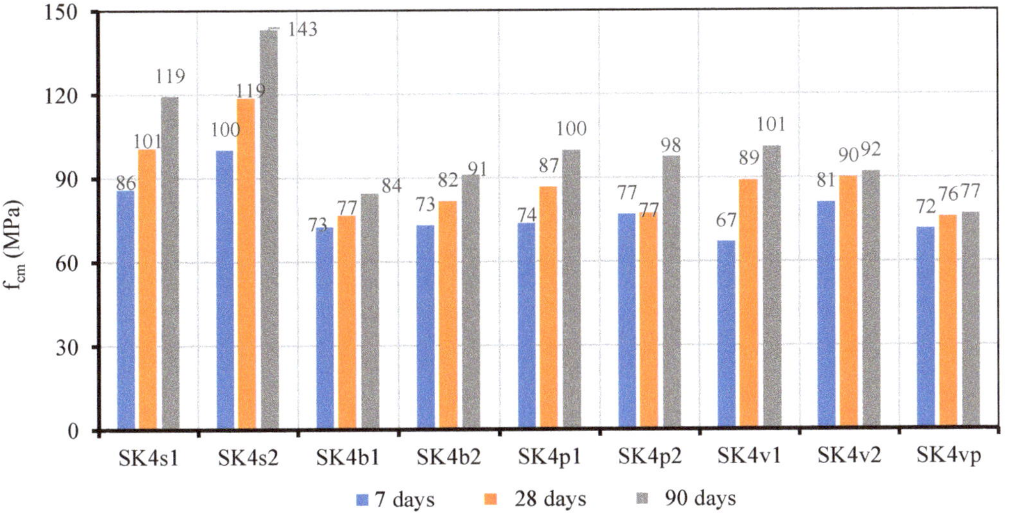

Figure 6. Average values of eco-UHDFRC compressive strength at 7, 28, and 90 days.

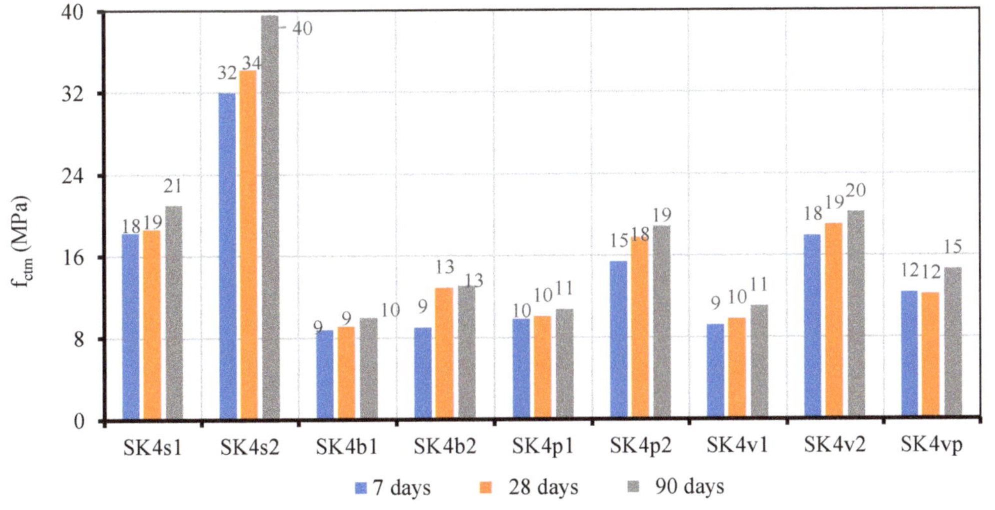

Figure 7. Average values of eco-UHDFRC tensile strength at 7, 28, and 90 days.

The results presented in Figures 6 and 7 demonstrate the high-performance potential of steel fibers compared to basalt, glass, and polymeric fibers. The eco-UHDC with steel fibers present compressive strength at 28 days of age 31 to 45% higher than those of eco-UHDC with basalt fibers, and between 12 and 32% higher than those of eco-UHDC with glass fibers, these differences being dependent on the fiber addition rate (1 and 2%, respectively). Concerning tensile strength, the eco-UHDC with steel fibers at 28 days present higher values compared to eco-UHDC with basalt fibers, around 104% in mixtures with 1% fiber and 166% in mixtures with 2% fiber, and in relation to eco-UHDC with glass fibers, the difference is 89% in the mixtures with 1% of fibers and 80% in the mixtures with 2% of fibers. The eco-UHDC with 2% of polymeric fibers at 28 days presents a compressive strength 14 and 27% lower than the eco-UHDC with 1% and 2% of steel fibers, respectively.

Regarding tensile strength, the eco-UHDC with 1% and 2% steel fibers present, respectively, values 85% and 240% higher than eco-UHDC with 2% polymer fibers. Compared to the eco-UHDC reference, SK4, steel fibers caused an increase of approximately 7 to 27% in compressive strength at 28 days and 50 to 176% in tensile strength, depending on the fiber addition rate.

Comparing the performance of eco-UHDC with basalt, glass, and polymeric reinforcement fibers, it is noted that, given the parameters analyzed, there are no significant improvements compared to the reference eco-UHDC, SK4, and there are even mixtures that show strength losses. Basalt and glass fibers have high strength but are ultra-fine fibers and have a high density toward the number of fibers in the matrix, which impairs workability and the air release in the matrix, resulting in less compact mixtures, less rigidity, and less compressive strength. The failure is brittle, not promoting any increase of ductility (Figure 8). In the case of polymeric fibers, larger dimensions allow the addition of a greater volume of fibers without a significant loss of workability; however, on the other hand, it promotes a reduction of the homogeneity of the matrix, with a consequent loss of compressive strength. The flexural strength of UHDC with polymeric fibers does not increase as much as with steel fiber reinforcement but provides good ductility (Figure 8).

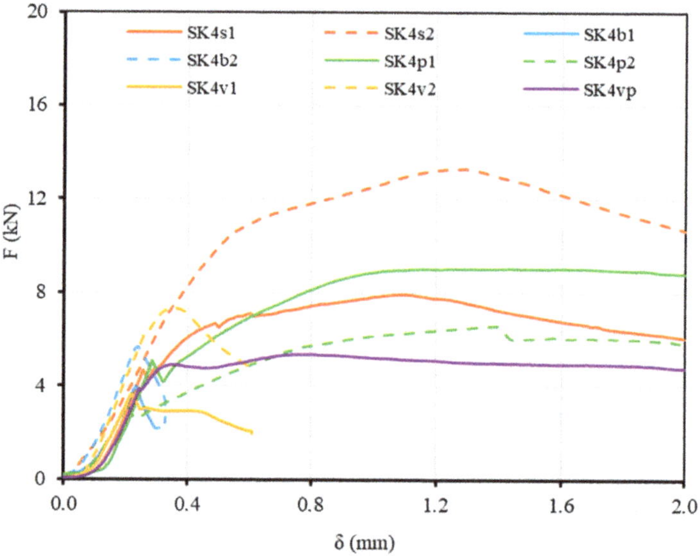

Figure 8. Load-deflection curves of the flexural test of eco-UHDFRC at 28 days.

3.2. *Optimized Eco-UHDC Mixtures*

Based on the mechanical properties of the results presented in the previous section, four eco-UHDC mixtures, SK3, SK4, SK7, and SK8, and two eco-UHDFRC mixtures, SK4s2 and SK4p2, were selected. These mixtures were characterized in more detail, namely in the remaining relevant mechanical properties (Young's modulus, flexural fracture energy), time-dependent properties (shrinkage and creep), and durability performance (capillary absorption; resistances to chloride and to carbonation; electrical resistivity).

SK3 was chosen to evaluate the reduction of the water/cement ratio in the main properties, remembering that this concrete differs from SK4 only in this ratio, and in small adjustments of the superplasticizer and of the volume of sand. The SK8 was chosen to evaluate the influence of the fly ash increase, differing from the SK4 only in the amount of fly ash, which is 10% higher. The SK7 mixture was chosen to evaluate the effect of Cape Verde pozzolan, compared to the SK4 concrete with fly ash. The SK4s2, eco-UHDC with 2% steel fibers, was selected because it presented better results compared to the other UHDFRC and presented significant improvements in mechanical performance compared to the reference matrix without fibers (SK4). Regarding tensile strength, the mixture with steel fibers reached a strength level that was almost triple the reference mixture at 28 days and with a 27% increase of the compressive strength. On the other hand, SK4p2, UHDC with 4% polymeric fibers, was selected due to the ductility showed after cracking (Figure 8).

3.2.1. Young's Modulus

The Young's modulus of concrete depends on the stiffness of its main components, i.e., the binder matrix, aggregates, and fibers. All the optimized mixtures were submitted to this test to evaluate the influence of the binder matrix variation, the water/cement ratio, and the type and dosage of additions (comparing SK4 with SK3, SK7, and SK8) on the Young's modulus. Comparing SK4, SK4s2, and SK4p2, it is possible to evaluate the influence of steel and polymeric fibers with the maximum dosages used on the mentioned property. Figure 9 shows the values obtained experimentally for the Young's modulus (E_{cm}) of the optimized eco-UHDC at 28 days of age.

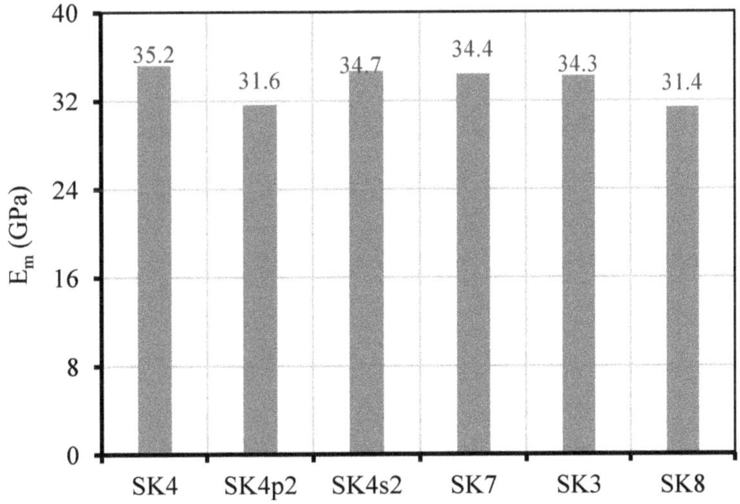

Figure 9. Young's modulus of eco-UHDC and eco-UHDFRC mixtures.

From the analysis of the above graph, it can be seen that: (1) The eco-UHDC with fibers (Sk4p2 and Sk4s2) had a lower Young's modulus; comparatively to the reference mixture, SK4, the difference is negligible in the mixture with steel fibers due to the small rise of air content. The biggest difference occurs with the eco-UHDFRC with polymeric fibers, Sk4p2, about 4 GPa less than the reference eco-UHDC, and the cause is the reduced stiffness of these fibers and the increase of air content (Table 4), being required to increase the superplasticizer dosage. (2) The SK3 mixture has a lower Young's modulus than the SK4 mixture, similarly to what happens with the compressive strength, contrary to the initial expectation, since SK3 has greater compactness. This result is probably due to the fact that this mixture does not have enough water to hydrate the cement and/or is due to the high dosage of superplasticizer used to obtain a fluidity similar to that of SK4. (3) The replacement of fly ash per pozzolan from Cape Verde, resulting in the mixture designated SK7, implied only a 2% decrease in the Young's modulus, leading to the conclusion that the two additions have approximately the same effect on the eco-UHDC mixture concerning Young's modulus, (4) The 10% increase of fly ash and the consequent reduction of limestone filler resulted in a 10% decrease of the Young's modulus (comparing SK4 and SK8), similarly to what happened with the compressive strength. This is probably because the limestone filler greatly contributes to the hardening reaction with the fly ash, accelerating the process. Therefore, the increase of fly ash and the decrease of filler may have implied the loss of this beneficial effect since the fly ash only reacts much later. For more advanced ages, the difference tends to decrease, as can be seen in the compressive strength evolution (Figure 4).

3.2.2. Flexural Fracture Energy

The flexural fracture energy of the selected eco-UHDFRC was experimentally determined, and the force-displacement relation is presented in Figure 10. As expected, the cracks started in the notch and developed a multi-stage, elastic, post-crack, and post-peak hardening behavior with significant high deformation capacity.

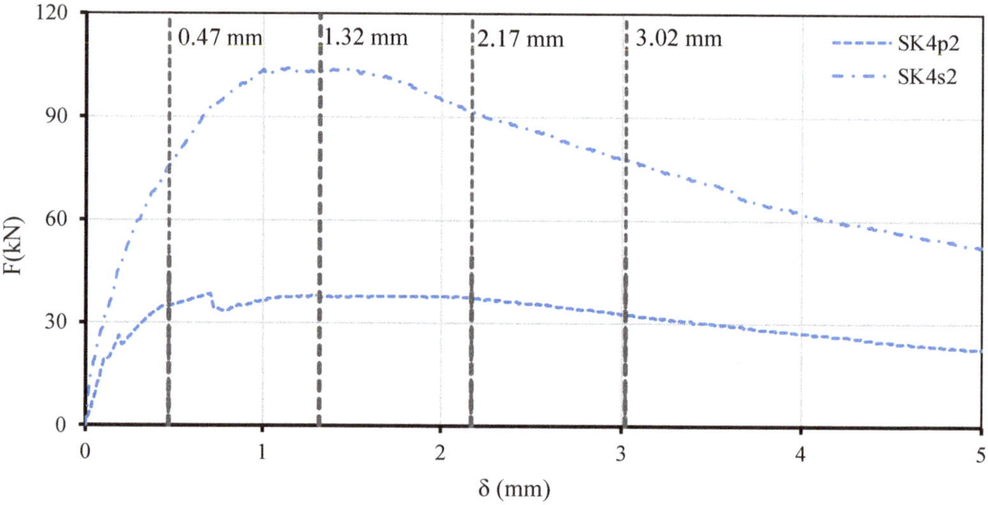

Figure 10. Load-deflection curves of the eco-UHDFRC residual flexural tensile strength test.

The residual flexural tensile strength that characterizes the behavior of the material up to a certain deflection (f_{Ri}), tensile stress under service conditions, and tensile stress used for checking ultimate limit states of structural elements produced with fiber concrete, f_{Fts} and f_{Ftu}, were determined from Equations (1)–(3), respectively. The results are presented in Table 5.

$$f_{Ri} = \frac{3}{2} \times F_i \times 10^3 \times \frac{L_R}{b_R \times {h_R}^2} \quad (1)$$

$$f_{Fts} = 0.45 f_{R1} \quad (2)$$

$$f_{Ftu} = f_{Fts} - \frac{w_u}{CMOD_3}(f_{Fts} - 0.5 f_{R3} + 0.2 f_{R1}) \geq 0 \quad (3)$$

where F_i—force; L_R—distance between supports; b_R—width of specimen; h_R—distance between top of the notch and top of the cross section. The force F_i is related to the deflection at mid-span, as follows: F1—force corresponding to a deflection of 0.47 mm; F2—force corresponding to a deflection of 1.32 mm; F3—force corresponding to a deflection of 2.17 mm; F4—force corresponding to a deflection of 3.02 mm; $CMOD_3$—crack mouth opening displacement corresponding to a deflection of 2.7 mm; and w_u equal to δ_3.

Table 5. Residual flexural tensile strength of eco-UHDFRC mixtures.

	Residual Flexural Tensile Strength (MPa)					
Mixtures	f_{R1}	f_{R2}	f_{R3}	f_{R4}	f_{Fts}	f_{Ftu}
SK4s2	24	33	29	25	10.9	9.7
SK4p2	11	12	12	11	5	4

3.3. Shrinkage and Creep

3.3.1. Shrinkage

The total shrinkage is the dimensional variation mainly caused by the combined effects of the drying shrinkage and the autogenous shrinkage. In UHPC, the autogenous shrinkage is the main component because this concrete has a very low water/binder ratio with extremely fine additions and because, generally, they do not include coarse aggregates [39,47,48].

The evolution of the shrinkage, ε_{cs}, measured in specimens produced with the select eco-UHDC is shown in Figure 11. Analyzing the graph in Figure 11, it is noted that, at very young ages, all mixtures have similar shrinkage. However, after 14 days, the differences became significant. The SK7, a mixture with pozzolan from Cape Verde, stands out for having the highest shrinkage with a difference approximately 15% higher than SK4, at 364 days. The SK8, with only 10% more fly ash than SK4, has the opposite behavior, presenting the lowest shrinkage because, as expected, the fly ash promotes the reduction of shrinkage. SK3, which differs from SK4 only in the lower W/B ratio, has a lower shrinkage than SK4 in certain phases, but then tends to stabilize and approach the values of SK4. A similar trend was observed by Eppers and Müller ([49] when the binder has a higher amount of C3A.

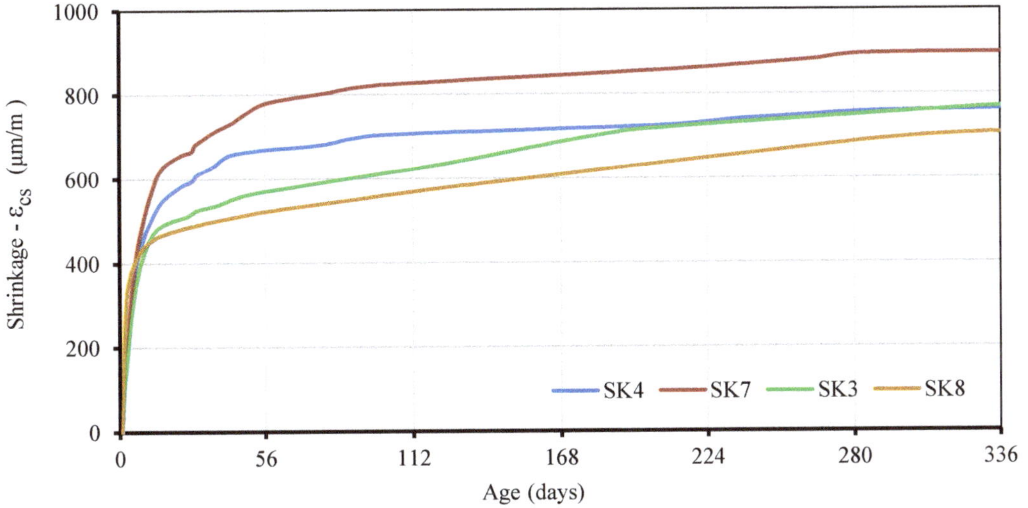

Figure 11. Evolution of total shrinkage of eco-UHDC with age.

Based on the results, it can be concluded that the partial replacement of cement by fly ash reduces the total shrinkage of the concrete, contrary to what happens with the replacement by pozzolan from Cape Verde. However, the results obtained in all characterized eco-UHDC are in line with the expected, since, according to Russel and Graybeal [50], the total shrinkage in UHPC is usually greater than 900 µm/m, depending on the curing method. Eppers and Müller [49] also reported autogenous shrinkage values of 600 to 900 µm/m at 28 days.

3.3.2. Creep

The obtained results for the creep characterization are presented through the evolution over time of the creep coefficient, $\varphi_c(t)$ (Figure 12), having the load applied at 28 days of age to assure a proper maturity of concrete. The $\varphi_c(t)$ is defined as the quotient between creep strain and elastic strain.

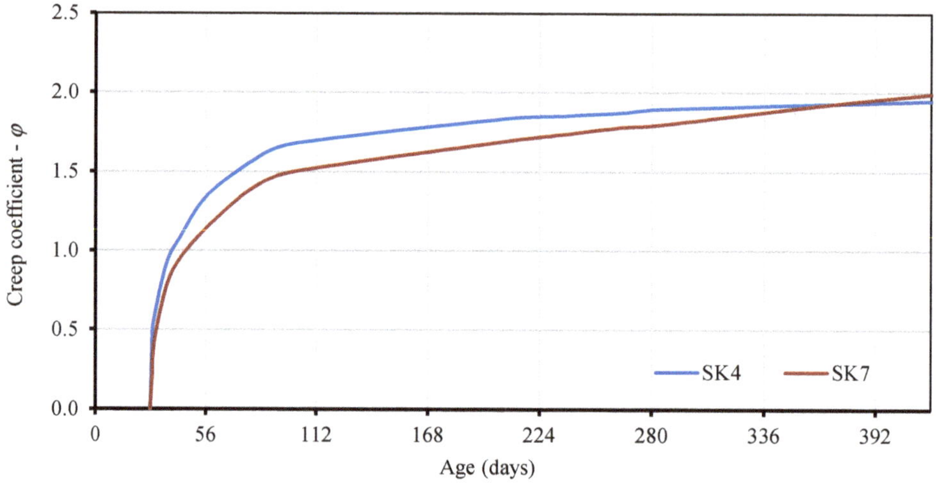

Figure 12. Evolution with age of creep coefficient: eco-UHDC with Cape Verde pozzolan (SK7) and eco-UHDC with fly ash (SK4).

Through the analysis of the curves presented in Figure 12, it can be concluded that the eco-UHDC with pozzolan from Cape Verde (SK7) initially has a creep coefficient about 13% lower than the eco-UHDC with fly ash (SK4). However, after one year, both mixtures show a similar creep coefficient and it is less than 2. It is also noted that the creep coefficient of SK4 seems to be stabilizing while the creep coefficient of SK7 appears to still evolve, which leads to the conclusion that pozzolan from Cape Verde tends to have higher creep coefficients than fly ash. The different maturity evolution of mixtures with these two pozzolanic additions is probably the reason for this effect. The reactivity of the Cape Verde pozzolan seems to be higher at younger ages and the fly ash has more pronounced evolution with age.

3.4. Durability—Exposed to Laboratory Environment

3.4.1. Carbonatation Resistance

The accelerated carbonation tests were carried out on SK4, SK7, Sk4p2, and SK4s2, and the average carbonation depth, C_d, is presented in Table 6. It is verified that, after 7 and 28 days of exposure, the carbonation in the eco-UHDC was null. The beginning of carbonation in the eco-UHDC started only at 56 days, being almost imperceptible. The two mixtures, SK4 and SK7, with fly ash and Cape Verde pozzolan, respectively, behaved similarly in all exposure periods.

Table 6. Carbonation depth in the eco-UHDC for various exposure periods.

Mixtures	$C_{7\ Days}$ (mm)	$C_{28\ Days}$ (mm)	$C_{56\ Days}$ (mm)	$C_{90\ Days}$ (mm)
SK4	0	0	2	2
SK7	0	0	2	2
SK4s2	0	0	0	2
SK4p2	0	0	0	0

The incorporation of fibers indirectly improved the carbonation resistance of the eco-UHDC matrix, with greater emphasis of the polymeric fibers since the concrete with this type of fibers had no carbonation in any of the analyzed periods. Even in the mixture with steel fibers (SK4s2), the carbonation depth was only observed after 90 days of exposure. The concrete carbonation resistance mainly depends on the porosity of the exposed surface, as well as on the distribution of the connected pores, and this porosity is related to the cementitious matrix and its compactness. In general, this type of concrete has very low porosity. However, the addition of fibers in this type of matrices, in addition to the characteristics provided in the hardened state, promotes a more efficient mixing process. This is due to the combined action of the fibers with the paste viscosity, resulting in macro and micro mixing shear forces that make the fresh mixture more flowable, releasing more air content. For this reason, fiber-reinforced UHDC mixtures commonly exhibit slightly lower porosity than unreinforced UHDC mixtures, being more stable and better controlling the exudation of water in the fresh state. In the latter, the water exuded from the mixture can originate connected pores that slightly promote the penetration of aggressive agents, whereas, in the former, the pores are less connected. Moreover, the addition of fibers provides a reduction of micro-cracks caused by drying shrinkage.

3.4.2. Carbonatation Resistance

The results of capillary absorption with time t_i, $Sa(t_i)$, resulting from the average of three specimens for each eco-UHDC tested, are shown in Figure 13. The height of capillary in each specimen, $h(t_i)$, was obtained through the arithmetic mean of the four faces (Figure 13).

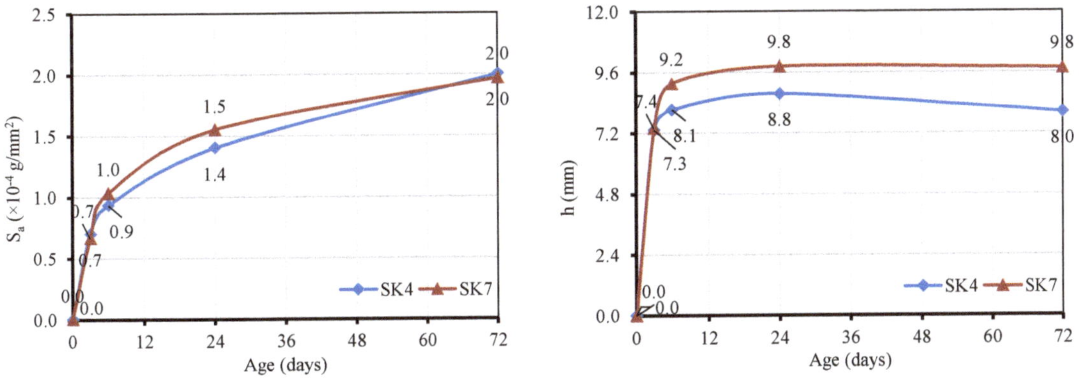

Figure 13. Results of capillary water absorption: capillary absorption, Sa (**left**); capillary ascension height (**right**).

The two eco-UHDCs clearly show a more pronounced capillary rise in the early stages, with a continuous increase up to 72 h. The eco-UHDC with pozzolan from Cape Verde presents, after 3 h, a capillary absorption slope slightly higher than that of the eco-UHDC with fly ash; however, after 72 h, the two concrete mixtures present similar results.

Figure 14 shows the relation between the capillary absorption, Sa, and the square root of time, obtained for each eco-UHDC as well as the trend line corresponding to the best fit of the points.

The trend lines of the two eco-UHDCs show correlation coefficients greater than 0.90. The slope of these lines represents the capillary absorption coefficient, and this coefficient is similar in the two concretes. This graph also shows the limit line corresponding to high-quality concrete, Sa = 0.1 mg/(mm^2·min$^{1/2}$), according to Browne [51]. This author proposed the following classification for the quality of concrete as a function of the Sa coefficient: above 0.2 mg/(mm^2·min$^{1/2}$) is "low-quality concrete"; between 0.1 and 0.2 mg/(mm^2·min$^{1/2}$) is "medium quality concrete"; and below 0.1 mg/(mm^2·min$^{1/2}$)

is "high-quality concrete". The two concretes can be classified as high-quality concretes. The results are far below the limit defined by the line Sa = 0.1 mg/(mm^2·min$^{1/2}$), proving through this durability indicator that both additions can be used as a partial substitute of cement without affecting this performance.

Capillarity water absorption (g/mm²)		
Time (min$^{1/2}$)	SK7	SK4
13.4	0.00007	0.00007
19.0	0.00010	0.00009
37.9	0.00015	0.00014
65.7	0.00020	0.00020

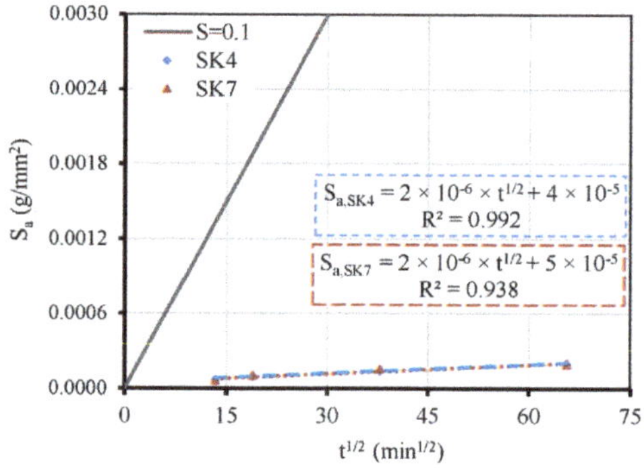

Figure 14. Capillarity water absorption in function of square root of time.

3.4.3. Chloride Migration Coefficient

The penetration depth of chlorides in specimens after the chloride migration test in a non-stationary regime was performed as shown in Figure 15.

Based on values of penetration depth, the non-steady-state chloride migration coefficient, D_{nssm}, was calculated using Equation (4).

$$\text{Dnssm} = \frac{0.0239\,(273+\text{T})\text{L}}{(\text{U}-2)\text{t}} \times \left(x_d - 0.0238\sqrt{\frac{(273+\text{T})\ \text{L}\ x_d}{\text{U}-2}} \right) \tag{4}$$

where U—absolute value of the applied voltage, V; T—average value of the initial and final temperatures in the anodic solution, °C; L—thickness of the specimen, mm; x_d—average value of the penetration depths, mm; and *t*-test—duration, hour.

Figure 16 shows the results of non-steady-state chloride migration coefficient, D_{nssm}, of the eco-UHDC mixtures obtained by the average of three specimens for each age. The same graphic shows the limits proposed by Luping [52] to classify concrete as a function of the chloride diffusion coefficient measured in the migration test. According to Luping [52], concrete is classified regarding the resistance to migration of chloride ions as: (i) very good strength concrete if D_{nssm} is less than 2×10^{-12} m^2/s; (ii) good resistance concrete if the D_{nssm} is between 2×10^{-12} m^2/s and 8×10^{-12} m^2/s; (iii) moderate resistance concrete if D_{nssm} is between 8×10^{-12} m^2/s and 16×10^{-12} m^2/s; and (iv) concrete not suitable for aggressive environments if D_{nssm} value is greater than 16×10^{-12} m^2/s. Therefore, the lower the D_{nssm}, the greater the resistance to chloride migration and, consequently, the greater the durability. All eco-UHDC mixtures correspond to good resistance concrete except eco-UHDC with Cape Verde pozzolan (SK7), which proves to be a very good resistance concrete at all ages. Figures 15 and 16 prove that mixtures improve resistance to chloride migration with age due to the continuous pozzolanic effect.

The mixture with pozzolan from Cape Verde (SK7) presents very low D_{nssm} results compared to the other mixtures, with differences ranging between 27% and 56%. Comparing the SK4 and SK7, which differ only in the addition type, the D_{nssm} results of SK7 are much lower, with differences of 56% at 28 and 56 days of age and 27% at 90 days (Figure 17a). This fact proves that the pozzolanic activity of fly ash is lower and slower than that of pozzolan from Cape Verde.

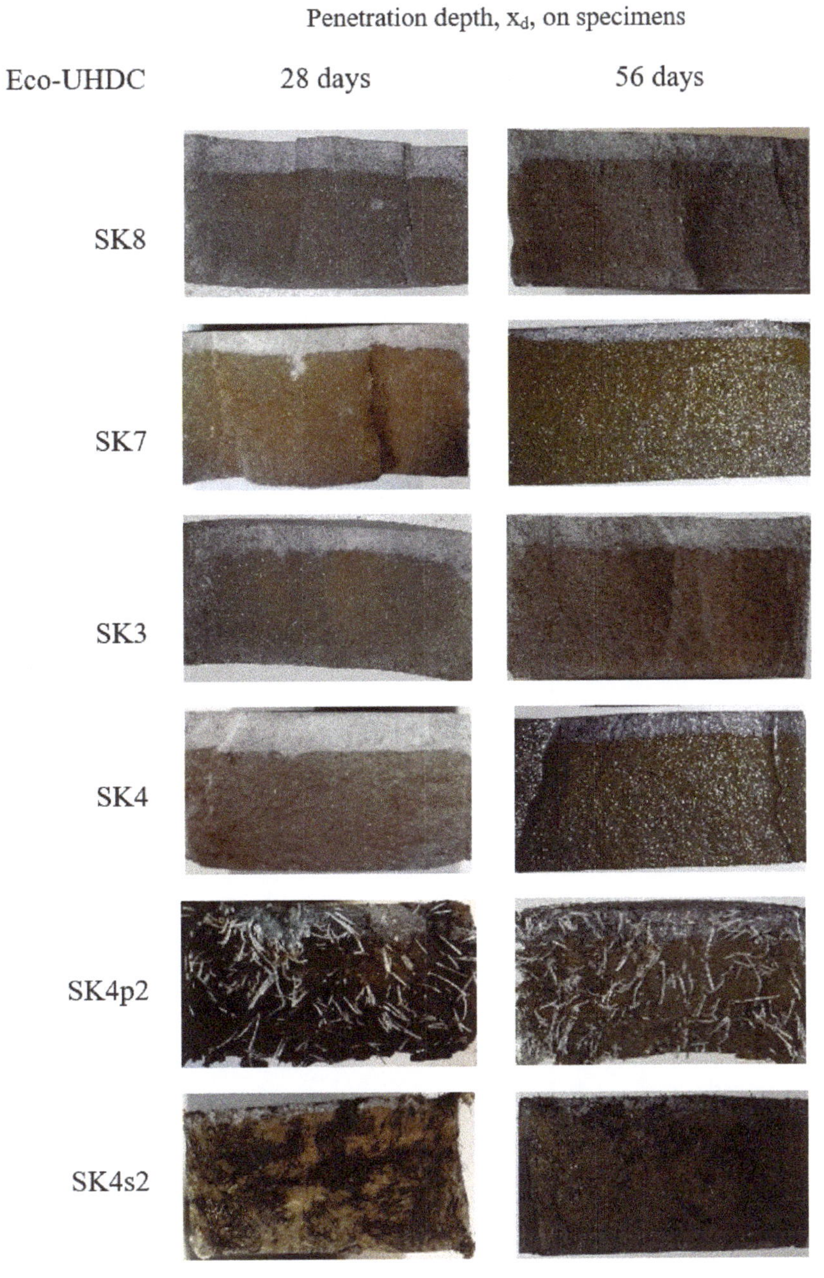

Figure 15. Depth of chloride penetration.

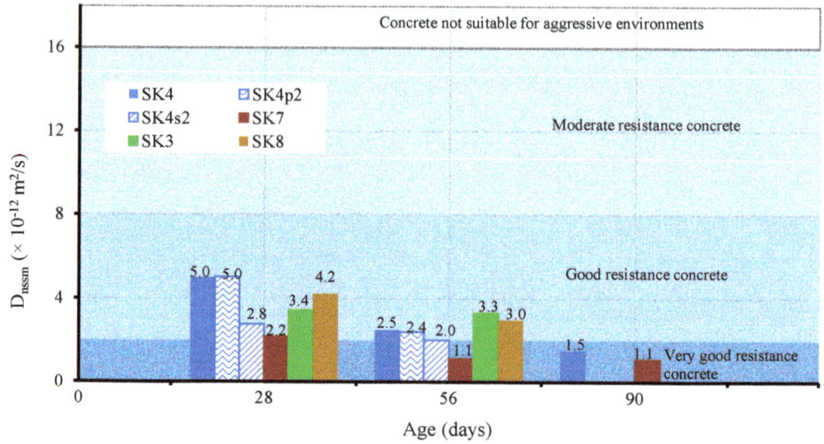

Figure 16. Non-steady state chloride migration coefficient, D_{nssm}, of the eco-UHDC mixtures at 28, 56, and 90 days.

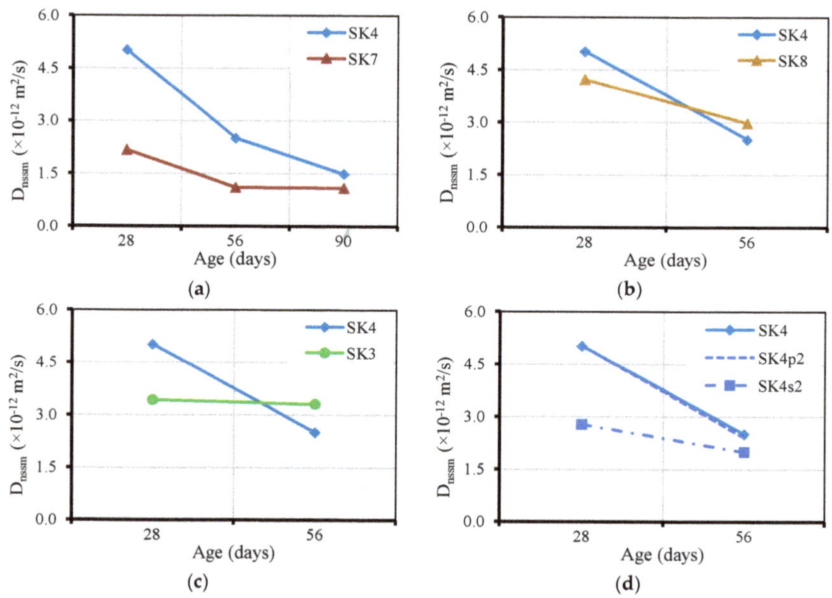

Figure 17. Variation of non-steady-state chloride migration coefficient with the age of eco-UHDC: (**a**) influence of Cape Verde pozzolan; (**b**) influence of the fly ash content, (**c**) influence of the W/B ratio, and (**d**) influence of the incorporation of fibers in the matrix.

Analyzing the influence of increasing the fly ash dosage, comparing the performance of SK4 and SK8 mixtures (Figure 17b), it is observed that the mixture with the highest dosage of fly ash has lower D_{nssm} values at 28 days. However, at 56 days of age, the trend is reversed. This trend was not expected since the diffusion coefficient tends to decrease with the increase in the percentage of addition of fly ash and with the increase of the concrete age [53]. Real et al. [54] found that the diffusion coefficient increases with an increasing percentage of fly ash, but only at young ages. These results were justified by the short period of wet curing and by the young age of the concrete when the tests were carried out (28 days), considering that it is a binder matrix with pozzolanic additions. The same authors recommended that chloride penetration resistance tests should be carried out for more advanced ages in concretes with additions whose pozzolanic reactions develop more slowly to correctly evaluate the real contribution of this type of addition. Several researchers report this fact [55,56], at least up to a certain percentage of replacement. According to some authors [57–59], there is an optimal content of fly ash that should be used to replace cement. This optimum content is related to the amount of C-H available in the mixture to satisfy the pozzolanic reaction developed by the fly ash. The diffusion coefficient for a given W/B ratio increases with the percentage of cement replacement by fly ash; however, the contribution of fly ash, namely in the refinement of the microstructure, is only relevant when the development of pozzolanic reactions is effective [58].

Comparing SK4 with SK3 (concrete with a lower water/binder ratio compared to SK4), observing Figure 17c, it can also be seen that, at 56 days of age, SK4 has a D_{nssm} 20% lower than SK3. Normally, the diffusion coefficient tends to increase in less compact concretes, regardless of the type and dosage of the binder [58]. However, the observed trend may be explained by the higher water/binder ratios, which facilitate faster cement hydration, producing more C-H and providing a faster activation of fly ash.

The mixture with polymeric fibers (SK4p2) does not show significant improvement compared to the reference eco-UHDC (SK4), unlike the mixture with steel fibers (SK4s2) which has lower D_{nssm} values of about 40% at 28 days and 20% at 56 days, compared to SK4 at the same age (Figure 17d).

3.4.4. Electrical resistivity

The resistance to chloride penetration was also evaluated through the electrical resistivity measured on the surface of the eco-UHDC to indirectly assess the ability of chloride ions to diffuse inside the concrete. Figure 18 shows the electrical resistivity results of the eco-UHDC mixtures, using the average of four readings per specimen and per age. As expected, the electrical resistivity increases with age due to the continuous hydration of cementitious materials, mainly by the pozzolanic reaction, producing a finer pore network with less connectivity.

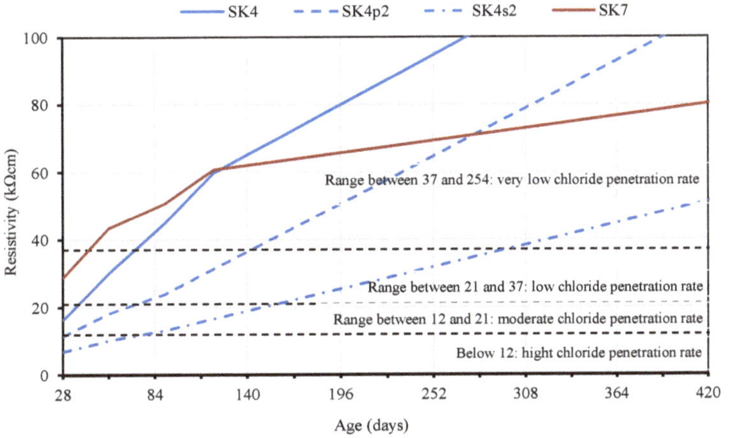

Figure 18. Variation of electrical resistivity of eco UHDC with age.

Comparing the electrical resistivity results of the developed eco-UHDCs, with the reference values proposed by AASHTO T358-15 [60], presented in Figure 18, it appears that the eco-UHDCs have a very low chloride penetration rate, but only after a long curing period (about 90 days for fiber-free concrete, 150 days for polymer fiber concrete, and 250 days for steel fiber concrete). This late effect is certainly due to the pozzolanic effect in the development of new C-S-H compounds.

The average electrical resistivity of SK7 (UHDC with Cape Verde pozzolan) is higher than that of SK4 (UHDC with fly ash) up to 90 days. From this age onward, there was a contrary trend, confirming the trend recorded in the study with LCC-low carbon concrete [61]. Comparing these values to those obtained in the chloride migration test in a non-stationary regime, this trend was not registered at the corresponding ages, although the graph presented in Figure 17a shows a convergence of results of the two concretes. This occurred probably because the prolonged curing period of the specimens submitted to the resistivity test allows the fly ash to have the necessary time to develop the pozzolanic reactions and, thus, improve the concrete performance.

The results obtained also show that the eco-UHDC with fibers, mainly with steel fibers, present a higher chloride penetration rate than the reference eco-UHDC, contradicting the results obtained in the chloride migration test in a non-stationary regime. These results corroborate what was reported in previous studies, that is, the presence of steel fibers influences the electric field generated by the resistivity device [62,63]. Therefore, when interpreting the results, it is necessary to bear in mind that it is not appropriate to directly compare the results of concrete with and without steel fibers.

3.4.5. Chloride Content at the Eco-UHDC Surface

Figure 19 shows the mean values of chloride content, Cs, per mass of cement and addition type II at 28, 56, and 90 days of age. These values were measured, initially, in concrete mass using Equation (5) and later calculated relatively to the mass of cement and type II addition using Equation (6) to be compared with the critical value of EN 206-1 [64]. One must consider that there is, in 100 g of SK4 mass, 35.8 g of cement plus fly ash, and in the case of SK7, 36.1 g of cement plus pozzolan from Cape Verde.

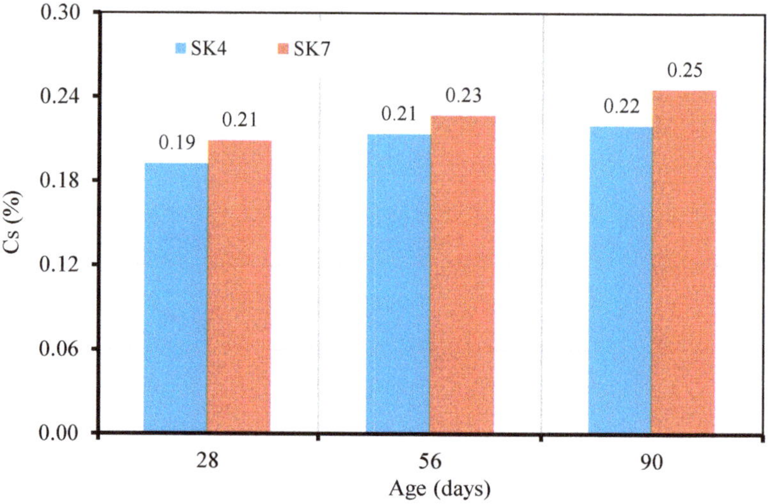

Figure 19. Chloride content at the eco-UHDC surface at 28, 56, and 90 days of age.

$$Cs = 3.545 \times f_{AgNO_3} \frac{V_{NH\,SCN}^{blanc} - V_{NHSCN}^{concrete}}{m_{concrete}} \quad (5)$$

f_{AgNO_3}—is molarity of silver nitrate solution;
$V_{NH\,SCN}^{blanc}$—volume of the ammonium thiocyanate solution used in the blank titration (mL);
$V_{NHSCN}^{concrete}$—volume of the ammonium thiocyanate solution used in the titration (mL);
$m_{concrete}$—mass of the concrete sample (g).

$$Cs_{m_{cement+addition\ type\ II}} = \frac{Cs}{m_{cement+type\ II\ addition}} \times 100 \quad (6)$$

The results show that, in general, the eco-UHDC with pozzolans from Cape Verde has a higher content of chlorides on the surface. This result was already expected since this concrete has a lower diffusion coefficient of chloride ions, thus demonstrating that pozzolan provides a greater ability to fix chloride ions in concrete compared to fly ash. So, this ability potentiates a higher concentration of Cs because this resistance creates a barrier to chloride diffusion. However, the two concrete mixtures present values that range from 38 to 52% lower than the critical chloride content (% CR), whose value is 0.4% according to EN 206-1 [64] Considering these values, it is concluded that the probability of corrosion occurring at the end of the periods of exposure to chlorides to which these concretes were subjected is very low (according to LNEC specification E 463 [65]).

3.5. Durability—Exposed to Maritime Environment

3.5.1. Chloride Content at the Eco-UHDC Surface

Figure 20 shows the average values of Cs per mass of cement and the type II addition of eco-UHDC exposed to the environmental conditions of Cape Verde in the tidal zone for circa 30 months.

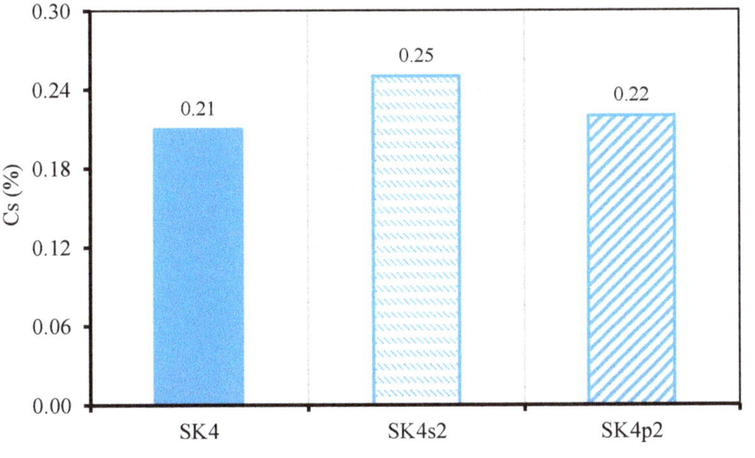

Figure 20. Chloride content per mass of cement on the surface of SK4, SK4s2, and SK4p2 concrete exposed for circa 30 months in the tidal zone.

The chloride content measured in SK4 and tested in a laboratory, when compared with the results of specimens exposed in the tidal zone, shows similar results. After circa 30 months of exposure to chlorides in the tidal zone, the results show that the probability of corrosion in the reinforcement of structures produced with the developed UHDC is almost nil, at least during the analyzed period since the Cs values are lower than the critical chloride content (% CR), which is, according to EN 206-1 [64], 0.4% (concrete with steel reinforcements or other embedded metals). The specimens with fibers present Cs values slightly higher than the reference concrete (SK4), approximately 18% higher in the SK4s2 (with steel fibers), and 6% higher in the SK4p2 (with polymeric fibers).

3.5.2. Depth of Chloride Ion Penetration

Figure 21 and Table 7 show the depth of chloride ion penetration obtained on each of the faces of the specimens. The depth of chloride ion penetration of each edge corresponds to the average of measurements at three points. There are some differences comparatively to the values measured in tests carried out following the LNEC E 463 [65]. The eco-UHDC without fibers (SK4) has the lowest depth of chloride ion penetration, followed by the concrete with steel fibers (SK4s2), and, finally, the concrete with polymeric fibers (SK4p2). These results contradict the results obtained in the tests under laboratory conditions, where the eco-UHDC with steel fibers was the concrete with the lowest depth of chloride penetration. These differences may be related to sea salinity that are different from the laboratory conditions and to the wet/dry cycles of the tidal zone that are not reproduced in the laboratory.

Figure 21. Depth of chloride ions penetration in eco-UHDC concrete exposed for circa 30 months in the tidal zone.

Table 7. Depth of chloride ions penetration, in mm, in UHDC concrete exposed for circa 30 months in the tidal zone.

Eco-UHDC	SK4	SK4s2	SK4p2
		(mm)	
Edge A	8	10	8
Edge B	7	7	15
Edge C	5	7	7
Edge D	4	7	5
Average	6	8	9

3.6. *Lifetime Expected and Minimum Cover Required*

The lifetime expected for a concrete structure is defined during the design phase and corresponds to the period during which the structure meets the safety, functionality, and aesthetic requirements without unforeseen maintenance costs [66]. These requirements are influenced not only by the mechanical and deferred properties of the concrete but also by the durability parameters. Most reinforced concrete structures are designed for a service lifetime of 50 or 100 years regarding the risk of rebar corrosion.

This section presents the expected service lifetime of reinforced concrete structures produced with the developed eco-UHDC concretes determined considering the resistances to carbonation and to the chlorides diffusion and the minimum cover defined in main codes. The estimated minimum concrete cover, $c_{min, dur}$, necessary to guarantee the proper corrosion resistance during the defined service lifetime for structures exposed to environmental conditions XC (corrosion induced by carbonation) and XS (induced by chloride ions present in seawater) are also presented. The results presented were obtained using the Tuutti model [67] for the degradation of reinforced concrete caused by steel corrosion, the recommendations described in the LNEC E465 [68], and considering two types of structures: (i) current structures with an intended service lifetime (t_g) equal to 50-years-old with a reliability class RC2 (service lifetime category or structural class S4), which corresponds to a safety factor γ equal to 2.3 [68]; (ii) special structures with a t_g equal to 100-years-old with a reliability class RC3 (lifetime category or structural class S5), which corresponds to a safety factor γ equal to 2.8 [68].

3.6.1. Prediction of the Service Lifetime of Reinforced Concrete Structures under the Environmental Exposure Classes XC

The prediction service lifetime for current and special structures produced with eco-UHDC, SK4, SK7, SK4s2, and SK4p2 exposed to XC environmental conditions are presented in Table 8. The developed concretes present resistance to carbonation, more than enough for the XC exposure class, and in all cases the estimated service life is over 400 years.

Table 8. Service lifetime (years) for current and special structures produced with the studied concrete under environmental conditions XC.

		Current Structures (Class S4, RC2, 50 Years)				Special Structures (Class S5, RC3, 100 Years)			
Exposure Class		**XC2**	**XC3**	**XC4 ***	**XC4 ****	**XC2**	**XC3**	**XC4 ***	**XC4 ****
Eco-UHDC	**$C_{min,dur}$ (mm) EC 2**	25		30		30		35	
SK4	tg (years)	>>400	>>400	>>400	>>400	>>400	>>400	>>400	>>400
SK7		>>400	>>400	>>400	>>400	>>400	>>400	>>400	>>400
SK4s2		>>400	>>400	>>400	>>400	>>400	>>400	>>400	>>400
SK4p2		>>400	>>400	>>400	>>400	>>400	>>400	>>400	>>400

XC4 *—dry region; and XC4 **—Wet region.

3.6.2. Determination of the Minimum Cover to Resist Carbonation-Induced Corrosion

The minimum cover required to guarantee the resistance against steel corrosion in reinforced concrete structures under the environmental exposure classes XC are presented in Table 9. These covers are much lower than the minimum values recommended by Eurocode 2 (EC2) [69] and EN 206-1 [64], with differences greater than 20 mm. In other words, the results showed that, for any structural class, the structures produced with eco-UHDC can have a cover less than 5 mm for any class of XC environmental exposure, considering only the criterion of durability. These results have a significant impact on the cost of the structures since it will be possible to produce elements with smaller cross-sections. It can also be seen that the eco-UHDC with fly ash (SK4) and the eco-UHDC with pozzolan from Cape Verde (SK7) have similar minimum covers. The introduction of fibers in eco-UHDC (SK4s2 and SK4p2) allows reducing the cover, the highest difference was obtained with the concrete with polymeric fibers (SK4p2); the minimum cover was reduced by half.

Table 9. Cover for S4 and S5 class structures under environmental conditions XC: minimum standards cover and minimum covers for the studied concrete.

			Current Structures (Class S4, RC2, 50 Years)				Special Structures (Class S5, RC3, 100 Years)			
Exposure Class			**XC2**	**XC3**	**XC4 ***	**XC4 ****	**XC2**	**XC3**	**XC4 ***	**XC4 ****
Minimum cover $C_{min,dur}$ (mm)		EC2	25	25	30		30	30	35	
	Eco-UHDC	SK4	1.3	1.9	2.6	2.9	1.7	3.0	4.0	4.2
		SK7	1.3	1.9	2.6	2.9	1.7	3.0	4.0	4.2
		SK4s2	1.2	1.7	2.4	2.6	1.6	2.7	3.6	3.8
		SK4p2	0.6	0.9	1.2	1.3	0.8	1.3	1.8	1.9

XC4 *– dry region; and XC4 **– Wet region.

3.6.3. Prediction of the Service Lifetime of Reinforced Concrete Structures under the Environmental Exposure Classes XS

The prediction service lifetime of current and special structures exposed to the XS environmental conditions, produced with the developed eco-UHDC, are presented in Table 10. The results show that, adopting the minimum cover recommended by Eurocode 2 (EC2) [69] and EN 206-1 [64], the service lifetime of structures produced with the eco-UHDC developed is adequate and significant when exposed to the environmental conditions XS1 and XS2. It is recalled that current structures and special structures must have a service lifetime of at least 50 and 100 years, respectively. These results may indicate some long-term economic and environmental benefits because the need for maintenance will occur later than expected. The SK7 mixture proves the higher resistance to chloride diffusion, fulfilling the required service life for all exposure classes (including XS3) and both structural classes.

Table 10. Service lifetime (years) for current and special structures produced with the developed concrete under environmental conditions XS.

			Current Structures (Class S4, RC2, 50 Years)				Special Structures (Class S5, RC3, 100 Years)			
Exposure Class			XS1	XS2		XS3	XS1	XS2		XS3
				1 m	1.4–25 m			1 m	1.4–25 m	
Eco-UHDC	SK3	t_g (years)	573	115	91	41	852	174	145	54
	SK4		246	77	66	18	365	127	112	23
	SK7		1581	211	158	113	2357	296	229	148
	SK8		362	91	75	26	539	145	124	34
	SK4s2		915	150	115	65	1361	218	175	86
	SK4p2		246	77	66	17	365	127	112	23

3.6.4. Minimum Cover to Resist Corrosion Induced by Chloride Ions Present in Seawater

The minimum cover required to guarantee the resistance against steel corrosion induced by chlorides in reinforced structures produced with the eco-UHDC are presented in Table 11. Comparing the minimum cover required in the structures produced with the eco-UHDC with the minimum covers recommended by Eurocode 2 (EC2) [69] it is possible to conclude that the developed eco-UHDCs provide good corrosion protection because the minimum cover required is lower than the values recommended by codes for exposure classes XS1 and XS2, with differences ranging from 12% to 55%, depending on the type concrete and exposure class. However, for current structures and exposure class XS3 (tidal zones), only the eco-UHDC with pozzolan from Cape Verde (SK7) and the eco-UHDC with steel fibers (SK4s2) proved to be suitable, since they are the only ones with a required cover below the minimum recommended by codes, with differences of 8 mm and 4 mm, respectively. It is concluded, once again, that the concrete with pozzolan from Cape Verde is more suitable to resist the penetration of chloride ions than the concrete with fly ash.

Table 11. Cover for S4 and S5 class structures under environmental conditions XS: minimum standards cover and minimum covers for the studied concrete.

			Current Structures (Class S4, RC2, 50 Years)				Special Structures (Class S5, RC3, 100 Years)			
Exposure Class [1]			XS1 [2]	XS2 [3]		XS3 [4]	XS1 [2]	XS2 [3]		XS3 [4]
				1 m	1.4–25 m			1 m	1.4–25 m	
	EC2		35	40		45	40	45		50
Minimum cover $C_{min,dur}$ (mm)	Eco-UHDC	SK3	20	23	26	47	25	29	33	58
		SK4	24	28	31	57	30	36	39	70
		SK7	16	18	20	37	20	23	26	46
		SK8	22	26	29	52	27	33	36	64
		SK4s2	18	21	23	42	22	26	29	52
		SK4p2	24	28	31	57	30	36	39	70

[1] In this analysis, regarding the distance to the coastline, the worst-case scenario was considered (structures are located next to the coastline); [2] Structures exposed to the air with sea salts; [3] Structures permanently submerged; [4] Structures in the tidal zone.

4. Conclusions

The UHDC matrix was optimized using a combination of filler material (limestone or quartz flour) and reactive pozzolanic additions as a partial replacement of Portland cement, aiming to improve its eco-efficiency. Various types of fibers (steel, polymeric, glass, and basalt) were also used to enhance the mechanical and durability performance. Based on the results experimentally obtained, the following conclusion are drawn:

(1) The optimization of the unreinforced UHDC matrix makes it possible to produce concrete with only 60% of cement in relation to the total binder, maintaining good workability and the desired mechanical characteristics (compressive strength higher than 100 MPa and flexural strength higher than 12 MPa at 56 days of age). The eco-UHDC matrix with fly ash develops greater compressive strength, up to 25%, compared to the eco-UHDC with other additions;

(2) The eco-UHDC with pozzolan from Cape Verde and fly ash show similar carbonation resistances and creep coefficients (lower than 2.0), but the shrinkage of the former is 15% higher than that of the latter at 364 days. It must also be highlighted that pozzolan from Cape Verde has a huge effect in reducing the diffusion coefficient of chloride ions, reaching 57% lower than that of the eco-UHDC with fly ash, but the difference decreases with the age of concrete due to the slower hardening provided by the fly ash;

(3) The shrinkage of all developed eco-UHDC varies between 700 and 900 μm/m after stabilization. This is an expected outcome since the total shrinkage of UHPC is normally higher than 900 μm/m and replacing cement with fly ash decreases the total shrinkage of concrete. The eco-UHDC with the replacement of cement by natural pozzolan from Cape Verde has an opposite trend;

(4) The addition of steel fibers and polymeric fibers provides greater ductility comparatively to the addition of glass and basalt fibers. Steel fibers obviously increase the flexural tensile strength, between 50 and 176%, depending on the fiber addition rate. The other types of fibers also have an influence on tensile strength, but their dosage must be equal to or higher than 2% volume. The glass and polymeric fibers have much more of a relevant effect than basalt fibers in the mechanical characterization;

(5) The introduction of fibers in eco-UHDC affects in different ways some mechanical properties. While the addition of steel fibers can increase up to 27% of the compressive strength, the basalt and glass fibers can decrease to circa 10% in both the compressive strength and the Young's modulus. These losses are due to the negative effect of those fibers on the workability and air release of the matrix, resulting in less compact mixtures and, thus, with higher air content, less stiffness, and less strength;

(6) The addition of steel and polymeric fibers also improves the durability performance of eco-UHDC since there is an increase in resistance to both carbonation and chloride

ions. Polymeric fibers are the most effective in terms of carbonation resistance since the UHDC with this type of fibers exhibited no carbonation in any of the analyzed periods. The UHDC reinforced with steel fibers is more suitable to be used in an environment under the action of chloride ions because this mixture reduces the migration coefficient, D_{nssm}, by 40% at 28 days and 20% at 56 days in comparison with the eco-UHDC used as reference (SK4);

(7) Considering only the requirements related to durability, the structures produced with the developed eco-UHDC, exposed to environmental conditions XC, XS1, and XS2, require a cover lower than the values recommended by EC2, with differences that can reach 55%. From all the developed eco-UHDC, the one with pozzolana from Cape Verde is the most suitable to be used in any structural class, submitted to any XC and XS exposure classes.

These results prove that it is possible to produce structural elements with eco-UHDC, with reduced cross-sections and high durability, and consequently, with economic and environmental benefits. The topic of concrete sustainability associated with the cost and energy consumption required to produce this type of concrete will be addressed in future research.

Author Contributions: K.R.: Conceptualization, methodology, formal analysis, writing—original draft preparation. H.C.: Conceptualization, methodology, validation, supervision, writing—review and editing. R.C.: Conceptualization, validation, supervision, formal analysis, writing—review and editing. E.J.: Conceptualization, supervision, writing—review and editing. All authors have read and agreed to the published version of the manuscript.

Funding: This research was funded by [Calouste Gulbenkian Foundation] grant number [SBG35-2016] and by [FCT] in the framework of project [UIDB/04625/2020].

Institutional Review Board Statement: Not applicable.

Informed Consent Statement: Not applicable.

Data Availability Statement: Data will be made available on request.

Acknowledgments: The authors acknowledge the Calouste Gulbenkian Foundation for the Doctoral Grant with the reference SBG35-2016; the ISEC—Polytechnic Institute of Coimbra for providing the facilities and all necessary resources to perform the study; and the companies Secil, Omya, Sarendur and BASF for their support. This work is part of the research activity carried out at Civil Engineering Research and Innovation for Sustainability (CERIS) and has been funded by Fundação para a Ciência e a Tecnologia (FCT) in the framework of project UIDB/04625/2020.

Conflicts of Interest: The authors declare no conflict of interest.

References

1. Zhang, X.; Wu, X.; Zhang, X.; Wang, L.; Tang, Y.; Qiu, F. Bond behaviors of pre- and post-yield deformed rebar embedded in ultra-high performance concrete. *Constr. Build. Mater.* **2022**, *341*, 127839.
2. Ferdosian, I. Ultra-high Durable Concrete: A Way towards Safe and Durable. In Proceedings of the 7th International Conference on Safety and Durability of Structures (ICOSADOS 2016), Vila Real, Portugal, 10–12 May 2016.
3. Ghafari, E.; Costa, H.; Júlio, E. Critical review on eco-efficient ultra high performance concrete enhanced with nano-materials. *Constr. Build. Mater.* **2015**, *101*, 201–208.
4. Ghafari, E.; Costa, H.; Júlio, E.; Portugal, A.; Durães, L. Enhanced durability of ultra high performance concrete by incorporing supplementary cementitious materials. In Proceedings of the Second International Conference on Microestructural-related Durability of Cementitious Composites, Amsterdam, The Netherlands, 11–13 April 2012; pp. 11–13.
5. Graybeal, B.; Hartmann, J.L. Strength and Durability of Ultra-High Performance Concrete Materials and Structures. In Proceedings of the 2003 Concrete Bridge Conference, Brisbane, Australia, 19–22 October 2003; pp. 1–20.
6. Hakeem, I.Y.; Althoey, F.; Hosen, A. Mechanical and durability performance of ultra-high-performance concrete incorporating SCMs. *Constr. Build. Mater.* **2022**, *359*, 129430. [CrossRef]
7. Li, J.; Wu, Z.; Shi, C.; Yuan, Q.; Zhang, Z. Durability of ultra-high performance concrete—A review. *Constr. Build. Mater.* **2020**, *255*, 119296.

8. Nasrin, S.; Ibrahim, A. Flexural response of Ultra-High-Performance Concrete (UHPC) hybrid bridge deck connections made with local materials. *Constr. Build. Mater.* **2021**, *270*, 121451. [CrossRef]
9. Shah, H.A.; Yuan, Q.; Photwichai, N. Use of materials to lower the cost of ultra-high-performance concrete—A review. *Constr. Build. Mater.* **2022**, *327*, 127045.
10. Yu, R.; Spiesz, P.; Brouwers, H.J.H. Development of an eco-friendly Ultra-High Performance Concrete (UHPC) with efficient cement and mineral admixtures uses. *Cem. Concr. Compos.* **2015**, *55*, 383–394.
11. Ghafari, E.; Costa, H.; Júlio, E. Statistical mixture design approach for eco-efficient UHPC. *Cem. Concr. Compos.* **2015**, *55*, 17–25.
12. De Larrard, F. Concrete optimisation with regard to packing density and rheology. In Proceedings of the 3rd RILEM International Symposium on Rheology of Cement Suspensions Such as Fresh Concrete, Reykjavik, Iceland, 19–21 August 2009.
13. Dingqiang, F.; Yu, R.; Kangning, L.; Junhui, T.; Zhonghe, S.; Chunfeng, W.; Shuo, W.; Zhenfeng, G.; Zhengdong, H.; Qiqi, S. Optimized design of steel fibres reinforced ultra-high performance concrete (UHPC) composites: Towards to dense structure and efficient fibre application. *Constr. Build. Mater.* **2021**, *273*, 121698.
14. Fennis, S.; Walraven, J.; Uijl, J. The use of particle packing models to design ecological concrete. *Heron* **2009**, *54*, 185–204.
15. Ferdosian, I.; Camões, A. Effective low-energy mixing procedure to develop high-fluidity cementitious pastes. *Matéria* **2016**, *21*, 11–17. [CrossRef]
16. Ferdosian, I.; Camões, A. Eco-efficient ultra-high performance concrete development by means of response surface methodology. *Cem. Concr. Compos.* **2017**, *84*, 146–156. [CrossRef]
17. Ghafari, E.; Costa, H.; Júlio, E. RSM-based model to predict the performance of self-compacting UHPC reinforced with hybrid steel micro-fibers. *Constr. Build. Mater.* **2014**, *66*, 375–383.
18. Sun, Z.; Yang, S.; Hang, M.; Wang, J.; Yang, T. Optimization design of ultrahigh-performance concrete based on interaction analysis of multiple factors. *Case Stud. Constr. Mater.* **2022**, *16*, e00858.
19. Yu, R.; Spiesz, P.; Brouwers, H.J.H. Mix design and properties assessment of Ultra-High Performance Fibre Reinforced Concrete (UHPFRC). *Cem. Concr. Res.* **2014**, *56*, 29–39.
20. Chan, Y.-W.; Chu, S.-H. Effect of silica fume on steel fiber bond characteristics in reactive powder concrete. *Cem. Concr. Res.* **2004**, *34*, 1167–1172. [CrossRef]
21. Droll, K. Influence of additions on ultra high performance concretes—Grain size optimisation. In Proceedings of the The International Symposium on Ultra-High Performance Concrete, Kassel, Germany, 13–15 September 2004; pp. 285–301.
22. Van Tuan, N.; Ye, G.; van Breugel, K.; Fraaij, A.L.A.; Bui, D.D. The study of using rice husk ash to produce ultra high performance concrete. *Constr. Build. Mater.* **2011**, *25*, 2030–2035. [CrossRef]
23. Zhao, S.; Sun, W. Nano-mechanical behavior of a green ultra-high performance concrete. *Constr. Build. Mater.* **2014**, *63*, 150–160.
24. Ferdosian, I.; Camões, A.; Ribeiro, M. High-volume fly ash paste for developing ultra-high performance concrete (UHPC). *Cienc. e Tecnol. dos Mater.* **2017**, *29*, e157–e161.
25. Faried, A.S.; Mostafa, S.A.; Tayeh, B.A.; Tawfik, T.A. Mechanical and durability properties of ultra-high performance concrete incorporated with various nano waste materials under different curing conditions. *J. Build. Eng.* **2021**, *43*, 102569.
26. Ghafari, E.; Costa, H.; Júlio, E.; Portugal, A.; Durães, L. The effect of nanosilica addition on flowability, strength and transport properties of ultra high performance concrete. *Mater. Des.* **2014**, *59*, 1–9.
27. Mostafa, S.A.; Faried, A.S.; Farghali, A.A.; El-Deeb, M.M.; Tawfik, T.A.; Majer, S.; Elrahman, M.A. Influence of nanoparticles from waste materials on mechanical properties, durability and microstructure of uhpc. *Materials* **2020**, *13*, 4530. [PubMed]
28. Qing, Y.; Zenan, Z.; Deyu, K.; Rongshen, C. Influence of nano-SiO_2 addition on properties of hardened cement paste as compared with silica fume. *Constr. Build. Mater.* **2007**, *21*, 539–545. [CrossRef]
29. Yu, R.; Spiesz, P.; Brouwers, H.J.H. Effect of nano-silica on the hydration and microstructure development of Ultra-High Performance Concrete (UHPC) with a low binder amount. *Constr. Build. Mater.* **2014**, *65*, 140–150.
30. Bajaber, M.A.; Hakeem, I.Y. UHPC evolution, development, and utilization in construction: A review. *J. Mater. Res. Technol.* **2021**, *10*, 1058–1074.
31. *EN 1015-3*; Methods of Test for Mortar for Masonry—Part 3: Determination of Consistence of Fresh Mortar (by Flow Table). CEN: Brussels, Belgium, 2001.
32. *EN 1015-6*; Methods of Test for Mortar for Masonry—Part 6: Determination of Bulk Density of Fresh Mortar. CEN: Brussels, Belgium, 1998.
33. *EN 1015-7*; Methods of Test for Mortar for Masonry—Part 7: Determination of Air Content of Fresh Mortar. CEN: Brussels, Belgium, 1998.
34. *EN 1015-11*; Methods of Test for Mortar for Masonry—Part 11: Determination of Flexural and Compressive Strength of Hardened Mortar. CEN: Brussels, Belgium, 1999.
35. *LNEC E 397*; Hardened Concrete—Determination of the Modulus of Elasticity of Concrete in Compression. Based on RILEM CPC 8—Modulus of Elasticity of Concrete in Compression. LNEC: Lisboa, Portugal, 1975. (In Portuguese)
36. *EN 14651*; Measuring the Flexural Tensile Strength (Limit of Proportionality (Lop), Residual). CEN: Brussels, Belgium, 2007.
37. *EN 12808-4*; Grouts for tiles—Part 4: Determination of Shrinkage. CEN: Brussels, Belgium, 2009.
38. Charron, J.-P.; Androuët, C. Shrinkage Mitigation of an Ultra-High Performance Concrete Submitted to Various Mixing and Curing Conditions. *Materials* **2021**, *14*, 3982.

39. Matos, A.; Nunes, S.; Costa, C.; Aguiar, J. Autogenous shrinkage in ultra-high performance concrete. In Proceedings of the National Meeting of Structural Concrete—BE2016, Coimbra, Portugal, 2–4 November 2016; pp. 1–10.
40. Zhang, J.; Li, V.C. Influences of Fibers on Drying Shrinkage of Fiber-Reinforced Cementitious Composite. *J. Eng. Mech.* **2001**, *127*, 37–44. [CrossRef]
41. *LNEC E 399*; Concrete Chraracterization of Creep in Compressive Test. National Laboratory of Civil Engineering (LNEC): Lisboa, Portugal, 1993. (In Portuguese)
42. Bažant, Z.P. Prediction of concrete creep and shrinkage: Past, present and future. *Nucl. Eng. Des.* **2001**, *203*, 27–38. [CrossRef]
43. Lopez, M. Creep and Shrinkage of High Performance Lightweight Concrete: A Multi-scale Investigation. PhD Thesis, Georgia Institute of Technology, Atlanta, GA, USA, 2005.
44. Neville, A.M. *Properties of Concrete*, 5th ed.; Pearson: London, UK, 2012.
45. Klausen, A.E.; Kanstad, T.; Bjøntegaard, Ø.; Sellevold, E. Comparison of tensile and compressive creep of fly ash concretes in the hardening phase. *Cem. Concr. Res.* **2017**, *95*, 188–194. [CrossRef]
46. Neves, R.; Gonçalves, A. Steel fiber reinforced concrete. Properties related to its durability. In Proceedings of the National Meeting of Structural Concrete—BE2000, Porto, Portugal, 22–24 November 2000; pp. 135–142. (In Portuguese).
47. Ahmad, S.; Zubair, A.; Maslehuddin, M. Effect of the key mixture parameters on shrinkage of reactive powder concrete. *Sci. World J.* **2014**, *2014*, 426921.
48. Gu, C.P.; Ye, G.; Sun, W. Ultrahigh performance concrete-properties, applications and perspectives. *Sci. China Technol. Sci.* **2015**, *58*, 587–599.
49. Eppers, S.; Müller, C. Autogenous shrinkage strain of ultra-high performance concrete (UHPC). In Proceedings of the Second International Symposium on Ultra High Performance Concrete, Kassel, Germany, 5–7 March 2008; pp. 433–441.
50. Russel, G.H.; Graybeal, B.A. *Ultra-High Performance Concrete: A State-of-the-Art Report for the Bridge Community*; US Department of Transportation: Washington, DC, USA, 2013.
51. Browne, R. *Field Investigations: Site & Laboratory Tests: Maintenance Repair and Rehabilitation of Concrete Structures*; CEEC: Lisboa, Portugal, 1991.
52. Luping, T. *Guideline for Practical Use of Methods for Testing the Resistance of Concrete to Chloride Ingress*; SP Swedish National Testing and Research Institute: Boras, Sweden, 2005.
53. Yoo, J.; Oh, B.; Park, S.; Kim, S. Chloride penetration resistance properties of concrete containing mineral admixtures. In Proceedings of the Fracture Mechanics of Concrete and Concrete Structures Assessment. Durability, Monitoring and Retrofitting of Concrete Structures, Jeju, Republic of Korea, 23–28 May 2010; pp. 1008–1016.
54. Real, S.; Bogas, J.A.; Pontes, J. Chloride migration in structural lightweight aggregate concrete produced with different binders. *Constr. Build. Mater.* **2015**, *98*, 425–436. [CrossRef]
55. Saraswathy, V.; Song, H.W. Effectiveness of fly ash activation on the corrosion performance of steel embedded in concrete. *Mag. Concr. Res.* **2007**, *59*, 651–661. [CrossRef]
56. Uysal, M.; Yilmaz, K.; Ipek, M. The effect of mineral admixtures on mechanical properties, chloride ion permeability and impermeability of self-compacting concrete. *Constr. Build. Mater.* **2012**, *27*, 263–270. [CrossRef]
57. Camões, A. High performance concrete with fly ash incorporation. In Proceedings of the Third International Conference on Unsaturated Soils, UNSAT 2002, Recife, Brazil, 10–13 March 2002; p. 487. (In Portuguese).
58. Carrinho, D. *Resistance to Chloride Penetration in Different Types of Concrete Subjected to Prolonged Immersion Tests*; Instituto Superior Técnico-Universidade de Lisboa: Lisboa, Portugal, 2018. (In Portuguese)
59. Nath, P.; Sarker, P. Effect of Fly Ash on the Durability Properties of High Strength Concrete. *Procedia Eng.* **2011**, *14*, 1149–1156. [CrossRef]
60. *AASHTO T 358-15*; Standard Method of Test for Surface Surface Resistivity Indication of Concrete's Ability to Resist Chloride Ion Penetration. AASHTO (American Association of State Highway and Transportation Officials): Washington, DC, USA, 2015.
61. Robalo, K.; Costa, H.; do Carmo, R.; Júlio, E. Enhanced mechanical and durability performances of low cement concrete with natural pozzolan addition. *J. Adv. Concr. Technol.* **2021**, *19*, 519–535. [CrossRef]
62. Camões, A.; Costeira, A. Durability of steel fiber reinforced concrete. In Proceedings of the National Meeting of Structural Concrete—BE2012, Porto, Portugal, 24–26 October 2012; pp. 1–10. (In Portuguese).
63. Lataste, J.F.; Behloul, M.; Breysse, D. Characterisation of fibres distribution in a steel fibre reinforced concrete with electrical resistivity measurements. *NDT E Int.* **2008**, *41*, 638–647. [CrossRef]
64. *EN 206-1*; Concrete—Part 1: Specification, Performance, Production and Conformity. CEN: Brussels, Belgium, 2007.
65. *LNEC E 463*; Determination of Diffusion Coefficient of Chlorides from Non- Steady-State Migration Test. National Laboratory of Civil Engineering (LNEC): Lisboa, Portugal, 2004. (In Portuguese)
66. Coutinho, J. *Improving the Durability of Concrete by Treating Formwork*, 1st ed.; FEUP: Porto, Portugal, 2005. (In Portuguese)
67. Tuutti, K. *Corrosion of Steel in Concrete*; The Swedish Cement and Concrete Research Institute: Boras, Sweden, 1982.

68. *LNEC E 465*; Concrete Methodology for Estimating the Concrete Performance Properties Allowing to Comply with the Design Working Life of The Reinforced or Prestressed Concrete Structures under the Environmental Exposures XC and XS. National Laboratory of Civil Engineering (LNEC): Lisboa, Portugal, 2007. (In Portuguese)
69. *EN 1992-1-1*; Eurocode 2–Design of Concrete Structures Part 1-1: General Rules and Rules for Buildings. CEN: Brussels, Belgium, 2010.

sustainability

MDPI

Article

Aquatic Weed for Concrete Sustainability

Mitiku Damtie Yehualaw [1], Mihiret Alemu [2], Behailu Zerihun Hailemariam [1], Duy-Hai Vo [3] and Woubishet Zewdu Taffese [4,*]

1 Faculty of Civil and Water Resource Engineering, Bahir Dar Institute of Technology, Bahir Dar University, Bahir Dar 6000, Ethiopia
2 Department of Construction Technology and Management, Woldia Institute of Technology, Woldia University, Woldia 7220, Ethiopia
3 Department of Civil Engineering, University of Technology and Education, The University of Danang, 48 Cao Thang Street, Hai Chau District, Da Nang 550000, Vietnam
4 School of Research and Graduate Studies, Arcada University of Applied Sciences, Jan-Magnus Jansson Aukio 1, 00560 Helsinki, Finland
* Correspondence: woubishet.taffese@arcada.fi

Abstract: Ordinary Portland cement (OPC) is the primary binder of concrete, accounting for approximately 5% to 7% of greenhouse gas (GHG) and carbon dioxide (CO_2) emissions with an annual production rate of more than 4 billion tons. It is critical to reduce the carbon footprint of concrete without sacrificing its performance. To this end, this study focuses on the use of water hyacinth ash (WHA) as a pozzolanic binder in the production of concrete as a partial replacement for cement. Four mixes are designed to achieve C-25-grade concrete with varying proportions of cement replacement with WHA of 0%, 5%, 10%, and 15% of the cement weight. Extensive experiments are performed to examine the workability, strength, durability, and microstructure of concrete specimens. The test results confirm that incorporating WHA in concrete improved its workability, strength, and durability. The optimal results are obtained at the maximum OPC replacement level, with 10% WHA. The use of WHA as a partial replacement for cement greatly reduces the energy required for cement production and preserves natural resources. More research is needed to use WHA on a large scale to achieve greater sustainability in the concrete industry.

Keywords: water hyacinth ash; concrete; fresh properties; compressive properties; microstructure properties; durability

check for updates

Citation: Yehualaw, M.D.; Alemu, M.; Hailemariam, B.Z.; Vo, D.-H.; Taffese, W.Z. Aquatic Weed for Concrete Sustainability. *Sustainability* **2022**, *14*, 15501. https://doi.org/10.3390/su142315501

Academic Editor: Constantin Chalioris

Received: 28 October 2022
Accepted: 19 November 2022
Published: 22 November 2022

1. Introduction

The development of any country is primarily characterized and measured by improvements in living conditions and the growth of its built infrastructure. Concrete is one of the most prevalent and widely consumed building materials utilized in the construction of megastructures and infrastructures, and it is important to the world that it is a highly versatile, all-around material, one that is used every year in development [1–5]. Despite its significant contribution to built infrastructure, concrete is responsible for the extensive use of natural resources in its production. Cement is a well-known, essential component in the production of concrete. The manufacturing of cement affects nonrenewable natural resources, the atmospheric environment, and costs, as well as engineering and durability behavior aspects [6–10]. Manufacturing 1 ton of Portland cement requires quarrying 1.5 tons of limestone and clay [6,11,12]. It has been reported that, throughout the world, 4.40 billion tons of cement are produced yearly, contributing to 5%–7% of GHG and CO_2 emissions [6,13–16].

On the other hand, generating 1 ton of clinker requires an enormous amount of energy (850 Kcal/kg) and raw materials (1.7 tons on average) [17–19]. The world has seen global climate changes as a result of these interrelated issues, which ultimately led to the depletion of the ozone layer. Because of the loss of ozone, the sun's infrared rays will reach the ground directly, drying up plants and exposing people and animals to several dangerous diseases.

Creating a sustainable system loop that can transform resources that are sent to landfills as waste into useful products for the construction sector would be the ideal goal to partially address the aforementioned situation. This would save natural resources, similar to how finding locally accessible, sustainable, eco-efficient, appropriate, cheap, and low-energy alternative agricultural and industrial waste materials to partially or completely replace cement has become mandatory and necessary to ease this global issue. Since plant extracts contain a variety of organic components, there has been a lot of research on replacing some of the cement with various plant wastes [6]. Numerous studies on cement auxiliary materials, such as silica fume, fly ash, rice husk ash, coffee husk ash, ground nut ash, bamboo leaves ash, banana leaves ash, sugarcane bagasse, corncob ash, blast furnace slag, animal bone powder, tobacco waste, and eggshell, have been conducted in the last two decades. Unexpectedly, these materials proved successful in meeting the needs of cement concrete solid development.

Water hyacinth (Eichhornia crassipes) is a free-gliding and -floating aquatic weed that grows in still or moderately mobile crisp water bodies, and it inhabits bodies of water in such quantities that these bodies either start to interfere with human usage or become a health hazard by serving as insect breeding areas [20]. It poses the risk of oxygen depletion, water quality degradation, biodiversity loss, water loss, and waterway blockage, all of which impede agriculture, fisheries, recreation, and hydropower. Water hyacinth has been referred to as a noxious weed and an invasive aquatic plant [12]. In less than two weeks, it grows dynamically and richly to produce large biomass and double its population. It consists of long and fibrous roots that may be up to 3 m in length and has a fibrous stem. The average length of the fiber is 1.604 mm and the average diameter is 5.5 microns [1,6]. It is a completely free source of biomass that is currently being underutilized as a supplementary cementitious material [1,6,21,22].

Ethiopia is one of the nations fighting against the presence of water hyacinth in its water bodies despite having abundant water resources and wetland ecosystems [23–26]. Due to the rapid growth of the water hyacinth, there is a risk that these waters will be infested by it if the situation persists. Water hyacinth, also known as "Imboch" in the local Amharic language, was first discovered in Ethiopia in 1956 in Koka Lake and the Awash River (Stroud, 1994), referenced in [17]. From late 1956 to late 2011, the weed's presence in water bodies decreased [27–32]. The lack of natural enemies for the weed has been largely blamed for the water hyacinth's rapid proliferation in Ethiopia over the past 30 years, according to the United Nations Environmental Programme (UNEP). Due to its ability to quickly blanket entire streams, water hyacinth is one of the most invasive and widely dispersed plants in Ethiopia [25]. It invaded Lake Tana in Bahir Dar, Ethiopia, affecting not only the aquatic life in the lake but also the socioeconomic activities of the local populace, whose livelihoods rely directly or indirectly on the lake's ecosystem services [33]. Dersseh et al. [33] stated that Lake Tana's invaded area was 20,000 ha, 50,000 ha, and 34,000 ha, respectively, in 2012, 2014, and 2015, compared with official agency estimates of less than 5000 ha during the peak growing season. According to a survey conducted from 4 to 14 October 2018 and excluding the infestation region in the floodplain, 2279.4 hectares of Lake Tana were projected to be covered by water hyacinth [33]. It cost an average of about USD 100,000 to control and remove water hyacinth from the invaded areas in Ethiopia between 2000 and 2013 [28]. Figure 1 shows the removal of invasive water hyacinth from Lake Tana. As it covers a large portion of the lake, anyone could imagine the adverse impact it will have on water quality, biodiversity, waterway flow, agriculture, fisheries, and recreation.

Previous studies have been conducted on the possible use of water hyacinth ash (WHA) as a partial replacement cement in concrete production [1,6–8,15,16], and they proved that workability improved with the addition of WHA; the concrete compressive strength and split tensile strengths decreased with 20% WHA content, and up to 10% WHA could replace cement. Damtoft et al. [7] investigated the mechanical performance of concrete that was increased with 15% WHA, including microsilica admixtures; the workability improved incrementally with the addition of WHA. Turner and Collins [16] identified that WHA has a lower specific gravity than cement (2.44 times lower than cement), and its fineness is twice as high as that of cement; additionally, consistency, slump, and setting time improved with the addition of WHA compared with normal cement; 10% WHA was the most optimal dosage for the partial replacement of cement. Vieira et al. [15] discovered that the water absorption of concrete containing 10% WHA was lower, and the chloride resistance was 3.5% higher when compared with control concrete, indicating greater durability.

Figure 1. Removal of an invasive water hyacinth from Lake Tana, Bahir Dar, Ethiopia.

In the present experimental research work, cement is substituted with WHA by 5%, 10%, and 15%, and it is utilized in concrete production to evaluate the workability, mechanical strength, and microstructures with thermogravimetric and Fourier-transformed infrared spectroscopy analyses. Finally, the effect of using WHA as a supplementary replacement material for cement in concrete durability properties is investigated.

2. Materials and Methods

2.1. Materials Used

2.1.1. Binders

In this work, ordinary Portland cement (OPC) and water hyacinth ash were used as binders. The adopted cement was OPC 42.5R, produced by Derba MIDROC Cement PLC. The cement quality was tested as per ASTM C1084–19 [34]. The grading and physical properties met the requirements of ASTM C 150–19 [35].

2.1.2. Coarse Aggregates

Locally available crushed stones, sieved and retained on sieves with sizes of 4.75 mm and above, were used as coarse aggregates. The coarse aggregate was also tested for various properties, such as unit weight, specific gravity, moisture content, water absorption capacity, and nominal aggregate size to ensure its suitability for the experiment. All physical tests were performed as per ASTM standards and conformed to all necessary standard requirements. Table 1 lists the appropriate test methods used and the corresponding physical properties gained as a result of the coarse aggregate.

2.1.3. Fine Aggregates

In this work, natural river sand from the city of Bahir Dar that passed through a 4.75 mm sieve was used as a fine aggregate. Table 2 shows the adopted test methods and physical properties for sand, and based on the result attained, it conforms to all ASTM standard requirements.

Table 1. Coarse aggregate test results.

No.	Property	Test Method	Result	Unit
1	Unit weight	ASTM C 29	1594.56	kg/m^3
3	Specific gravity	ASTM C 128	2.75	-
4	Moisture content	ASTM C128	1.01	%
5	Water absorption capacity	ASTM C128	1.41	%
6	Nominal aggregate size	-	25	mm

Table 2. Fine aggregate test results.

No.	Property	Test Method	Result	Unit
1	Fineness modulus	ASTM C117	2.68	-
2	Specific gravity	ASTM C 136	2.71	-
3	Moisture content	ASTM C 128	1.8	%
4	Water absorption capacity	ASTM C128	2.45	%

2.1.4. Water Hyacinth

Figure 2 shows the main steps of the WHA production process. Firstly, water hyacinth was manually collected from Lake Tana in Bahir Dar, Amhara National Regional State, Ethiopia, because the use of machinery was prohibitively expensive. It is important to note that the manual collection method is a time-consuming and labor-intensive activity. Then, the collected samples were washed and cleaned with potable water to remove dirt and impurities. Thereafter, the washed samples were cut uniformly into smaller pieces and sundried for over a week. Finally, the WHA was found after the dried water hyacinth samples were burned in a Muffle Furnace at 800 °C for 6 h and cooled at room temperature for 24 h to convert the organic matter into an inorganic substance. The samples were then ground using a milling machine and sieved down to 150 μm. WHA samples passed through a 150 μm sieve were used as a cement replacement material [1].

2.2. Mix Proportions and Sample Preparation

The mix design was performed based on the physical test results of aggregates, as in the ACI 211.1 [36] mix design procedure for normal concrete. All concrete mixes were designed to achieve C-25-grade concrete with a target mean strength of 33.5 MPa. A cement content of 360 kg/m^3, a constant water–cement ratio of 0.49, and a slump of 25 to 50 mm were used to achieve the specified target mean strength. The designed mix proportion of the concrete was 1:2:3 (binder:fine aggregates:coarse aggregates). All mixes were batched with 0%, 5%, 10%, and 15% WHA by weight to replace cement. The mix proportion of cement to WHA and the mix designation are provided in Table 3.

Table 3. Mix designation.

No.	Mix Designation	Percentage of Cement	Percentage of WHA
1	0% WHA	100	0
2	5% WHA	95	5
3	10% WHA	90	10
4	15% WHA	85	15

Figure 2. Main steps of the WHA production process.

A mechanical mixer was used to obtain a uniform mixture of ingredients. The mixing process entailed first dry mixing the binder and aggregates for three minutes followed by the slow addition of water as mixing continued. After all of the water was added, the mixing was continued for an additional three minutes in order to achieve a homogeneous mixture. Immediately after mixing, the slump of the mixture was evaluated, followed by pouring the fresh mixtures into pre-oiled molds to evaluate mechanical and durability properties.

2.3. *Test Methods*

Following various tests on the material qualities of cement, fine aggregates, coarse aggregates, and WHA, the fresh property of all concrete samples was determined in terms of the concrete's workability. The ASTM C143-12 [37] slump test was used to examine the workability of concrete in a new concrete mix. The short-term water absorption capacity test came before the compressive strength test. The amount of water absorption—measured as residual water left in concrete interfaces—that determines how durable or fragile it is. Following a 72 h water curing period and a drying phase in an oven at 105 °C, the weights of the specimens were properly measured to determine the percentage of water absorption. The specimens were subsequently submerged in cold water for an additional hour before being properly weighed [33].

The water absorption of each concrete mixture was evaluated after 7 and 28 days of curing in accordance with ASTM C 642 [38]. For the compressive strength test, concrete cube specimens were cast in 150 mm × 150 mm × 150 mm and cured for 3, 7, 28, 56, and 91 days. After casting, the cubes were wrapped in plastic sheets and kept at room temperature for 24 h. The cubes were removed from the molds after 24 h of casting and submerged in water for curing until the time of the test. The compressive strength tests were performed on the samples in accordance with ASTM C 109 [39].

To study the behavior of WHA concrete in different environments, e.g., dampness, salts, acids, etc., different chemical solutions were prepared. The experimental work for chemical resistance was carried out according to ASTM C722 [40]. The resistance to sulfate attack was conducted on the compressive strength of the concrete cubes by immersing the samples in a 5% Na_2SO_4 solution. Fourier-transformed infrared (FTIR) spectroscopy tests were conducted on the selected mixes to identify the chemical compounds present in the hardened WHA samples using a DIGILAB FTS3500 with a transmittance wavelength range of 500–4000 cm^{-1} [41–45]. The consumption and amount of calcium hydroxide in the sample were determined by thermogravimetric analysis (TGA). The TGA enables the determination of fixed calcium hydroxide. To perform the TGA tests, the samples were prepared for 5% and 10% WHA at curing ages of 7 and 28 days.

3. Results and Discussion

3.1. Fresh Properties of WHA

The slump of concrete containing varying percentages of WHA is shown in Figure 3. As can be seen in Figure 3, as the percentage of WHA increases, the workability of the concrete improves. The slump of fresh concrete with 0% WHA was 35, whereas the slump with 5% WHA was 48, representing a 37.14% improvement in workability. The higher the percentage of WHA in the cement replacement, the better the workability of the concrete. Because WHA has a smooth surface and spherical shape, it lowers the specific surface area more than ordinary Portland cement [42]. The same effects were observed by [1,7,16].

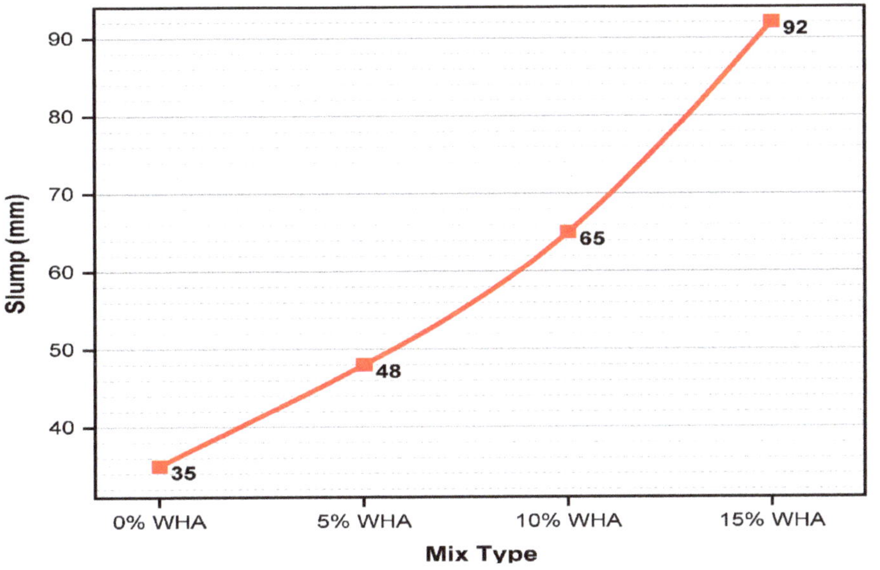

Figure 3. Workability of concrete for various percentages of WHA.

3.2. Water Absorption Capacity

The water absorption of each concrete mixture was evaluated after 7 and 28 days of curing according to ASTM C 642 [38]. As shown in Figure 4, the water absorption of concrete significantly reduced as the age of the concrete increased. On the other hand, as the percentage of WHA cement replacement increases, the concrete absorbs slightly less water. This demonstrates that voids in WHA-replaced concrete are fewer and nonporous when compared with the control-mix concrete. The increase in mass expressed as a percentage of dry mass was used to analyze water absorption. It is evident that the cover concrete rapidly lost water while curing due to hydration. For concrete with a control mix, surface water absorption was 7.48% and 4.5% after 7 and 28 days, respectively. This higher water absorption could be attributed to the higher porosity of the concrete preserved on the surface due to air [38]. The amorphous phase of WHA may facilitate the hydration reaction and result in fewer unhydrated WHA particles, decreasing water absorption in the concrete samples [42].

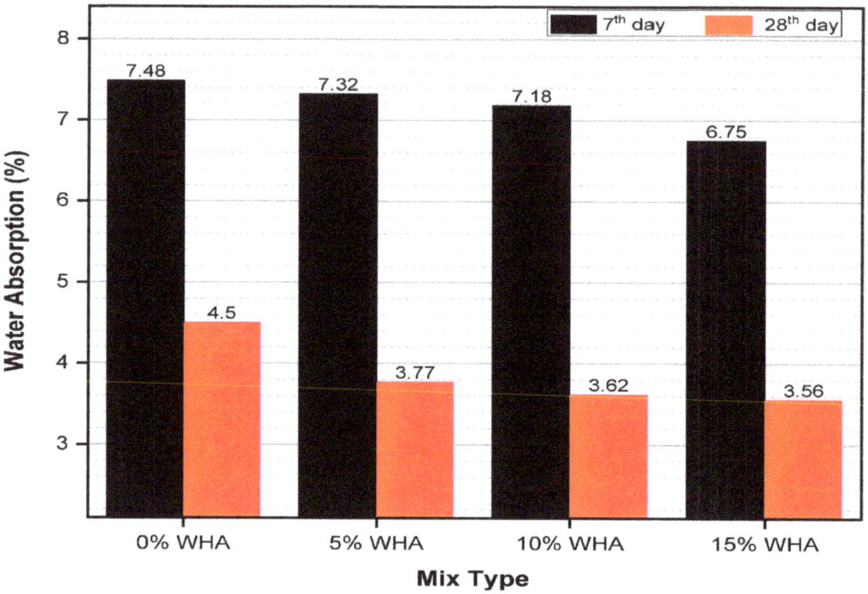

Figure 4. Water absorption of concrete with different percentages of WHA.

3.3. Compressive Strength Development

To test the compressive strength of the hardened concrete, first, the concrete cube samples were immersed in a water tank containing clean water at a normal temperature of 25 ± 2 °C for curing ages of 3, 7, 28, 56, and 91 days, then they were dried for 24 h in the open air. The result of the test for each sample in all curing periods is shown in Figure 5. From the figure, it can be seen that, except for the 15% WHA-containing samples, the compressive strength of all concrete cubes increased incrementally with the WHA replacement level with respect to the normal (0% WHA) concrete. The figure also reveals that all samples, including the 15% WHA, achieved both the designed (25 MPA) and mean target (33.5 MPa) compressive strength at the later age of 28 days, and the result is supported by [1,13,25]. Compared with the 0% WHA, at 28 days, the strengths were improved by 1.4 MPa (4.23%) and 2.6 MPa (7.85%) for 5% WHA and 10% WHA, respectively, whereas the 15% WHA samples decreased by 1.10 MPa (3.23%). It can also be seen that the strength increased with curing age for all of the samples. Among the 56-day strength enhancements obtained above for concrete containing 10% WHA, a remarkable enhancement of (42%) was obtained with respect to the reference control mix. This enhancement could be related to a higher amount of calcium–silicate–hydrate (C–S–H) formed as a consequence of WHA addition. The main reason for the improvement in concrete compressive strength is the pozzolanic reaction from WHA and calcium hydroxide, which promotes the formation of hydrated calcium silicate. However, concrete without WHA can only rely on cement to hydrate into a small amount of C–S–H. Hydrated calcium silicate is one of the most important elements that provide strength. Therefore, the compressive strength of concrete with 10% WHA achieved maximum strength and was the optimum replacement level; the result is in agreement with previous findings [8]. According to [1,16], the incorporation of more than 10% WHA in cement concrete continuously decreased the compressive strength because increasing WHA content would greatly accelerate the rate of hydration in the concrete, resulting in the early evaporation of water from the mix before the full reaction could occur, resulting in a reduction in compressive strength.

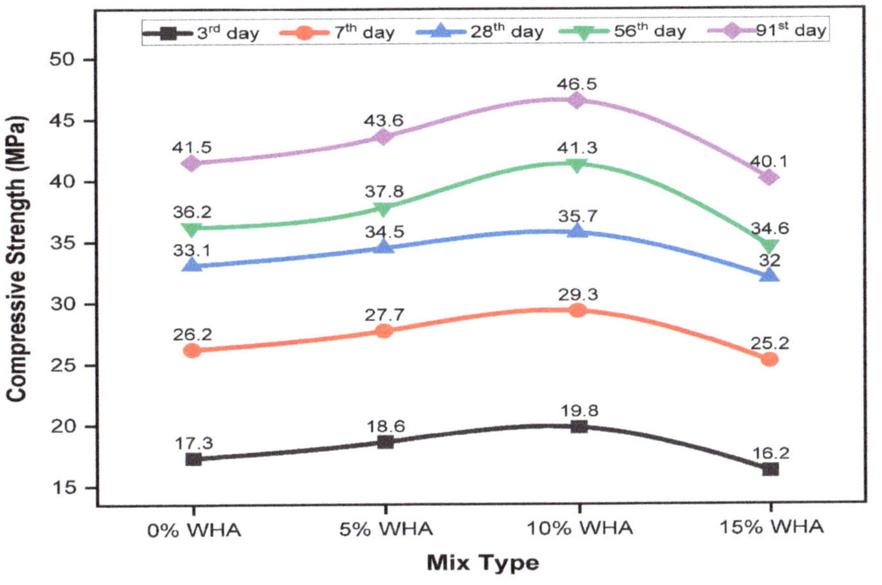

Figure 5. Compressive strength of concrete for different curing ages.

3.4. Thermogravimetric Analysis

In the TGA technique, the sample is first heated, and the mass change due to heating is recorded as a derivative thermogravimetric (DTG) curve. In studies on cementitious samples, the most significant mass changes are the mass loss due to H_2O release (dehydration, dehydroxylation) until they reach 550 °C and above 550 °C due to CO_2 release (decarbonation). In a sample that was treated with isopropanol and left in a vacuum to remove the free water, the mass loss until it reached 550 °C corresponded to the amount of bound water in the hydrate phases [46]. Between 450 and 550 °C, the main thermal transition (endothermic) was seen. The mass loss at this temperature range can be used to determine the amount of calcium hydroxide in the sample [46]. This peak indicates that $Ca(OH)_2$ or C–H breakdown occurs during heating, resulting in a considerable mass loss in the sample. Between 550 and 850 °C, a slight second peak was noticed, which is thought to be caused by the dissolution of $CaCO_3$ and the loss of water from the C–S–H gel. A low Ca/Si ratio suggests that C–A–S–H gel crystallinity reduces with increasing Al concentration and that increased Al-absorbed content results in more C–A–S–H phases. The following list of chemical reactions can be used to characterize the breakdown processes [46].

Dehydration of C–H:

$$\mathrm{Ca(OH)_2} \rightarrow \mathrm{CaO} + \mathrm{H_2O} \uparrow \tag{1}$$

Carbonation of $CaCO_3$:

$$\mathrm{CaCO_3} \rightarrow \mathrm{CaO} + \mathrm{CO_2} \uparrow \tag{2}$$

Dehydroxylation of C–S–H:

$$\mathrm{C-S-H} \rightarrow \mathrm{C-A-S-H} + \mathrm{HO_2} \uparrow \tag{3}$$

In this study, TGA investigations were performed for the concrete samples containing 5% WHA after the 7th day of curing and for the 10% WHA samples after the 7th and 28th days of curing (Figures 6–8). As shown in the figures, to level the weight loss, we subtracted the upper dotted line *Y*-axis value from the lower dotted line *Y*-axis value to obtain the weight loss (%) value for each portion. Figures 6 and 7 show the TGA results for 5% WHA and 10% WHA seventh-day curing samples, respectively. On day 7, the 5% WHA samples decomposed into an average of 27% C–H from the concrete mixes after an approximate heating temperature of 450 °C; it lost 10% after 50 °C (500 °C), and no decompositions occurred until the temperature reached 600 °C (Figure 6). Beyond 600 °C, water decomposed from C–S–H, and decarbonation (the release of CO_2 from $CaCO_3$) occurred. Together they lost 23% of their mass at 750 °C; after which, it remained constant. The thermal stability of the sample linearly decreased in terms of dehydration, dehydroxylation, and decarbonation for the 10% WHA content (Figure 7); compared with the 5% WHA mixes, it decomposed an equivalent amount of water and CO_2 (23%); a continuous heating temperature of more than 750 °C was required. This corroborates that the 5% WHA samples complete dehydration at a 750 °C heating temperature. As shown in Figure 8, a TGA examination of the 28th-day sample of concrete incorporating 10% WHA revealed that the $Ca(OH)_2$ dehydrated and lost about 0.30% of its mass at a temperature of 500 °C, and the figure indicates that there was no mass loss for C–H and other components of the concrete until the heating temperature rose above 600 °C. Then, after reaching 600 °C, the C–S–H gel and $CaCO_3$ released water and CO_2, respectively; they lost about 11% of their mass at a temperature of 700 °C, and further continuous reductions were observed up to 800 °C. This suggests that adding more WHA to the cement results in a pozzolanic reaction in the C–S–H gel produced by the cement and silica after the initial addition of WHA, and this also increases the heat of hydration; hence, the water within the C–S–H and the CO_2 present in the $CaCO_3$ evaporated rapidly, resulting in a large amount of dehydroxylation and decarbonation. This implies that the 5% WHA concrete sample contained more water than the 10% WHA concrete sample, and it takes more time (curing age) to achieve strength at early ages, as evidenced by the compressive strength result.

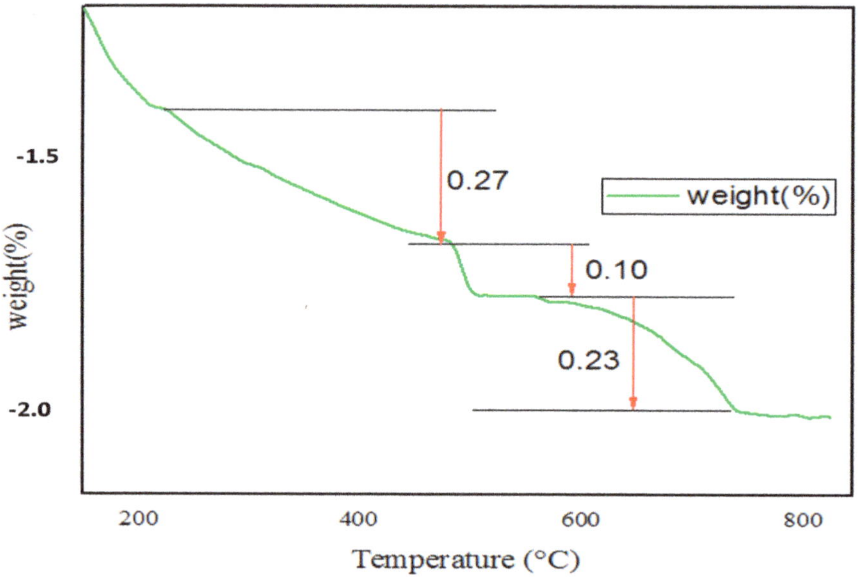

Figure 6. TGA curve for 5% WHA for 7 days of curing.

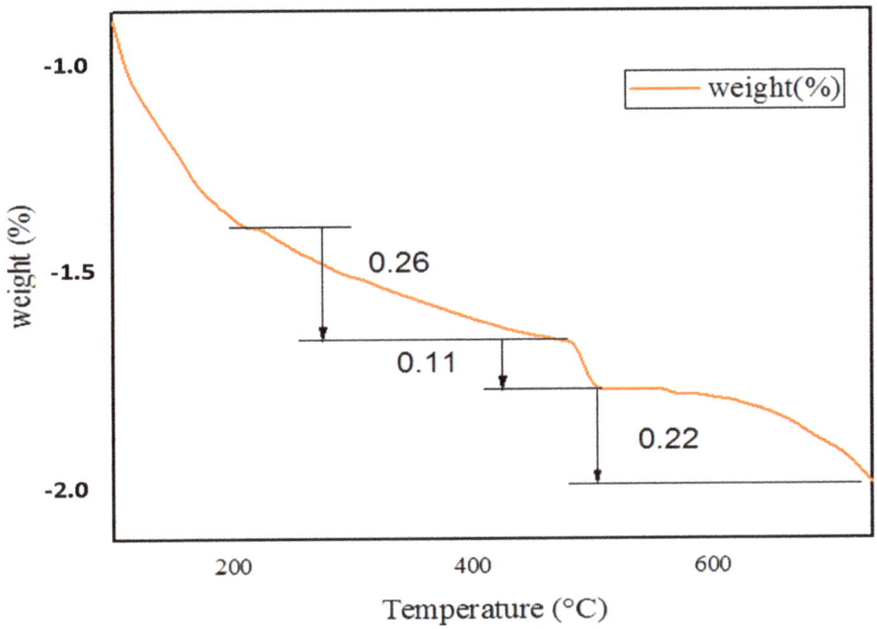

Figure 7. TGA curve for 10% WHA for 7 days of curing.

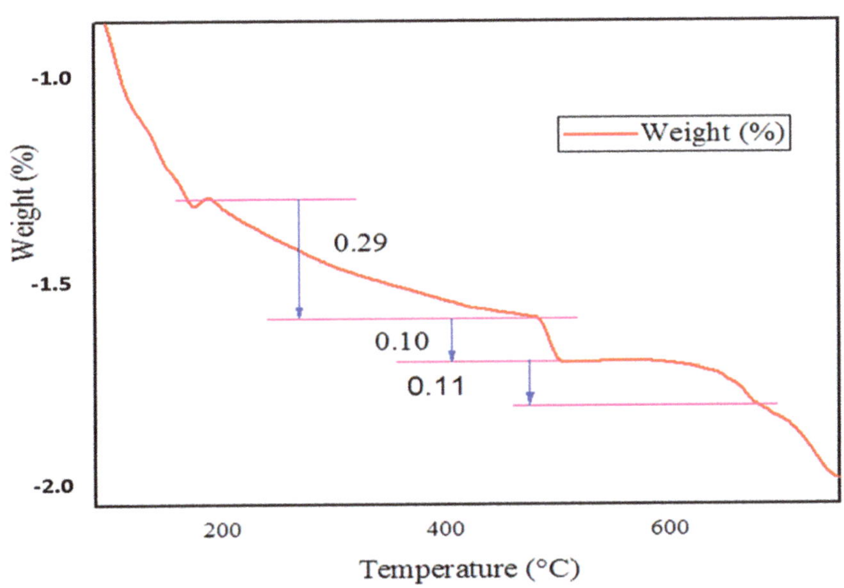

Figure 8. TGA curve for 10% WHA for 28 days of curing.

3.5. Fourier Transformed Infrared Spectroscopy Analysis

Figures 9 and 10 show the infrared spectra of selected WHA mixes (0% WHA, 5% WHA, 10% WHA, 15% WHA) at curing ages of 7 and 28 days (with main bands marked). The results show that the wavenumbers and curves of the 7- and 28-day pastes were essentially the same, indicating that the hydration products did not change with the development of the hydration process, with the exception of the 7th-day samples, in which the number of hydrated quartz particles was higher than that of the 28th-day samples, where the wavelength stretched for the 10% WHA mixes. In the sample with the highest percentage of 10% WHA samples, the unreacted particles displayed a sharply intensified peak at wave numbers 470, 600, and 770 cm^{-1}. This indicates an incomplete activation reaction, which increased the amount of partially reacted and unreacted particles in the sample and decreased its mechanical strength. This is consistent with the earlier discoveries of [45].

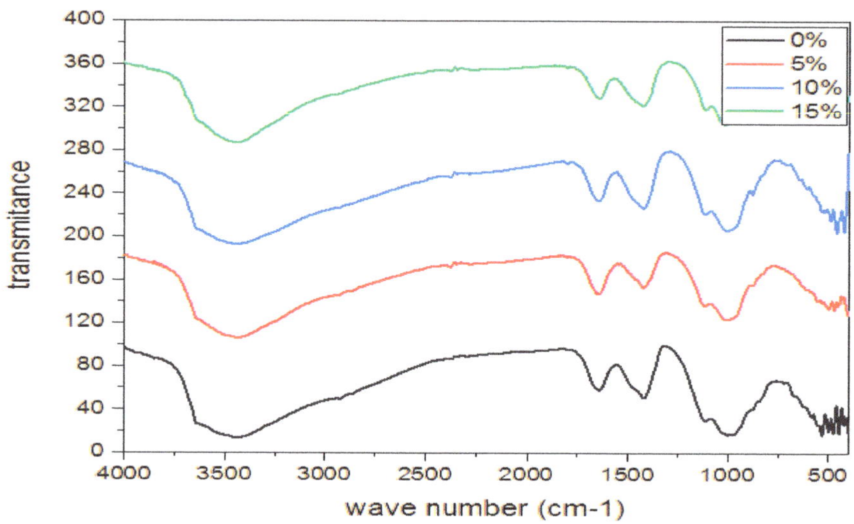

Figure 9. FTIR spectra of concrete with WHA for the 7th day.

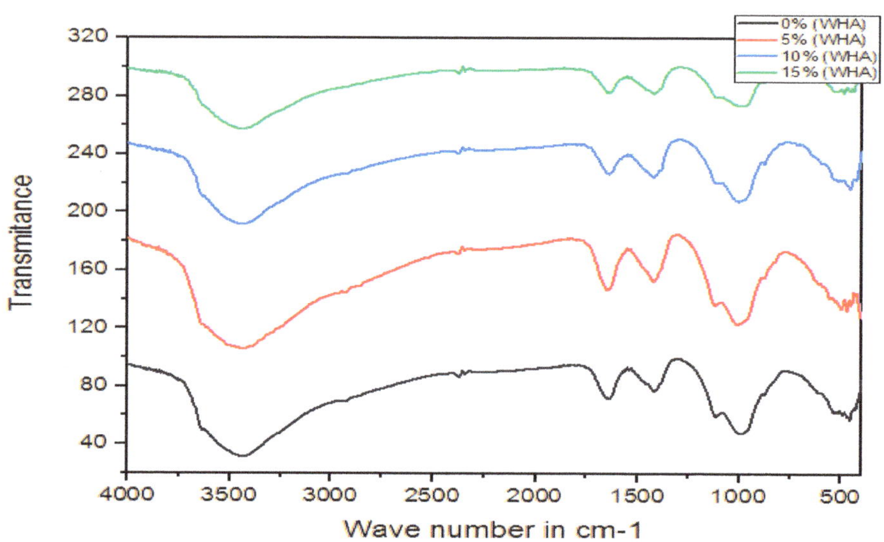

Figure 10. FTIR spectra of concrete with WHA for the 28th day.

Figure 9 shows the infrared spectra of all 7-day samples, with spectra ranging from 500 to 4000 cm^{-1} for the characteristic wavenumber and the associated functional groups of cement paste. From the result of the FTIR spectrum analysis of WHA, the major peaks were as follows: 3650, 3720, 2350, 2290, 1500, 920, 889, and 680 cm^{-1}. Due to the presence of calcium in the sample, a low wavenumber band can be observed at 684 cm^{-1}. H–O–H band vibrations were found at the peak value of 1500 cm^{-1}. The peak value of 684.15 cm^{-1} indicates asymmetric stretching vibrations in Al–O bonds due to anhydrous WHA. Other researchers have observed T–O bond-stretching at the peak values of 966 cm^{-1} and 1635 cm^{-1}, which correspond to H–O–H bonds and Al–O bonds at a wavenumber of 688 cm^{-1} [47].

At the age of 28 days (Figure 10), the material became sharper at a peak of 960.80 cm^{-1} due to the presence of activated alkali; this shift represents the gel component of the WHA material reacting with alkali activators. Because of this reaction, a new product, aluminosilicate hydrate gel, is formed. This study demonstrates the formation of chain links such as Si–O–Al. The extending vibrations related to O–H twisting vibrations and H–O–H extending vibrations (3600 to 1600 cm^{-1}) represent free water and provided proof of a soluble base initiation response in the concrete [38], revealing that the expansion of H–O–H extending vibrations could be observed at 3500 cm^{-1}. During the hydration reaction, 1030 to 1170 cm^{-1} is attributed to the Ca-modified silica gel (C–S–H). The band area detail of 4000–3500 cm^{-1} indicates the loss of $Ca(OH)_2$, 3500–1600 cm^{-1} indicates the stretching of the –OH bond and bending of the H–O–H vibrations, and 1000–800 cm^{-1} indicates the loss of the $CaCO_3$ gel [15].

3.6. Sulphate Attack Resistance

Several past studies have revealed that mortar and concrete produced with pozzolans with OPC performed better against sulfate attacks than plain OPC mortar and concrete. The deterioration of concrete due to sulfate attack is mostly measured in terms of weight loss, strength change, spalling, cracking, and expansion [48]. The changes in compressive strength loss in the percentage of hardened concrete made with 5%, 10%, and 15% WHA compared with the control mix (0% WHA) immersed in 5% Na_2SO_4 solution for 3, 7, 28, 56, and 91 days are plotted as a function of immersed curing time in Figure 11. A higher negative relative percentage indicates greater resistance to sulfate attack and higher positive results for lower resistance and vice versa [49]. The result in the figure shows that, for 5% and 10% WHA replacement levels for OPC, the concrete samples had better resistance to sulfate attack for all immersed curing times. Notably, after 56 days of curing time, both the 5% and 10% WHA samples outperformed 100% OPC concrete samples in terms of sulfate resistance, with 7.31% and 3.51%, respectively. On the other hand, they achieved 8.37% and 4.54% after 91 days of immersed time. However, a deterioration in compressive strength due to sulfate attack was observed when the cement was replaced with 15% WHA for all immersing periods. For instance, its compressive strength decreased by 6.21% (1 MPa) and 4.1% (1.7 MPa) for 3- and 91-day immersions, respectively. In addition, for all concrete samples, the sulfate attack resistances decreased when the samples were kept in sulfate solution for up to 28 days and gradually improved at the later age of 91 days. The ettringite and expansive gypsum that are formed as a result of sulfate attack are most likely the cause of degradation, cracking, spalling, and expansion in 100% OPC concrete. When OPC paste is submerged in sulfate solutions, the stiffness of the material reduces, and the water absorption values of ettringite increase, resulting in the formation of gypsum and ettringite [47,50]. Another problem is that by lowering cohesion in the hydrated OPC paste, as well as reducing adhesion among the aggregate particles, the expansion and cracking caused by sulfate attack enhances the compressive strength loss of concrete specimens [50]. Therefore, WHA plays a major role in OPC concrete by improving its sulfate resistance; 5% WHA is the maximum replacement level to resist any effects on the concrete and provides better resistance to sulfate attacks.

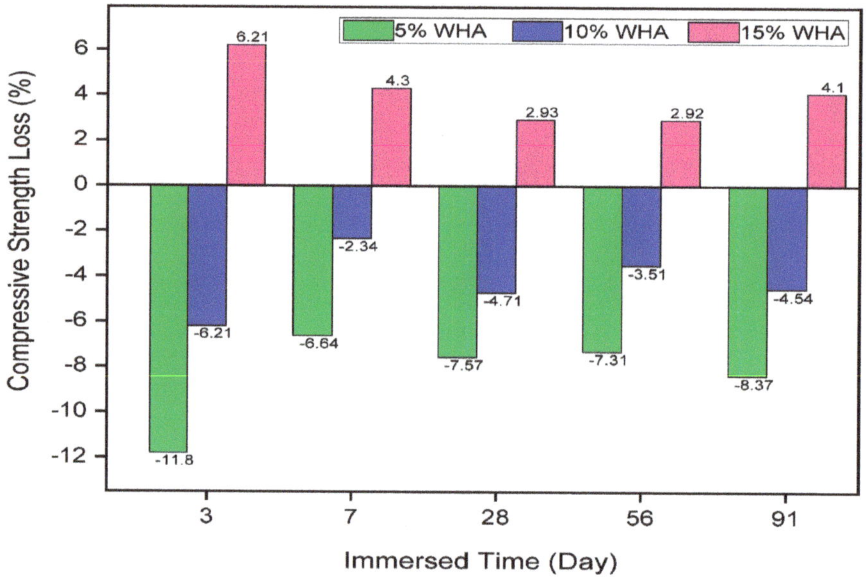

Figure 11. Resistance of concrete to sulfate attacks.

All of the experimental results confirmed that replacing cement with WHA of up to 10% improves the engineering and microstructure properties of concrete. The use of WHA has significant environmental and socioeconomic implications. It helps to reduce the amount of ordinary Portland cement used in the production of concrete, which eventually reduces the energy required for cement production and preserves natural resources. Aside from improving the engineering and microstructural properties of concrete, the use of WHA in concrete could also reduce the negative impact of this invasive aquatic weed on human health, water quality, biodiversity, waterway flow, agriculture, fisheries, and recreation. All of these have significant economic and health consequences for society.

The use of WHA with only OPC is a limitation of this work. It is critical to conduct several experimental studies with WHA and other supplementary cementitious materials, with or without an alkali-activator, to examine its impact on several engineering and durability properties of concrete. Such experiments enable the identification of the best combinations to yield better mechanical and durability properties, as well as sustainability. In the future, the authors intend to conduct relevant experimental research with the inclusion of ground-granulated blast-furnace slag (GGBFS) in geopolymers, concrete, or mortar.

This study examined the impact of WHA on the workability, strength, durability, and microstructure properties of concrete to make use of an invasive aquatic weed, water hyacinth, removed from Lake Tana. The results corroborated that water hyacinth in the form of ash improved the performance of concrete. To achieve greater sustainability in the concrete industry, a thorough study of how to scale up the use of WHA is required. Indeed, because water hyacinth has several negative effects on natural waterbodies, the research should focus on planting and harvesting it in controlled artificial waterbodies to avoid the associated environmental and socioeconomic impacts. As a result, the authors encourage agricultural and environmental researchers to investigate the safe cultivation of water hyacinth. A lifecycle assessment of WHA will then be used to determine the long-term impact on concrete sustainability.

4. Conclusions

This work investigated the fresh, mechanical, and durability properties, as well as the microstructural performance, of concrete with WHA partially replacing ordinary Portland cement. An extensive literature review on this additive (WHA) was conducted, experiments were carried out, and test results were discussed. Based on the obtained results, the following conclusions can be drawn:

- Burning water hyacinth at 800 °C for 6 h yields a good pozzolanic property, suitable for use as a supplementary cementitious material in concrete.
- The slump values of concrete are highly dependent on the proportion of WHA. High levels of WHA increase slump results and, hence, the workability of the concrete mix.
- The concrete specimens exhibited a reduction in water absorption with curing time, which can be attributed to the minimal voids and pores in the concrete samples.
- The concrete specimens with the highest percentage of WHA exhibited the greatest durability.
- Adding 5% to 10% WHA to concrete improves its compressive strength. However, further increasing the WHA replacement level results in a decrease in the compressive strength of the concrete, with a 10% replacement of cement with WHA yielding the optimal results.
- A microstructure analysis confirmed that the structure of C–S–H gels changes in samples when WHA is added, which contributes to compressive strength. The decomposition of $Ca(OH)_2$ and the formation of CaO, as well as the release of H_2O, reduce the porosity of concrete.
- Replacing OPC with WHA in concrete by up to 10% would improve the sulfate attack resistance of the concrete.
- Further research is recommended to improve the use of WHA in concrete on a large scale in order to achieve greater sustainability without adverse environmental and socioeconomic impacts.

Author Contributions: Conceptualization, M.A.; methodology, M.A.; software, M.A., B.Z.H., M.D.Y. and W.Z.T.; validation, M.A. and M.D.Y.; formal analysis, M.A. and M.D.Y.; investigation, M.A. and M.D.Y.; data curation, M.A., B.Z.H., D.-H.V., M.D.Y. and W.Z.T.; writing—original draft preparation, M.A.; writing—review and editing, B.Z.H., D.-H.V., M.D.Y. and W.Z.T.; visualization, M.D.Y. and W.Z.T.; supervision, M.D.Y.; project administration, M.D.Y. All authors have read and agreed to the published version of the manuscript.

Funding: This research received no external funding.

Institutional Review Board Statement: Not applicable.

Informed Consent Statement: Not applicable.

Data Availability Statement: The data supporting the findings of this study are available upon request.

Conflicts of Interest: The authors declare no conflict of interest.

References

1. Krishnan, S.; Murugesh, V. Experimental Study on Strength of Water Hyacinth Ash As Partial Replacement of Cement in Concrete. *Int. J. Res. Appl. Sci. Eng. Technol.* **2018**, *9*, 93–96.
2. Divyah, N.; Thenmozhi, R.; Neelamegam, M.; Prakash, R. Characterization and behavior of basalt fiber-reinforced lightweight concrete. *Struct. Concr.* **2019**, *22*, 422–430. [CrossRef]
3. Prakash, R.; Thenmozhi, R.; Raman, S.N. Mechanical characterisation and flexural performance of eco-friendly concrete produced with fly ash as cement replacement and coconut shell coarse aggregate. *Int. J. Environ. Sustain. Dev.* **2019**, *18*, 131–148. [CrossRef]
4. Prakash, R.; Raman, S.N.; Subramanian, C.; Divyah, N. Eco-friendly fiber-reinforced concretes. In *Handbook of Sustainable Concrete and Industrial Waste Management*; Woodhead Publishing: Sawston, UK, 2022; pp. 109–145. [CrossRef]
5. Srividhya, M.N.; Vidjeapriya, S.R. Enhancing the performance of hyposludge concrete beams using basalt fiber and latex under cyclic loading. *Comput. Concr.* **2021**, *28*, 93–105. [CrossRef]
6. Balasundaram, N.; Murugesh, V. Experimental behaviour of water hyacinth ash as partial replacement of cement on short column. *Int. J. Recent Technol. Eng.* **2019**, *7*, 648–650.

7. Damtoft, J.S.; Lukasik, J.; Herfort, D.; Sorrentino, D.; Gartner, E.M. Sustainable development and climate change initiatives. *Cem. Concr. Res.* **2008**, *38*, 115–127. [CrossRef]
8. El-Gamal, S.; El-Hosiny, F.; Amin, M.; Sayed, D. Ceramic waste as an efficient material for enhancing the fire resistance and mechanical properties of hardened Portland cement pastes. *Constr. Build. Mater.* **2017**, *154*, 1062–1078. [CrossRef]
9. Ilki, A.; Kumbasar, N. Behavior of damaged and undamaged concrete strengthened by carbon fiber composite sheets. *Struct. Eng. Mech.* **2002**, *13*, 75–90. [CrossRef]
10. Prakash, R.; Thenmozhi, R.; Raman, S.N.; Subramanian, C. Characterization of eco-friendly steel fiber-reinforced concrete containing waste coconut shell as coarse aggregates and fly ash as partial cement replacement. *Struct. Concr.* **2019**, *21*, 437–447. [CrossRef]
11. Prakash, R.; Thenmozhi, R.; Sudharshan, N.R.; Subramanian, C. Mechanical behaviour of polypropylene fibre reinforced concrete containing waste coconut shell as coarse aggregates and fly ash as partial cement replacement. *Rev. Fac. Ing. Univ. Antioq.* **2019**, *94*, 33–42. [CrossRef]
12. Prakash, R.; Divyah, N.; Srividhya, S.; Avudaiappan, S.; Amran, M.; Raman, S.N.; Guindos, P.; Vatin, N.I.; Fediuk, R. Effect of Steel Fiber on the Strength and Flexural Characteristics of Coconut Shell Concrete Partially Blended with Fly Ash. *Materials* **2022**, *15*, 4272. [CrossRef]
13. Torres, M.; Rodríguez-Puertas, C.; Alonso, M.; Puertas, F. Alkali activated slag cements using waste glass as alternative activators. *Rheol. Behav.* **2015**, *54*, 45–57. [CrossRef]
14. Huntzinger, D.N.; Eatmon, T.D. A life-cycle assessment of Portland cement manufacturing: Comparing the traditional process with alternative technologies. *J. Clean. Prod.* **2009**, *17*, 668–675. [CrossRef]
15. Vieira, D.R.; Calmon, J.L.; Coelho, F.Z. Life cycle assessment (LCA) applied to the manufacturing of common and ecological concrete: A review. *Constr. Build. Mater.* **2016**, *124*, 656–666. [CrossRef]
16. Turner, L.K.; Collins, F.G. Carbon dioxide equivalent (CO_2-e) emissions: A comparison between geopolymer and OPC cement concrete. *Constr. Build. Mater.* **2013**, *43*, 125–130. [CrossRef]
17. Puertas, F.; García-Díaz, I.; Barba, A.; Gazulla, M.; Palacios, M.; Gómez, M.; Martínez-Ramírez, S. Ceramic wastes as alternative raw materials for Portland cement clinker production. *Cem. Concr. Compos.* **2008**, *30*, 798–805. [CrossRef]
18. Prakash, R.; Thenmozhi, R.; Raman, S.N.; Subramanian, C.; Divyah, N. Mechanical characterisation of sustainable fibre-reinforced lightweight concrete incorporating waste coconut shell as coarse aggregate and sisal fibre. *Int. J. Environ. Sci. Technol.* **2020**, *18*, 1579–1590. [CrossRef]
19. Prakash, R.; Raman, S.N.; Divyah, N.; Subramanian, C.; Vijayaprabha, C.; Praveenkumar, S. Fresh and mechanical characteristics of roselle fibre reinforced self-compacting concrete incorporating fly ash and metakaolin. *Constr. Build. Mater.* **2021**, *290*, 123209. [CrossRef]
20. Ingwani, E.; Gumbo, T.; Gondo, T. The general information about the impact of water hyacinth on Aba Samuel Dam, Addis Ababa, Ethiopia: Implications for ecohydrologists. *Ecohydrol. Hydrobiol.* **2010**, *10*, 341–345. [CrossRef]
21. Murugesh, V. Experimental investigation on water hyacinth ash as the partial replacement of cement in concrete. *Int. J. Civ. Eng. Technol.* **2017**, *8*, 1013–1018.
22. Dinesh, M.V. Experimental Behavior of Water Hyacinth Ash as the Partial Replacement of Cement in Concrete. *Int. J. Sci. Res. Sci. Eng. Technol.* **2017**, *3*, 244–248.
23. Navarro, L.; George, P. Water Hyacinth in Africa and the Middle East: A survey of Problems and Solutions. 2000, p. 130. Available online: http://books.google.com/books?hl=es&lr=&id=hhXqXNQ0WSQC&pgis=1 (accessed on 13 March 2022).
24. IBC. *Ecosystems of Ethiopia*; Institute of Biodiversity and Conservation: Addis Ababa, Ethiopia, 2011.
25. Dechassa, N. Current Status of Water Hyacinth (*Eichhornia crassipes*) in Ethiopia: Achievements, Challenges and Prospects: A Review. *J. Environ. Earth Sci.* **2020**, *10*, 36–47. [CrossRef]
26. Gizaw, E.; Legesse, W.; Haddis, A.; Deboch, B. Assessment of factors contributing to eutrophication of Aba Samuel water reservior in Addis Ababa, Ethiopia. *Ethiop. J. Health Sci.* **2002**, *14*, 109–113.
27. Stroud, A. Water hyacinth (*Eichhornia crassipes* [Mart.] Solms) in Ethiopia. In Proceedings of the 9th Annual Conference of the Ethiopian Weed Science Committee, Addis Abeba, Ethiopia, 9–10 April 1991; EWSC: Addis Abeba, Ethiopia, 1994.
28. Zewdie, O.; Lemma, K.; Nikus, Y. *Prosopis juliflora*: Potentials and problems. *AREM*. 2005. Available online: https://agris.fao.org/agris-search/search.do?recordID=ET2006000002 (accessed on 13 March 2022).
29. Zelalem, W.; Kifle, D.; Mengistu, S.; Wondie, A. Weed-bed Macro-invertebrate Composition and Abundance in Relation to Water Hyacinth (*Eichhornia crassipes* (Mart) Solms) in the North-Eastern. *J. Agric. Environ. Sci.* **2017**, *2*, 50–64.
30. Firehun, Y.; Struik, P.C.; Lantinga, E.A.; Taye, T. Joint use of insects and fungal pathogens in the management of water hyacinth (*Eichhornia crassipes*): Perspectives for Ethiopia. *J. Aquat. Plant Manag.* **2013**, *51*, 109–121.
31. Yirefu, F.; Yirefu, F.; Struik, P.; Struik, P.; Lantinga, E.; Lantinga, E.; Tessema, T.; Tessema, T. Occurrence and diversity of fungal pathogens associated with water hyacinth and their potential as biocontrol agents in the Rift Valley of Ethiopia. *Int. J. Pest Manag.* **2017**, *63*, 355–363. [CrossRef]
32. Tegene, S.; Ayele, N. Prevalence and Intensity of Water Hyacinth Infestation in the Water Bodies of Rift Valley, Ethiopia. *J. Agric. Nat. Resour. Sci.* **2014**, *1*, 118–126.

33. Dersseh, M.G.; Kibret, A.A.; Tilahun, S.A.; Worqlul, A.W.; Moges, M.A.; Dagnew, D.C.; Abebe, W.B.; Melesse, A.M. Potential of Water Hyacinth Infestation on Lake Tana, Ethiopia: A Prediction Using a GIS-Based Multi-Criteria Technique. *Water* **2019**, *11*, 1921. [CrossRef]
34. *ASTM C1084*; Standard Test Method for Portland-Cement Content of Hardened Hydraulic-Cement. ASTM: West Conshohocken, PA, USA, 2015; Volume i, pp. 315–319. [CrossRef]
35. Time, I.S. ASTM C150-12 Common Reference Type I Portland Cement for Use in ASTM C989-12 Slag Activity Testing. 2012. Available online: http://www.ccrl.us/ (accessed on 13 March 2022).
36. American Concrete Institute. *Standard Practice for Selecting Proportions for Normal, Heavyweight, and Mass Concrete*; American Concrete Institute: Farmington Hills, MI, USA, 1991; pp. 1–38.
37. *ASTM C143/C143M*; Standard Test Method for Slump of Hydraulic-Cement Concrete. ASTM: West Conshohocken, PA, USA, 2015; pp. 1–4. [CrossRef]
38. *ASTM C 642-06*; Standard Test Method for Density, Absorption, and Voids in Hardened Concrete. ASTM International, United States. Annu. B. ASTM Stand.: West Conshohocken, PA, USA, 1997; pp. 1–3.
39. *ASTM C109/C109M-02*; Standard Test Method for Compressive Strength of Hydraulic Cement Mortars. Annu. B. ASTM Stand.: West Conshohocken, PA, USA, 2020; Volume 4, p. 9.
40. ASTM. Standard Specification for Ferromolybdenum. *Current* **2000**, *89*, 6–7. [CrossRef]
41. Hwang, C.-L.; Yehualaw, M.D.; Vo, D.-H.; Huynh, T.-P. Development of high-strength alkali-activated pastes containing high volumes of waste brick and ceramic powders. *Constr. Build. Mater.* **2019**, *218*, 519–529. [CrossRef]
42. Vo, D.-H.; Hwang, C.-L.; Thi, K.-D.T.; Yehualaw, M.D.; Chen, W.-C. Effect of Fly Ash and Reactive MgO on the Engineering Properties and Durability of High-Performance Concrete Produced with Alkali-Activated Slag and Recycled Aggregate. *J. Mater. Civ. Eng.* **2020**, *32*, 04020332. [CrossRef]
43. Frías, M.; Savastano, H.; Villar, E.; de Rojas, M.I.S.; Santos, S. Characterization and properties of blended cement matrices containing activated bamboo leaf wastes. *Cem. Concr. Compos.* **2012**, *34*, 1019–1023. [CrossRef]
44. Yehualaw, M.D.; Hwang, C.-L.; Vo, D.-H.; Koyenga, A. Effect of alkali activator concentration on waste brick powder-based ecofriendly mortar cured at ambient temperature. *J. Mater. Cycles Waste Manag.* **2021**, *23*, 727–740. [CrossRef]
45. Hwang, C.-L.; Yehualaw, M.D.; Vo, D.-H.; Huynh, T.-P.; Largo, A. Performance evaluation of alkali activated mortar containing high volume of waste brick powder blended with ground granulated blast furnace slag cured at ambient temperature. *Constr. Build. Mater.* **2019**, *223*, 657–667. [CrossRef]
46. Shah, V.; Scott, A. Hydration and microstructural characteristics of MgO in the presence of metakaolin and silica fume. *Cem. Concr. Compos.* **2021**, *121*, 104068. [CrossRef]
47. Balagam, S.R.K. Thermogravimetric Analysis of Concrete. *IJMTER* **2018**, *4*, 5.
48. Krishnan, E.; Solaiyan, V. Study on engineering and microstructural properties of ambient cured GP concrete produce from GGBS and M-sand. *J. Crit. Rev.* **2020**, *7*, 499–506.
49. Zhang, S.P.; Zong, L. Evaluation of Relationship between Water Absorption and Durability of Concrete Materials. *Adv. Mater. Sci. Eng.* **2014**, *2014*, 650373. [CrossRef]
50. Nie, Q.; Zhou, C.; Shu, X.; He, Q.; Huang, B. Chemical, Mechanical, and Durability Properties of Concrete with Local Mineral Admixtures under Sulfate Environment in Northwest China. *Materials* **2014**, *7*, 3772–3785. [CrossRef]

sustainability

MDPI

Article

External Sulfate Attack of Ambient-Cured One-Part Alkali-Activated Self-Consolidating Concrete

Dima Kanaan, Ahmed M. Soliman * and Amine el Mahdi Safhi

Department of Building, Civil and Environmental Engineering, Concordia University, Montreal, QC H3G 1M8, Canada
* Correspondence: ahmed.soliman@concordia.ca

Abstract: The mechanism of sulfate attack on alkali-activated materials, particularly the alkali-activated self-consolidating concrete (AASCC), is complex and contradictory. This could be due to the wide range of precursor and activator materials used in the production of AASCC mixtures, which has called into question the reliability and validity of existing evaluation procedures and practices. This paper presents a systematic research effort on AASCC mixtures, based on granulated blast-furnace slag, prone to various sulfate attack scenarios that are thought necessary to establish a proposed criterion. The conducted experimental design demonstrated that single-, binary-, and ternary-precursor AASCC samples, activated with 1:1 Na_2CO_3 and MetaNa_2SiO_3, partially submerged in sodium, magnesium, and mixed sulfate solutions could experience a dual sulfate attack scheme. Sulfate attack can occur in the immersed section in sulfate solutions, while physical sulfate attack can occur in the portion above the solution level. The influence of physical sulfate attack on the concrete's characteristics was not significant given that the damage was confined to the outer surface. However, the damage was primarily monitored by the AASCC different systems' pore structure, which resulted in the leaching of ions from samples to solutions. It was found that maintaining the pH in the sulfate solutions increased the rate of damage of AASCC mixtures. Furthermore, binary, and ternary precursor blends partially replacing slag with SF, or FA resulted in decreased porosity, surface scaling, and AASCC deterioration caused by an expansion in the volume of very small diameter pores. Finally, in all AASCC systems, gypsum and ettringite were the primary degradation products of sulfate attack.

Keywords: external sulfate attack; alkali-activated materials; self-consolidating concrete; chemical attacks; durability; granulated blast-furnace slag

check for updates

Citation: Kanaan, D.; Soliman, A.M.; Safhi, A.e.M. External Sulfate Attack of Ambient-Cured One-Part Alkali-Activated Self-Consolidating Concrete. *Sustainability* **2023**, *15*, 4127. https://doi.org/10.3390/su15054127

Academic Editors: Woubishet Zewdu Taffese and Sandra Barbosa Nunes

Received: 24 December 2022
Revised: 14 February 2023
Accepted: 19 February 2023
Published: 24 February 2023

1. Introduction

External sulfate attack (ESA), caused by the invasion of sulfuric ions in soils, underground, marine, or industrial wastewater, is a substantial means of deterioration of concrete in-service. Even though sulfate typically damages the cement-paste matrix, its adversity depends on the types of binder, nature, and concentrations of sulfate solutions, concrete quality, and surrounding conditions. Generally, cement paste pores are filled with a highly basic solution (i.e., pH > 12.5). Consequently, any medium with a lower pH value would be an aggressive condition for the cementitious matrix. Concrete exposed to ESA suffers from expansion, cracking, strength loss, and eventually, disintegration.

Different test methods are utilized to examine the ESA mechanisms and effects, e.g., ASTM C1012 "Standard Test Method for Length Change of Hydraulic-Cement Mortars Exposed to a Sulfate Solution" [1]. This commonly used standard was criticized due to its limitation (orientation-expansion effect). In 1997, Taylor reclaimed that laboratory research focuses on expansion and specimen cracking, but field experiences denoted a higher loss of adhesion and strength as a prevalent sign of deterioration [2]. Additionally, most of the existent standardized durability testing methods have especially been founded to test the long-term durability performance of Portland cement systems. The lack of standardized testing and evaluation criteria for alkali-activated materials (AAMs) symbolizes one of the main barriers facing its spread as no reliable long-term durability data are available. Various testing methods are thus vital to better predict the AAM's different damage mechanisms while being exposed to sulfate attacks (i.e., partial immersion exposure tests).

Thus, this research was dedicated to investigating the chemical and physio-chemical degradation mechanisms (physical or chemical) of alkali-activated self-consolidating concrete (AASCCs) partially exposed to aggressive aqueous solutions. This is crucial as the transport mechanism in partially exposed concrete structures does not appear to cause the same distress as in entirely exposed concrete components. Diffusion is the main transport mechanism that triggers a higher deterioration rate under saturated conditions. It is also known to be slow compared to other mechanisms, as an aqueous ionic transport mechanism, and is not apparent until a thermal or concentration gradient is generated. Hence, the "sulfate-related deterioration rate defies the expectations of the relevant models" [3].

2. Research Significance

The mechanism of sulfate attack on AAM, particularly AASCC, is complex and contradictory. This could be due to the wide range of precursor and activator materials used in the production of AASCC mixtures. The physical form of sulfate attack is not evident when testing a fully saturated specimen, the standard method of exposure recommended by the ASTM C1012 [1]. In addition, there is no standard test available for the partially immersed concrete in sulfate solutions. This has called into question the reliability and validity of existing evaluation procedures and practices [4–6]. This paper presents a systematic research effort on AASCC mixtures prone to various sulfate attack scenarios that are thought necessary to establish a proposed criterion. AASCCs were partially exposed to 10% single and mixed sulfate solutions in a controlled pH environment as part of accelerated degradation tests, a step to simulate harsh field conditions. This would result in a better understanding of the sulfate attack mechanism as well as the development of more reliable assessment tools for predicting AASCC performance and service life.

3. Experimental Program

3.1. Materials and Mixture Proportion

Ground granulated blast furnace slag (GGBFS) was employed as the main precursor to produce single, binary, and ternary AASCC. Fly ash Class-F (FA) and silica fume (SF) were added with various proportions replacing GGBFS. The physical and chemical properties of the precursor materials, along with the X-ray diffraction patterns, were previously published in a previous work [7]. Two dry-powder activators, MetaNa_2SiO_3 and Na_2CO_3 with a 1:1 ratio, were mutually utilized to activate the AASCC mixtures. The used Na_2SiO_3 had the density of 1.09 g/cm^3 and a molar ratio of 1.0. The Na_2CO_3 powder had $\geq$99.5% purity and the density of 2.53 g/cm^3. Natural siliceous sand with a fineness modulus of 2.5, a specific gravity value of 2.68, and water absorption of 1.5% was used. Furthermore, coarse aggregate with a maximum nominal size of 19 mm, a specific gravity value of 2.71, and water absorption of 0.4% was used. Their volumetric contents were set to 60% and 40%, respectively. The water-to-binder (*w/b*) ratio was fixed at 0.40 by the mass of the binder. It is important to note that no superplasticizers were used in the production of AASCC mixtures. AASCC mixtures were divided into three groups: single-, binary-, and ternary-precursor groups. The only difference amongst mixtures in the same group is the activator content (i.e., 16, 20, and 25%). Mixtures with appropriate performance, i.e., the ability to balance the restricted fresh characteristics, strength gain, and expected durability performance, were chosen for further long-term investigation. The precise proportions of the mixtures were disclosed in a previous research paper [7].

Five AASCC mixtures were prepared with different precursor blends and compositions but the same activator dosage (25%), as indicated in Table 1. The S-25 designation refers to a single precursor 100% GGBFS AASCC combination activated with 25% dosage. Both binary AASCC groups used 90% GGBFS and 10% SF (assigned code B1) or 70% GGBFS and 30% FA (assigned code B2). The AASCC ternary mixtures contained 60% GGBFS, 33% FA, and 7% SF (code T1) or 50% GGBFS, 45% FA, and 5% SF (code T2).

Table 1. Summary of mixture's proportions used for the designed AASCC mixtures (in kg/m^3).

Mixture ID	Slag	FA	SF	Sand	CA
S-25	525	0	0	646	830
B1-25	472.5	0	52.5	652	825
B2-25	367.5	157.5	0	637	840
T1-25	315	173	37	625	840
T2-25	289	210	26	624	840

Along with ASTM C1012 [1], samples were continuously exposed to aggressive solutions of sodium sulfate (Na_2SO_4) and magnesium sulfate ($MgSO_4$). The selection of the different cations (Na^+ and Mg^{2+}) accompanying SO_4^{2-} shows not only several attacking and destruction mechanisms but also various solubilities. Although sodium is not an aggressive component, it has a significant impact on the solubilization of cement species and the pH of the pore solution. The effect of magnesium sulfate is far more substantial than either sodium or calcium sulfate. Magnesium and sodium solubilities were 35.7 g and 28.1 g in 100 g water at 25 °C, respectively [8]. These values are significantly higher than that of calcium (i.e., 0.205 g in 100 g water at 25 °C). The sulfate salts used in assessing the sulfate resistance of AASCC mixtures were Na_2SO_4 and $MgSO_4$ of purity > 99%. Each solution was prepared by dissolving either 50 g or 100 g of the solutes in 1 L tap water to obtain (5% and 10%) of the sulfate solution.

3.2. *Testing Program*

Several mixtures were tested to evaluate their fresh and mechanical properties, according to EFNARC requirements [9]. Mixtures satisfying the limits were considered for additional investigation targeting a balance of the fresh characteristics, strength gain, and expected durability performance. A synthesis of AASCC, mixtures with optimum performance were regarded for further assessment in this paper [10,11]. After the mixing procedure, the fresh properties were evaluated, and numerous specimens were casted to study the hardened properties. Cylindrical specimens with dimensions of 100 Ø 200 mm were cast without any vibration. Instantly after casting, samples were covered to prevent the evaporation of water and left intact at room temperature (23 ± 2 °C). Past 24 ± 4 h, the samples were demolded and cured in an ambient curing condition until the age of testing.

3.2.1. Single Damage-Factor

In partly exposed sulfate structures, the lower portion may be saturated with seawater, but the upper remains at ambient temperature and humidity conditions. The deterioration commonly occurs when the sulfate is drawn into the concrete, permeated upwards, and then evaporated upon reaching the surface. Accordingly, various types of degradation, i.e., leaching, microcracking, paste and permeability characteristic alteration, efflorescence, and surface scaling, may appear. The degradation can be accelerated through the transfer of larger amounts of sulfate throughout the matrix (Exposure I scenario of the current work). For instance, Boyd and Mindess [12] partially exposed concrete cylinders to a 5% Na_2SO_4 solution and concluded that while scaling implies significant damage, the internal damage caused by sulfate attacks may be more severe. Up to now, this physical form of sulfate attack is not evident when testing a fully saturated sample. Additionally, there is no available standard for the partially immersed concrete in sulfate solutions.

The concrete deterioration scheme proposed for single damage factor tests was divided into six groups (Figure 1). Sulfate solutions were replenished every 4 weeks with fresh ones. The importance of monitoring the pH of the sulfate solutions, although not specified in ASTM C1012 [1], is associated with the real field conditions in which concrete would be exposed to a continual supply of sulfate ions [13,14]. Mixtures were evaluated based on their visual appearance, changes in mass, cross-section variation, and pH values of the sulfate solutions for 183 d at the end of each 30 d.

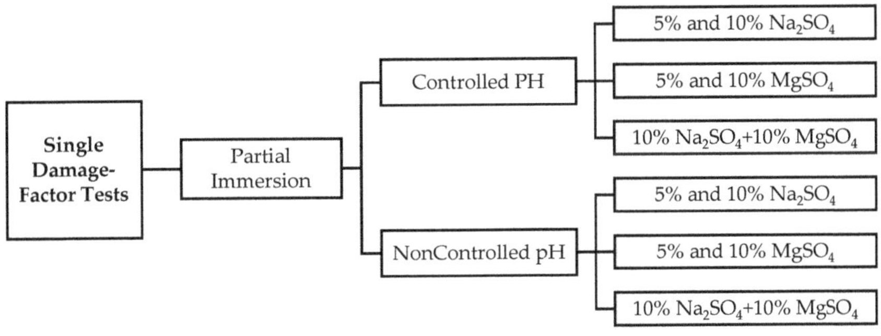

Figure 1. AASCC deterioration scheme.

A step to simulate harsh field conditions, AASCCs were partially exposed to 10% Na_2SO_4 and 10% $MgSO_4$ in addition to a mixed solution of 10% Na_2SO_4 +10% $MgSO_4$ in laboratory conditions. The attack intensity is well known to be mainly dependent on the pH of aggressive solutions. Attempting to control the pH is a simulation of field conditions in which concrete is exposed to an environment that usually contains mobile sulfate. Therefore, the effect of monitoring the pH of the solution on the leaching mechanisms of the AASCC mixtures was assessed. Alternatively, the resistance of AASCC mixtures to 5% Na_2SO_4 and 5% $MgSO_4$ was evaluated for comparative purposes. The initial pH value for sulfate solutions was kept in the range of 6.0–8.0, as recommended by the ASTM C1012 [1]. The exposure to the sulfate solutions lasted for 6 months (26 weeks) under ambient laboratory conditions. For specimens in controlled pH environments, the pH was maintained at the range of 6.0–8.0 by titration with dilute sulfuric acid solutions at constant time intervals. To guarantee a homogenous distribution of the solutions, regular stirring was accompanied by adding a pH regulator (i.e., sulfuric acid solutions).

3.2.2. Multiple Damage-Factor

Field concretes undergo wetting and drying cycles due to fluctuations in water levels caused by flooding, rainwater runoff, tides, and the cyclic migration of sulfate-containing groundwater through capillarity. Such activities induce the crystallization of sulfate salts under the concrete surface layer leading to cracking, scaling, and delamination. Existing test approaches cannot capture these complicated deterioration mechanisms (i.e., continuous immersion exposure). Besides, standard immersion tests have frequently presented different performances relative to the observed ones in real-field cases. A mandatory shift to performance-based specifications demands the elaboration of a performance-based evaluation that better depicts field conditions.

The question of whether wetting–drying cycles and the underlying mechanisms of this form of damage are physical or chemical is still controversial among researchers. For instance, the length change of concrete cylinders exposed to Na_2SO_4 solutions by 2.1% in a 20-year long-term research program was investigated [15]. Samples were exposed to wetting–drying cycles to accelerate the test. In each cycle, the samples were immersed in sulfate solutions for 16 h at room temperature, then air-dried for 8 h at 54 °C. A year after, it was reported that eight years of continuous immersion exposure triggered comparable damage to concrete samples under wetting–drying. De Almeida [16] proposed immersing samples in 16% Na_2SO_4 solutions for 2 h (wetting) followed by drying at 105 °C for 10 to 15 h. It was concluded that specimens failed under physical rather than chemical action. Likewise, mortar cylinders were partially immersed in 10% sodium and 10% magnesium sulfate solutions and exposed to 32–95% constant or cyclic relative humidity (RH) for up to 151 h [17]. This extensive study reported a thick efflorescence layer on the surface of partially immersed specimens in 10% Na_2SO_4 solution only, particularly with cyclic RH. The resistance of mortars to 5% sulfate solution accompanied with wetting–drying cycles were investigated [18]. Each cycle consisted of the exposure at room temperature for 6 d, followed by drying for 1 d at 100 °C. The latter exposure was observed to be more aggressive compared to continuous immersion exposure and has contributed to the complete disintegration of the samples within 17 weeks. Also, Haynes et al. [19] experimented on partially immersed concrete cylinders in (5% Na_2SO_4) and subjected them to several temperatures and RH. The conclusion was that specimens disintegrated after the exposure to cycles between 20 °C with 82% RH and 40 °C with 31% RH in two-week intervals.

In-depth research is still required on sulfate attack assessments, including wetting–drying cycles. Several questions remain unanswered, e.g., the real mechanisms of deterioration after these cycles and the degree and time of drying that can simulate the real field conditions. Variations in these parameters remain a challenge for researchers and standardization agencies. Besides, like partial immersion exposure tests, there is no available standard for concrete exposed to drying and wetting cycles. Further, the properties of AAM in various exposure conditions simulating field-like conditions need to be tackled. This section would provide an adequate understanding of the mechanisms of deterioration of AASCC mixes in the evaporation zones, resembling partially buried concrete elements, to develop reliable data on their durability. For multiple damage factor, mixes were evaluated while the exposure to different sulfate attack environments was considered with other concurrent damage mechanisms (i.e., cyclic environmental conditions).

After the curing process of AASCC mixtures, concrete cylinders were kept in ambient laboratory conditions (i.e., 23 ± 2 °C, RH of 50%) for 48 h to remove the excess moisture from surfaces. This technique would guarantee sufficiently dry samples for a uniform basis of analogy. After that, their preliminary physio-mechanical characteristics were recorded before the exposure to sulfate occurs. Specimens were partially immersed in highly concentrated solutions of (10%) sodium sulfate and (10%) magnesium sulfate solutions to evaluate the resistance AASCCs to combined effects of ESA with other ongoing deteriorating mechanisms (i.e., cyclic environmental conditions). Figure 2 illustrates the technique used to resemble the performance of concretes exposed to ESA [20]. The exposed part of the specimens, up to $^{1}/_{3}$ of the cylinder height, was fully immersed and sealed to minimize the evaporation of solutions and ensure that they were uptake by the cylinder. This section investigated the capability of holistic testing techniques in capturing the AASCC deterioration mechanisms within a reasonable testing duration. Therefore, sulfate solutions with a very high concentration were used to accelerate the deterioration process.

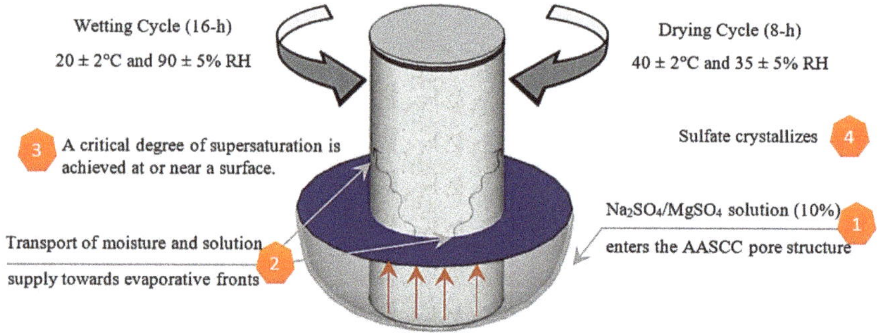

Figure 2. Schematic diagram of a specimen exposed to a field-like multi-damaging condition, inspired from [20].

The effect of interacting deterioration parameters, partial immersion in sulfate solutions under cyclic environmental conditions could depict the seasonal variations of atmospheric temperature (T) and RH found in many geographic locations, i.e., south of the USA. These unfavorable exposure conditions, an exposed portion of specimens to cycles of varied ambient temperature and RH, resembled a repeated crystallization deliquescence and hydration–dehydration processes that can cause rapid decay of AASCC mixtures. The repetitive crystal growth of sodium and magnesium sulfate from anhydrous to hydrous states was associated with a significant volume increase due to phase changes. Based on the stability phase diagram of the Na_2SO_4 solution, the direct precipitation of thenardite (above ~32.4 °C) and the rapid crystallization of mirabilite after thenardite dissolution causes a higher deterioration level in porous mixtures than mirabilite crystallization only [21]. Based on the available thermodynamic data and close to room ambient temperature, the crystalline phase of $MgSO_4$ (epsomite) would be stable. However, under dry conditions, epsomite would tend to dehydrate to form a monohydrate kieserite phase [22].

In this study, cylindrical specimens were subjected to alternating temperature and RH cycles for 183 d incessantly. Each cycle (24 h) includes two consecutive stages; an 8 h hot/dry stage (40 ± 2 °C and 35 ± 5% RH) followed by a 16 h temperate/humid stage (20 ± 2 °C and 90 ± 5% RH). The loss in the sulfate solutions was due to the continuous uptake, mainly during the dehydration (drying) cycles. They were frequently refilled up to one third of the cylinder's height. Moreover, sulfate solutions were renewed every 30 cycles. A conceptual framework is presented that considers four main visible deteriorating phenomena by ESA, i.e., spalling, delamination, cracking, and loss of cohesion (mass and strength loss). These complex physicochemical processes were monitored at two levels: macroscopic and integrated microscopic.

3.2.3. Mass Monitoring

To study the performance of AASCC mixtures in sulfate environments, specimens were partially immersed in sulfate solutions after a 90-d curing age. The masses of the samples were recorded monthly for 183 d. For mass change measurements, samples were weighed before the exposure to sulfate solutions at age 90 d and were taken as initial masses (M_i). After a specified time interval, samples were removed from the sulfate solutions and were brushed using a nylon brush to remove the attached debris. Under ambient laboratory conditions, samples were left to dry for 30 min before visual inspection, and measurements of masses. The change in weight (MC, %) was calculated as follows:

$$MC\ (\%) = \frac{M_i - M_j}{M_i} \times 100 \quad (1)$$

where M_i (g) is the average weight of two cylindrical samples; and M_j (g) is the average weight of two cylindrical samples exposed to sulfate solutions.

3.2.4. Cross-Section Variations Monitoring

During the sulfate attack, cross-sectional variations of samples can be monitored by certain geometrical parameters, such as sample volume, longitudinal expansion, and lateral variations of transverse sections (radius). The longitudinal expansion is the most questioned geometrical parameter under the ASTM C1012 [1]. Expansion and swelling of the cementitious matrix under sulfate attacks are frequently related to the formation of ettringite and gypsum due to crystal growth pressure [3]. Gypsum and ettringite formations require a calcium source that can be supplied from the leaching of portlandite and C-S-H. Portlandite leaching and progressive C-S-H decalcification led to a softened cement matrix and reduced strength. The formation of calcium sulfate-containing degradation products is related to Portland cement behavior. Expansion, however, does not capture certain degradation-induced changes, while volume and sample monitoring for the radius enables the evaluation of leaching activities and geometrical variations accordingly. The extent of leaching of elements is affected by the pH level, binder microstructure, permeability, and porosity of the material considered. The behavior of various AASCC mixture systems under sulfate attack is associated with the existence of C-S-H and C-A-S-H hydrotalcite gel; the former phase maybe highly resilient to sulfate attack.

The cylindrical samples' longitudinal and volume variations were directly evaluated from experimental data where the proposed lateral vs. longitudinal deformation pattern corresponds to the deformation path. This allows various binder materials to be classified according to the magnitude of the deformation in concrete mixtures. The lateral deformation is more sensitive to sulfate attacks than longitudinal expansion, making this approach useful in testing low expansion materials (i.e., AAMs). Equation (2) refers to the relative variation in the cross-sectional radius derived from the relative variation of the elongation x and the relative variation in volume z monitored during the test.

$$y = \sqrt{\frac{1+z}{1+x}} - 1 \tag{2}$$

where x is $\Delta L/Li$, y is $\Delta R/Ri$, and z is $\Delta V/Vi$ with Vi initial global volume, Li initial sample length, and Ri initial radius of a cross-section.

3.2.5. Microstructural Observations

After the visual inspection of test specimens, their degradation mechanisms were analyzed using X-ray diffraction (XRD), ICP-OES (inductively coupled plasma-optical emission spectrometry), ion chromatography, and differential scanning calorimetry (DSC) at a heating rate of 10 °C/min. X-ray diffraction (XRD) was conducted employing a Bruker D8 Advance diffractometer (CuKα radiation, 1.5406 Å) with an imaging plate detector to collect data in a range of 10–90° 2θ. XRD measurements were performed on powder samples, passing #200 sieve (75 μm), extracted from the superficial surfaces of the exposed samples (0–15 mm) after each exposure interval. For further medium analysis with the ICP and ion chromatography, the solutions' pH measurements and ion concentrations, i.e., Ca^{2+}, Na^{+}, SO_4^{2-}, and Mg^{2+}, was performed after the immersion of AASCC specimens.

4. Results and Discussions

4.1. Exposure I: Controlled pH

Sulfate penetration is often governed by diffusion through the pore structure and is consequently slow in the early stages of exposure. While exposed to a 10% sulfate solution in step 1, gypsum precipitation and ettringite production may reduce and offset the sulfate concentration in the pore solution. The presence of microcracks after a period of exposure raises the sulfate concentration in the pore solution in step 2 as sulfate can pass through the open cracks unhindered. For AASCC cylinders partially exposed to various concentrations of sulfate solutions, it was found that the crystallization pressure does not increase comparably with the supersaturation of the pore solution. Instead, changes in solution concentration due to the dual interactions between the paste and sulfate solutions may affect the crystallization pressure less in the Exposure I (high supersaturation) scenario than in the Exposure III (low supersaturation) environment. Then again, due to kinetic effects, the lower the concentration of sulfate in solutions, the slower the formation of expansive products, such as ettringite, resulting in lower crystallization pressure [23]. The sulfate concentration has a significant effect considering different exposure scenarios. The difference in crystallization pressure in AASCC cylinders have different causes. Using 10% sulfate solutions (Exposure I) and 5% solutions (Exposure III), the sulfate concentrations inside the cylinders were expected to be alike despite the variation in sulfate concentration between the container solution and samples. Still, the higher stress could be due to higher sulfate concentration (Exposure I), promoting cylinder supersaturation.

4.1.1. Visual Appearance

Single-precursor mixtures. The Malhotra et al. [24] rating system was used to assess the visual situation of each concrete cylinder after six months of partial immersion in sulfate solutions. The visual examination of AASCC single-precursor mixtures (S-25) revealed no signs of deterioration at the samples surface in the 10% Na_2SO_4 (Table 2) and 10% Na_2SO_4 + 10% $MgSO_4$ mixed solutions in a controlled pH environment. An exception was noticed when the concrete cylinders were exposed to a 10% $MgSO_4$ solution. The exposed areas were cracked, and surface delamination was observed.

Table 2. AASCC mixtures exposed to 10% sulfate solutions at a controlled pH environment.

Sulfate Solution	Exposure I				
	S-25	B1-25	B2-25	T1-25	T2-25
$MgSO_4$					
Na_2SO_4					
Mixed					

Binary-precursor mixtures. The samples of binary-1 mixes using 10% SF replacing slag and subjected to 10% sulfate solutions showed that magnesium sulfate greatly affected the concretes properties. The AASCC binary-1 mixture showed an expansion and cracking after exposure to 10% $MgSO_4$. The exposed area of the concrete mixture also suffered a massive loss (Table 2). Samples in mixed solutions exhibited a significant surface spalling above the exposed area to sulfate solution. While in Na_2SO_4 solution (Table 2), cracking along the length of the sample was more distinct. For binary-2 mixtures using 30% FA replacing slag, no visible signs of deterioration were detected after exposure to different sulfate solutions after 6 months. However, cracks were noticeable in the surface of samples exposed to 10% $MgSO_4$ and 10% mixed sulfate solutions.

Ternary-precursor mixtures. No visible signs emerged of deterioration while exposing the ternary mixtures to different sulfate solutions in a controlled pH environment. However, the ternary-1 set of mixtures yielded a greater surface scaling at the exposed area in the 10% $MgSO_4$ (Table 2) than in 10% Na_2SO_4 sulfate solution.

4.1.2. Mass and Cross-Section Variations

Single-precursor mixtures. Figure 3a shows the mass variation of single-precursor AASCC cylinders after the partial immersion in 10% $MgSO_4$, 10% Na_2SO_4, and 10% mixed controlled pH sulfate solutions. While Figure 3b shows the cross-section variation of these specimens after the partial immersion for 182 d. When exposed to a 10% $MgSO_4$, 10% mixed, and 10% Na_2SO_4 sulfate solutions, the cross-section decreases (0.5%, 1%, and 1.4%)

were negligible, while the mass remained to decrease but below 1%, probably due to alkalis and calcium leaching. The absence of significant damage in AASCC single-precursor specimens using 100% slag in the 10% sulfate solutions suggests the scarce formation of expansive products, such as ettringite and gypsum. The presence of aluminum in C-A-S-H and hydrotalcite gels, combined with the absence of portlandite, can result in high resistance to sulfate attack in single-precursor mixtures containing 100% slag [25]. The XRD analysis suggests that the supply of calcium and/or aluminum in single-precursor mixtures is insufficient to allow ettringite formation, as shown in Figure 4. However, traces of gypsum were detected in the samples exposed to 10% $MgSO_4$ solutions.

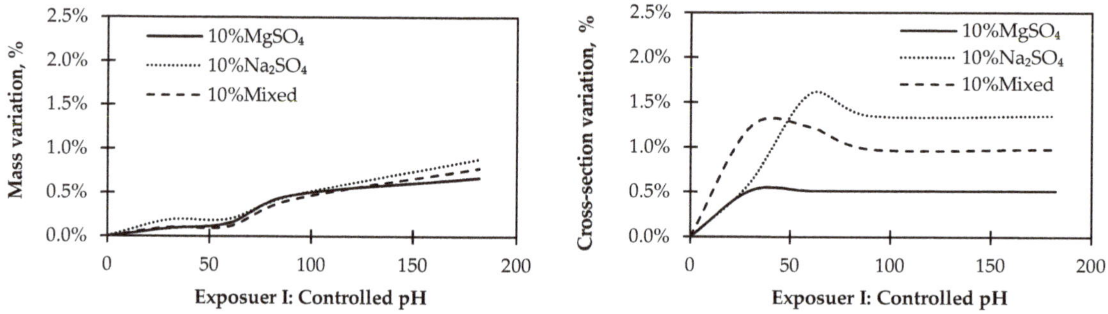

(a) (b)

Figure 3. Single-precursor AASCC mixtures exposed to controlled pH environment: (**a**) Time-dependent mass change; (**b**) Time-dependent cross-section change.

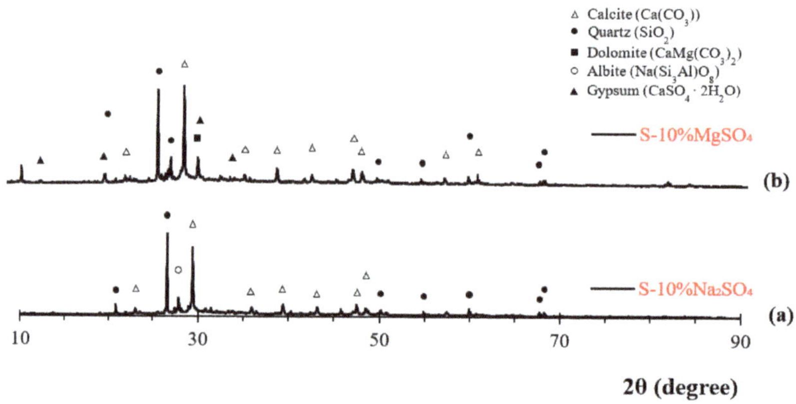

Figure 4. XRD spectra of AASCC single-precursor specimens exposed to sulfate attack: (**a**) In 10% Na_2SO_4 solution; (**b**) In 10% $MgSO_4$ solution.

Binary-precursor mixtures. The mass and cross-section variations of binary-precursor specimens after 6 months of partial immersion in 10% different sulfate solutions were presented in Figure 5. It can be adhered that the results in binary mixtures vary depending on the precursor type used, either SF in binary-1 or FA in binary-2. In the controlled 10% $MgSO_4$ solution, the binary-1 AASCC mixture made with 10% SF replacing slag yielded a maximum mass and cross-section loss of ~6.6% and 11.5%, respectively. This expected behavior can be attributed to the two-way ion diffusion between the $MgSO_4$ sulfate solution and mixtures. The high concentrations of Ca, Mg, and Na in the binary-1 AASCC mixture have been observed in ion chromatography data in Table 3, resembling the sulfate resistance performance of the binary-1 mixture. The binary-2 mixture made with 30% FA, on the other hand, had a mass and cross-section loss that was ~86% lower than the binary-1 mixture. When exposed to 10% Na_2SO_4 and mixed sulfate solutions, the mass loss was limited to 1% in binary-1 and 1.5% in binary-2 mixtures. For both binary sets, the minimal mass and cross-section variation was observed in 10% Na_2SO_4 due to the C-(N)-A-S-H gel changes, leading to the formation of microcracks within the binary network structure. However, the $MgSO_4$ attack mechanism can cause decalcification of calcium-containing systems such as C-(N)-A-S-H gels with high calcium content. This can lead to the formation of gypsum ($CaSO_4 \cdot 2H_2O$) gels that lack cementing properties [4–6,26].

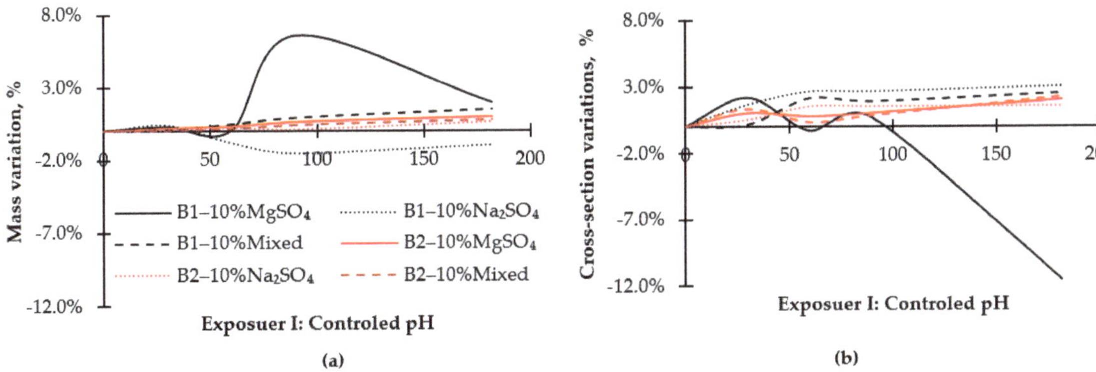

Figure 5. Time dependent (**a**) mass change and (**b**) cross-section change of binary-precursor AASCC mixtures exposed to 10% sulfate solutions at a controlled pH environment.

Table 3. Ion chromatography analysis and pH after 182 d in 10% sulfate solutions.

Sulfate Solution	Mixture ID	pH	Element Concentration (mg/L)			
			Na	Ca	Mg	SiO_3
10% $MgSO_4$	Single	10.1	3515	756	87	26
	Binary-1	9.40	32,485	34,820	581	48
	Binary-2	10.1	6468	604	257	396
	Ternary-1	9.90	6920	836	999	25
	Ternary-2	10.1	3516	13,679	1368	135
10% Na_2SO_4	Single	13.6	5739	42	247	22
	Binary-1	9.80	98,134	25,322	436	64
	Binary-2	11.3	4070	112	262	320
	Ternary-1	12.8	4439	202	25	16
	Ternary-2	12.8	5190	614	3	81
10% $MgSO_4$ + 10% Na_2SO_4	Single	10.9	6299	522	145	20
	Binary-1	9.80	88,184	22,405	1266	53
	Binary-2	10.3	5886	460	794	638
	Ternary-1	10.2	5800	1348	1678	25
	Ternary-2	10.3	6783	5846	1109	84

In the AASCC binary-1 mixtures with 10% SF substituting slag, XRD analysis revealed various degradation products in different sulfate solutions (Figure 6). Gypsum was the predominant expansive product freebie in $MgSO_4$ solution samples after 6 months of immersion. It is important to note that gypsum formation can result in softening and a loss of mass and strength over time. However, a trace of ettringite was found in samples subjected to Na_2SO_4 solution due to the interactions between sulfate ions and AASCC paste components. Likewise, ettringite was present in the samples exposed to a 10% $MgSO_4$ + 10% Na_2SO_4 mixed solution, and a significant amount of thenardite precipitates were identified. The damage above the solution level can be explained through the salt crystallization pressure theory. A supersaturated solution can create salt crystals, which exert pressure on the concrete pore walls, potentially damaging the matrix [27].

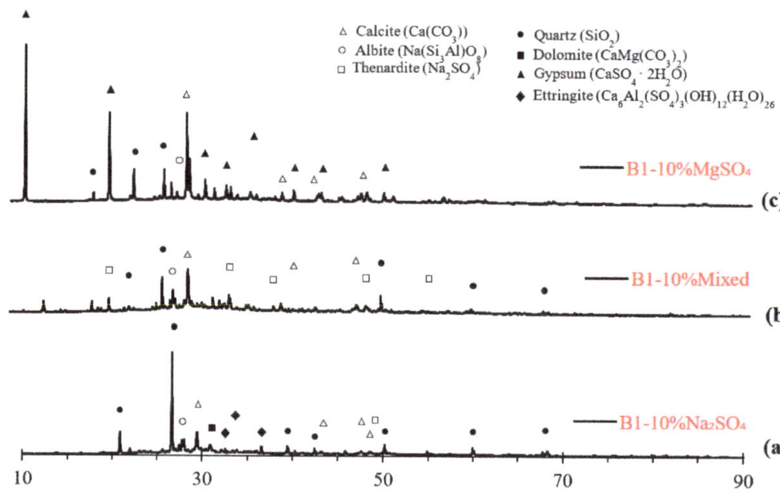

Figure 6. XRD spectra (2θ°) of binary-1 AASCC specimens exposed to sulfate attack: (**a**) In 10% Na_2SO_4 solution; (**b**) In 10% mixed solution; (**c**) In 10% $MgSO_4$ solution.

Overall, the average pore diameter determines the permeability, which affects the ion exchange rate after exposure to a sulfate environment. Therefore, increasing the densification level of the forming pore structure appears to enhance the performance of concrete exposed to aggressive environments. The positive substitution of slag with SF in binary-1 mixtures reduced their durability when exposed to external sulfate environments. This could be due to an increase in the small-diameter pores percentage in binary-1 mixtures and an increase in capillary suction. As a result, the specimen's upper surface area for evaporation increased, making the cylinder more vulnerable to physical attacks.

The only main component responsible for sulfate attack detected in the degraded sections above the solution level in all AASCC samples was gypsum (Figure 7). However, there was evidence of physical sulfate attack damage in the form of surface scaling above the solution level, notably in the samples subjected to 10% mixed solution. Thus, chemical sulfate attack resulting from dual interactions between sulfate solutions and samples is more likely to be the dominating mechanism.

Ternary-precursor mixtures. The mass and cross-section variations of the ternary-precursor mixtures were almost identical. Both sets lost ~2.3% of their mass but remained constant until the end of the test (Figure 8). At the same time, ternary-1 specimens resulted in ~2% and 4% cross-section loss in 10% $MgSO_4$ + 10% Na_2SO_4 and 10% Na_2SO_4 solutions, respectively. The mass loss in ternary blends can be credited to the occurrence of C-A-S-H gels in calcium-rich mixtures due to the inclusion of Al_2O_3 in their structure, which resembles the sulfate attack mechanism in OPC mixtures. Incorporating low or free-calcium precursors, on the other hand, increased the ion exchange reactions between the various sulfate solutions and the ternary-precursor matrix. However, because of the high competition between carbonate and sulfate ions, extensive ettringite formation may be overlooked and kept under control when a sodium carbonate activator is used.

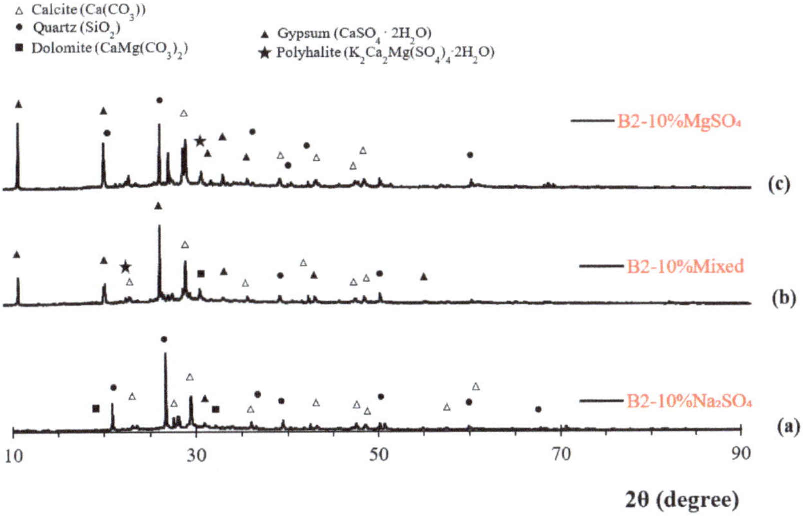

Figure 7. XRD spectra of binary-2 AASCC specimens exposed to sulfate attack: (**a**) In 10% Na_2SO_4 solution; (**b**) In 10% mixed solution; (**c**) In 10% $MgSO_4$ solution.

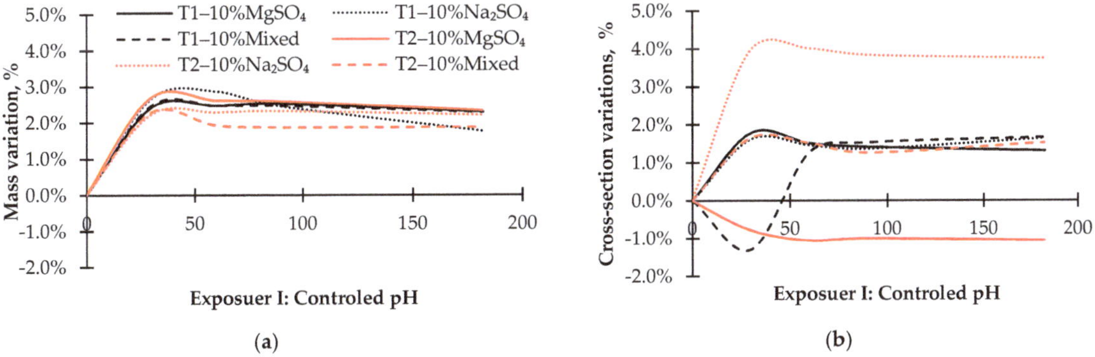

Figure 8. Time dependent (**a**) mass change and (**b**) cross-section change of ternary-precursor AASCC mixtures exposed to 10% sulfate solutions at a controlled pH environment.

XRD analyses (Figure 9) verified the structural changes in ternary-precursor blends, revealing thenardite in the damaged regions of samples exposed to a 10% Na_2SO_4 solution; gypsum was formed in specimens immersed in a 10% $MgSO_4$ solution. Overall, the formation of expansive gypsum and ettringite crystals is associated with exposure to external sources of sulfate attack. These crystals can colonize and accumulate in the concrete mixture's pores. Once filled, these crystals can initiate significant volumetric strain, leading to microcracking and concrete deterioration. Thus, crystallization of sulfate in macroscopic pores and cracks is desired. This is related to the formation of large-size expansive crystals, i.e., ettringite, which cannot exert a high crystallization pressure and thus are unlikely to be the source of the destruction. Since small crystals are prone to dissolution due to their unstable state, they are in equilibrium with higher sulfate concentrations.

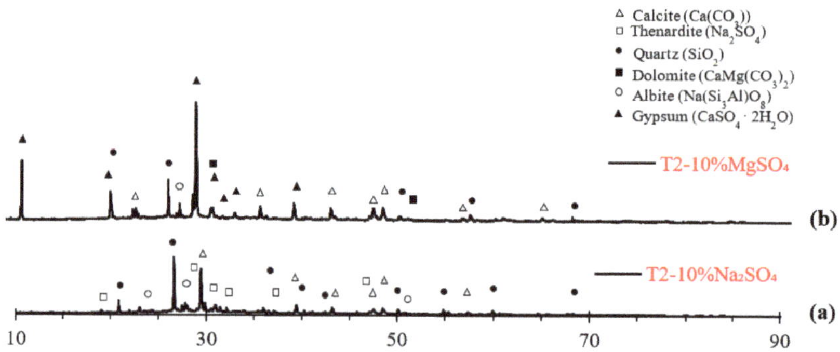

Figure 9. XRD spectra (2θ°) of ternary-2 AASCC specimens exposed to sulfate attack: (**a**) In 10% Na_2SO_4 solution; (**b**) In 10% $MgSO_4$ solution.

4.1.3. Ion Chromatography

Table 3 shows the ion chromatography analysis of different AASCC systems in controlled 10% sulfate solutions after 182 d of exposure. Na^+, Ca^{2+}, and Mg^{2+} concentrations in 10% $MgSO_4$, 10% Na_2SO_4, and 10% Na_2SO_4 + 10% $MgSO_4$ sulfate solutions were significantly increased during the testing period. For single-precursor mixtures, concrete cylinders with 100% slag showed high concentration levels of leached Ca when exposed to 10% $MgSO_4$ (756 mg/L) and 10% $MgSO_4$ + 10% Na_2SO_4 (522 mg/L) than in 10% Na_2SO_4 (42 mg/L) sulfate solutions. In contrast, the concentrations of Na and Ca in all sulfate solutions were extremely high, especially in binary-1 mixtures with 10% SF replacing slag, indicating that these elements were leached from the mixtures into solutions, which was in accord with the literature [4,28]. The decrease in pH of sulfate solutions $\leq$ 9.8 resulted in the decalcification of C-(A)-S-H gel, which was revealed by the increased Ca concentration in the sulfate solutions. For example, in $MgSO_4$ and mixed sulfate solutions, the Ca concentration was 34,820 mg/L and 25,322 mg/L, respectively compared with 22,405 mg/L in Na_2SO_4 solutions.

Exposure of binary-2 silicon-rich mixtures to various sulfate solutions resulted in extensive leaching of Si, Na, and Ca into the solutions. For instance, in 10% $MgSO_4$ and 10% Na_2SO_4 solutions, the concentration of SiO_3 was ~396 and 320 mg/L, respectively. While in 10% mixed sulfate solution, the concentration of SiO_3 recorded was 638 mg/L after 182 d. This can be caused by the leaching of the unreacted sodium silicate content present in the pore solution or the gel structure. However, due to the continuous monitoring of pH and the depletion of silicon sources present in the AASCC binary FA-slag mixtures, the amount of leached Si decreased over time. The high pH value of the sulfate solutions used to make binary-2 mixtures have been identified as a contributing factor to the Si leaching, which was shown earlier [29–31].

The Na^+, Ca^{2+}, and Mg^{2+} concentrations in ternary-precursor mixtures also increased over time, proving a more distinct impact of aggressive sulfate solutions on calcium-containing compounds, such as C-(N)-A-S-H gel in $MgSO_4$ and hybrid solutions. The test results of both ternary sets showed that higher pH values between 10 and 13 after 182 d of testing due to pore solution alkalis leaching or Ca release into the solutions due to the system's reaction with different sulfate compounds and ion exchange [4,28,32]. Ternary-2 and ternary-1, e.g., had Ca concentrations of 13,679 and 836 mg/L, 614 and 202 mg/L, and 5846 and 1348 mg/L in 10% $MgSO_4$, 10% Na_2SO_4, and mixed sulfate solutions, respectively.

4.2. Exposure II: NonControlled pH

4.2.1. Visual Appearance

Single-precursor mixtures. Table 4 shows the visual appearance of AASCC single-precursor mixtures exposed partially to an uncontrolled pH environment. It revealed no signs of deterioration in 10% $MgSO_4$ solution, few spalls of concrete at the exposed areas of specimens immersed in 10% Na_2SO_4, cracking and mass loss of concrete over the sulfate solution level of specimens immersed in 10% mixed solution. All specimens were surrounded with whitish precipitates, which indicated that the concrete had been affected.

Table 4. AASCC mixtures exposed to 10% sulfate solutions at uncontrolled pH environment.

Sulfate Solution	Exposure II				
	S-25	B1-25	B2-25	T1-25	T2-25
$MgSO_4$					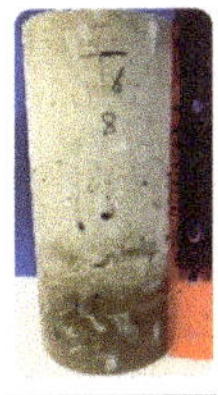
Na_2SO_4					
Mixed					

Binary-precursor mixtures. The uncontrolled pH exposure condition of binary-1 specimens to sulfate solutions for 6 months resulted in a higher level of deterioration, unexpectedly (Table 4). The severe loss of the concrete matrix at the bottom edge and above the exposed area was caused by the exposure to the $MgSO_4$ solution. In 10% Na_2SO_4 solution, broken portions above the partially exposed areas were noticeable, while white deposits were observed in 10% mixed solutions. For the binary-2 set of mixtures, in an uncontrolled pH sulfate environment, the surface erosion was minimal in all samples above the exposed area; however, it was significantly greater in 10% $MgSO_4$ (Table 4) and 10% $MgSO_4$ + 10% Na_2SO_4 sulfate solutions.

Ternary-precursor mixtures. After 6 months of exposure to sulfate in an uncontrolled pH environment, the scaling of ternary-2 (Table 4) concrete surfaces appeared higher than that of ternary-1 mixes, which appeared above the sulfate solution level.

4.2.2. Mass and Cross-Section Variations

Single-precursor mixtures. Figure 10 shows the time-dependent mass and cross-section variations of AASCC 100% slag specimens during the partial immersion in 10% $MgSO_4$, 10% Na_2SO_4, and 10% $MgSO_4$ + 10% Na_2SO_4 non-controlled pH sulfate solutions. White deposits were observed for all mixtures when exposed to all sulfate solutions, although mass loss remained constant $\leq$1%, most likely due to alkalis leaching in a near-neutral pH environment.

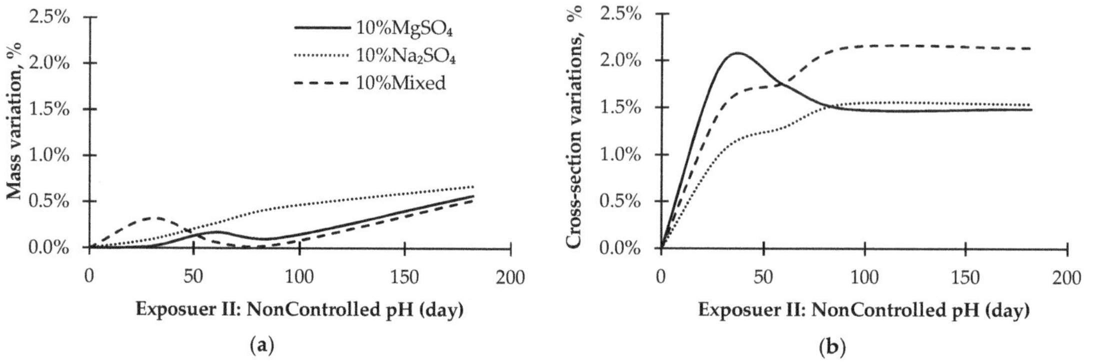

Figure 10. Single-precursor AASCC mixtures exposed to an uncontrolled pH environment time-dependent: (**a**) mass change; (**b**) cross-section change.

On the other hand, the reaction between sulfate and AASCC slag mixtures caused a change in the cross-section of specimens, especially before the renewal of solutions ~100 d. Exposure to 10% $MgSO_4$ solutions for 30 d produced a 2% change in cross-section, which is about double the increase in the other sulfate solutions, yet the change was minimal till the end of the test. In 10% Na_2SO_4 and mixed sulfate solutions, a progressive increase in cross-section loss was noticed until the end of the test, indicating the little possible development of expansive ettringite or gypsum.

Binary-precursor mixtures. The mass change of binary-precursor samples exposed to 10% $MgSO_4$, 10% Na_2SO_4, and 10% mixed solutions in an uncontrolled pH environment is illustrated in Figure 11. The mass loss noted for binary-1 samples subjected to $MgSO_4$ solution was 0.4% after 30 d of exposure, declining until the sulfate solution was renewed, yielding a mass gain of nearly 0.1% after 90 d. Samples continued to lose mass when the pH dropped between 6–8 at the end of the exposure period (2.1%) during the solution renewal. A different trend is observed for binary-2 samples in $MgSO_4$ solutions, where a marginal mass gain (0.1%) was noticed after 30 d, followed by a gradual mass loss reaching ~1%. However, neither the binary set of samples exposed to Na_2SO_4 and mixed sulfate solutions were significantly affected, with mass losses of (0.8% and 0.7%) for binary-1 and (1.1% and 0.9%) for binary-2, respectively. Figure 11b shows the cross-section changes that occurred due to the degradation and surface scaling of the AASCC two binary-precursor sets. The highest cross-section change was observed in the $MgSO_4$ solution for the binary-1 specimens (4.4%) made with 10% SF replacing slag, followed by cylinders exposed to Na_2SO_4 and mixed sulfate solutions with up to 2.4%. On the other hand, the binary-2 AASCC portion immersed into all sulfate solutions was mostly intact for all cylinders. Above the solution level, exposure to 10% mixed, 10% Na_2SO_4, and 10% $MgSO_4$ sulfate solutions for 182 d resulted in 2%, 1.7%, and 1.4% cross-section variation, respectively.

Ternary-precursor mixtures. After 182 d of exposure to sulfate solutions, the exposed surface of the specimens had no deposits and mostly showed no visible signs of deterioration in both sets of ternary mixtures. However, there were some mass changes in the specimens of <4% at the end of the testing period (Figure 12a). The ternary-1 samples prepared using a higher slag content lost a mass of 3.2% in the mixed sulfate, 2.5% in the Na_2SO_4, and 2.2% in the $MgSO_4$ sulfate solutions after 182 d of immersion. The extent of damage to ternary-1 specimens due to partial uncontrolled sulfate attack was not greatly associated with the cross-section changes. Ternary-2 cylinders gained 0.1% in the magnesium sulfate solution after 90 d of exposure and lost 0.5% at the end of the test. However, ternary-2 mixtures exposed to mixed and sodium sulfate solutions lost 1.8% and 0.5% at an uncontrolled pH environment. In the solutions of Na_2SO_4 + $MgSO_4$ and Na_2SO_4, ternary-2 samples had the most cross-section changes: 1.1% and 1.3% loss, while around 1.5% gain in the $MgSO_4$ solution (Figure 12b).

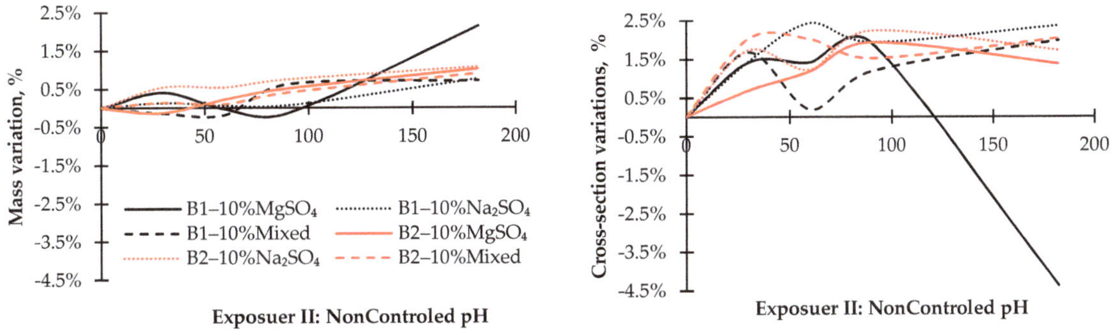

(a) (b)

Figure 11. Time dependent (**a**) mass change and (**b**) cross-section change of binary-precursor AASCC mixtures exposed to 10% sulfate solutions at an uncontrolled pH environment.

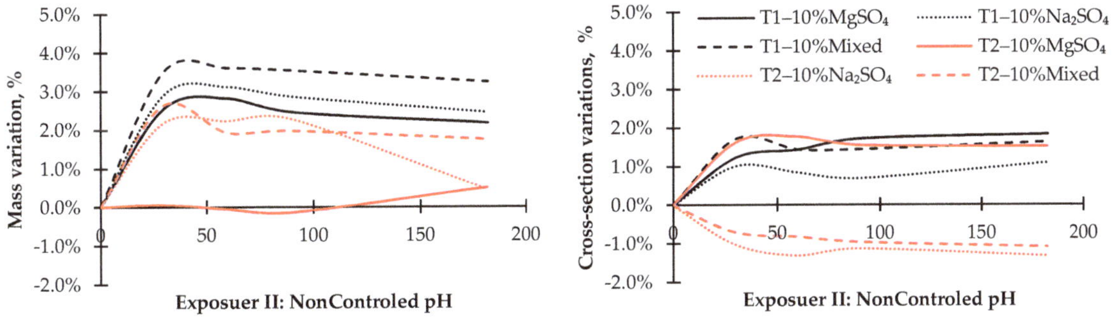

(a) (b)

Figure 12. Time dependent (**a**) mass change and (**b**) cross-section change of ternary-precursor AASCC mixtures exposed to 10% sulfate solutions at an uncontrolled pH environment.

4.3. Exposure III: 5% Sulfate Solution—Controlled pH

4.3.1. Visual Appearance

Single-precursor mixtures. No signs of deterioration were detected after the exposure of slag-AASCC mixtures to 5% $MgSO_4$, 5% Na_2SO_4, and 5% mixed sulfate solutions, as shown in Table 5. Overall, it can be noticed that all single-precursor samples stayed visually intact with visual integrity, except the ones subjected to 5% $MgSO_4$ solution.

Binary-precursor mixtures. For binary-1 mixtures using 10% SF replacing slag after 6 months of partial immersion in the 5% different sulfate solutions. The deteriorating conditions were most pronounced in 5% $MgSO_4$ (Table 5) and 5% Na_2SO_4 sulfate solutions. Cracks were detected along the length of the samples in all sulfate solutions. The soft whitish material above the exposed area also showed signs of damage. However, all binary-2 specimens were visually minimally intact, with some superficial cracks visible above the sulfate's solutions level.

Table 5. AASCC mixtures exposed to 5% sulfate solutions at controlled pH environment.

Sulfate Solution	Exposure III				
	S-25	**B1-25**	**B2-25**	**T1-25**	**T2-25**
$MgSO_4$					
Na_2SO_4	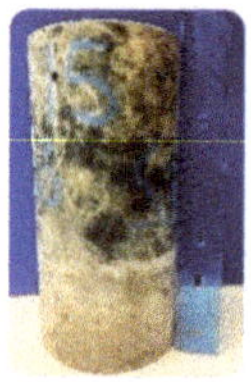				
Mixed					

Ternary-precursor mixtures. For both ternary-1 and ternary-2 AASCC mixtures using slag, FA, and SF, the immersed portions in the 5% sulfate solutions were mainly intact. Surface cracks and damage in the concrete's ternary-1and ternary-2 were confined to the concrete segments above the sulfate solutions level, the maximum exposure level.

4.3.2. Mass and Cross-Section Variations

Single-precursor mixtures. The mass change of samples made from 100% slag binder after 182 d of partial immersion in 5% of various sulfate solutions is illustrated in Figure 13. Exposing specimens to 5% $MgSO_4$ solution resulted in a 1% loss of mass, which is higher by 11%, and 67% than when exposed to 5% Na_2SO_4 and mixed sulfate solutions. However, the cross-section variation of the same set of samples immersed in 5% $MgSO_4$ solution was approximately 25% lower than that of the specimens exposed to 5% Na_2SO_4 and mixed sulfate solutions. The absence of significant deterioration in single-precursor slag specimens in 5% Na_2SO_4 and mixed solution could prove that expansive ettringite or gypsum formation is relatively low. On the other hand, specimen damage in $MgSO_4$ solutions could be related to the decalcification and transformation of C-(N)-A-S-H gels.

Binary-precursor mixtures. Figure 14a shows the mass and cross-section change of the binary-precursor cylinders in different 5% sulfate solutions. In the sodium sulfate solution, the binary-1 samples formulated with 10% SF replacing slag performed better than the binary-2 and had around 1.3% mass increase compared with 0.2% mass loss over the time of the experiment. Again, a similar trend was observed in the 5% mixed sulfate environment, with binary-1 gaining 0.2% mass after 182 d of exposure compared to losing 0.7% mass in binary-2. In contrast, in the binary-1 and binary-2 samples exposed to 5% $MgSO_4$ solutions, masses fluctuated and steadily increased from the initial value, i.e., 0.9% and 0.8%, respectively. The corresponding cross-section variations after 182 d of partial exposure are shown in Figure 14b. Some fluctuation of the cross-section was detected in all binary-precursor samples, especially in the magnesium and mixed sulfate solution. For example, the variations of the binary-precursor specimens finally equilibrated with a 5% and 1.2% decrease compared to the initial value in the binary-1 and binary-2 specimens, respectively. However, in the sodium sulfate solutions, 2.8% cross-section decline was measured in the binary-1 samples compared with ~1.7% in binary-2 samples. Almost similarly in the solution of sodium sulfate + magnesium sulfate, where 2.2% and 1.5% reduction was found in the binary-1 and binary-2 specimens, respectively.

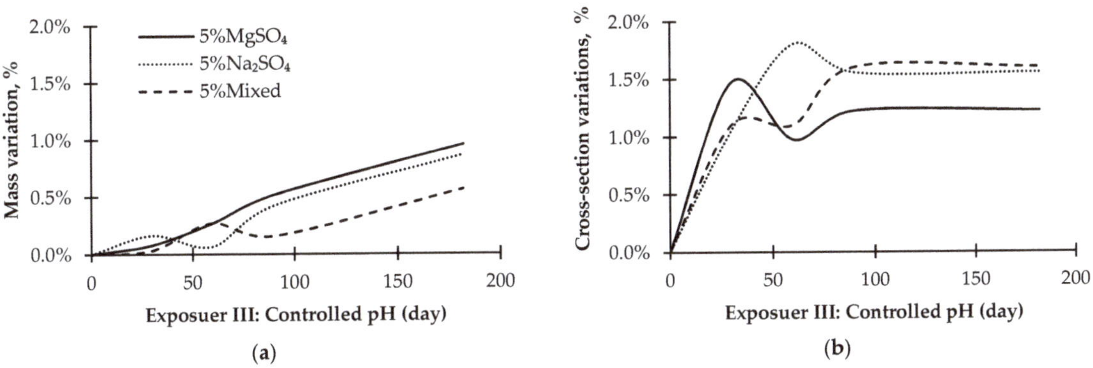

Figure 13. Single-precursor AASCC mixtures exposed to 5% sulfate solutions at a controlled pH environment: (**a**) Time-dependent mass change; (**b**) Time-dependent cross-section change.

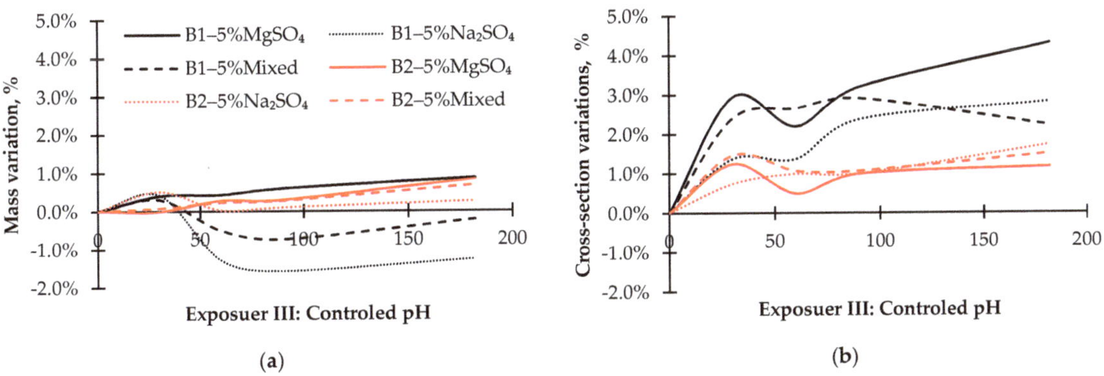

Figure 14. Time dependent (**a**) mass change and (**b**) cross-section change of binary-precursor AASCC mixtures exposed to 5% sulfate solutions at a controlled pH environment.

Overall, in binary mixes exposed to 5% solution, the damage was limited to the microstructural matrix in addition to few traces of damage above the embedded portion of cylinders, while the portion immersed in the solution was recovered in intact condition. The physical sulfate attack and leaching of ions were the proposed primary theories of damage. In contrast, a significant and steady paste transformation was observed after exposure to a 5% $MgSO_4$ solution, but little damage was observed in a 5% Na_2SO_4 solution. These results corroborate the findings of Ismail et al. [4], who discovered that when binary mixtures are exposed to $MgSO_4$ solution, the C-A (N)-S-H undergoes decalcification and silicate polymerization, but it remained essentially intact after exposure to Na_2SO_4.

Ternary-precursor mixtures. For ternary-precursor mixtures exposed to different 5% sulfate solutions, the mass and cross-section variations are shown in Figure 15. After 182 d of immersion, ternary-1 cylinders with a higher Ca content showed slightly more damage than ternary-2 cylinders. For example, after immersing ternary-1 samples in $MgSO_4$, Na_2SO_4, and mixed solutions, mass losses of 3.3%, 2.5%, and 2.7% were reported, respectively. In contrast, the mass variation of all ternary-2 samples decreased as the CaO amount decreased, resulting in 0.4%, 2.1%, and 2.3% mass loss in $MgSO_4$, Na_2SO_4, and mixed solutions, respectively. Contrary to expectations and mass change results of ternary-1 samples, cross-section variation in Na_2SO_4 and mixed sulfate solutions for ternary-2 mixtures was high at ~2.1% and 2.3%, respectively, then in $MgSO_4$ solution at ~0.4%.

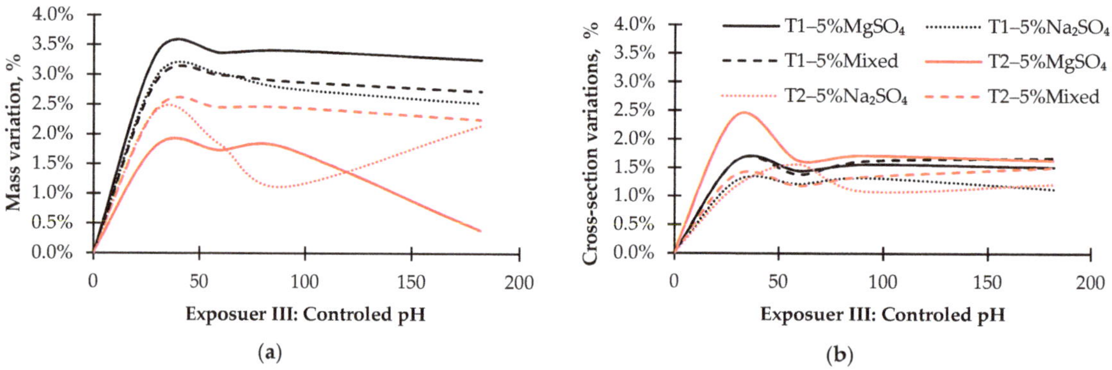

Figure 15. Time dependent (**a**) mass change and (**b**) cross-section change of ternary-precursor AASCC mixtures exposed to 5% sulfate solutions at a controlled pH environment.

4.3.3. Ion Chromatography

Table 6 shows the variation in pH and element concentration of solutions containing the three different series of AASCC samples, single-, binary-, and ternary-precursor samples and subjected to 5% $MgSO_4$, 5% Na_2SO_4, and 5% $MgSO_4$ + 5% Na_2SO_4 sulfate solutions. The pH of the solutions increased tremendously during the exposure to different sulfate solutions to maintain electroneutrality due to the migration and counter diffusion of alkali ions from the specimens to solutions. Prior to the immersion of specimens, the pH of all sulfate solutions was in range of 6–8, and after 182 d of exposure, the pH increased considerably. The deterioration levels are due to the concentrations of the elements leached up to 182 d, after which the damage of the cylinders in 50 g/L solutions takes off, indicating the onset of significant cracking as found by Ye et al. [33].

Table 6. Ion chromatography analysis and pH after 182 d in 5% sulfate solutions.

Sulfate Solution	Mixture ID	pH	Element Concentration in Different Sulfate Solutions (mg/L)			
			Na	Ca	Mg	SiO_3
5% $MgSO_4$	Single	12.1	4205	62	313	24
	Binary-1	9.80	51,829	1103	120	72
	Binary-2	10.5	1900	62	26	4598
	Ternary-1	9.90	1426	876	1441	101
	Ternary-2	11.1	18,826	10,123	1271	127
5% Na_2SO_4	Single	13.6	5081	38	193	23
	Binary-1	10.5	124,089	1795	67	72
	Binary-2	12.9	6170	50	356	552
	Ternary-1	12.8	4730	876	1832	49
	Ternary-2	12.7	7769	4596	1111	81
5% $MgSO_4$ + 5% Na_2SO_4	Single	13.0	5617	423	105	28
	Binary-1	13.0	72,134	7719	371	41
	Binary-2	10.9	5325	238	375	25
	Ternary-1	10.1	1950	588	933	16
	Ternary-2	10.5	8924	8322	4446	58

Despite the pH rise caused by interactions between sulfate solutions and AASCC various matrices, all elements were highly soluble at low sulfate concentrations, such as 5% solutions, as opposed to their status at higher sulfate concentrations as 10% solutions. In single-precursor systems, the extensive damage of the AASCC matrix was observed during immersion in 5% $MgSO_4$ + 5% Na_2SO_4 solutions where the leached Ca concentration was 423 mg/L compared to 7719 mg/L in binary-1 and 238 mg/L in binary-2 systems. In contrast, extensive deterioration and Ca leaching were observed when low-calcium systems, i.e., ternary-precursor systems, especially in ternary-1 samples (10,123 mg/L) in $MgSO_4$ sulfate solutions.

4.4. *Exposure IV: Drying and Wetting Cycles*

After 30 d of continuous exposure to daily wetting and drying cycles with drying for 8 h to a temperature of 40 ± 2 °C and RH of 35 ± 5%, followed by a wetting cycle for 16 h to a temperature of 20 ± 2 °C and RH of 90 ± 5%, efflorescence appeared on the concrete cylinders' drying surface above the solution level. In addition, salt precipitates were found on the immersed portion of the samples in sulfate solutions. According to previous studies [19–21,27,34–36], this exposure condition is favorable for the formation and precipitation of hydrous mirabilite from anhydrous thenardite and epsomite from kieserite, which can cause damage. In the Na_2SO_4-H_2O system, thenardite (Na_2SO_4) transformation into mirabilite ($Na_2SO_4 \cdot 10H_2O$) occurs at 20 °C and below 76% RH. This process generates an increase in the volume of 314% [37]. Meanwhile, three stable crystalline phases exist in the $MgSO_4$-H_2O system: epsomite ($MgSO_4 \cdot 7H_2O$), hexahydrite ($MgSO_4 \cdot 6H_2O$), and kieserite ($MgSO_4 \cdot H_2O$). When RH reaches > 80%, kieserite transforms into hexahydrite and epsomite, with the latter produced by hydration of hexahydrite [38]. The exposure lasted for six months (183 cycles of wetting and drying), and all samples were examined to diagnose the rate of damage. Higher surface scaling of concrete surfaces appeared on some cylinders above the sulfate solution level due to the crystallization of salts. Other samples, in contrast, revealed that the portion of concrete immersed in the solution was relatively ruined as the leaching process's kinetics increased with temperature [39]. It should be noted that the reduced permeability can be attributed to the denser structure of aluminosilicate gels, which reduced the manifestation of microcracking due to weathering effect. The schematic diagram (Figure 2) illustrates the relevant sulfate efflorescence phenomena.

4.4.1. Visual Appearance

Single-precursor mixtures. After six months of exposure, i.e., 183 cycles of wetting and drying, the specimens exhibited signs of damage. Surface scaling appeared on the surfaces of single-precursor cylinders exposed to 10% $MgSO_4$ and 10% $MgSO_4$ + 10% Na_2SO_4 mixed solutions (Table 7). The concrete mass loss and surface spalling were most pronounced in the 10% $MgSO_4$ solution. However, exposed cylinders to 10% Na_2SO_4 solution showed no signs of degradation. It is worthy to mention that the cyclic conversion of sulfate crystals can generate stresses in concrete pores. This phenomenon can lead to surface scaling and substantial deterioration levels [35].

Binary-precursor mixtures. For binary-1 mixtures, the mass loss due to the exposure to various sulfate solutions was caused in $MgSO_4$ and hybrid solutions. Instead, less degradation was noted with the Na_2SO_4 solution (Table 7) with less surface scaling and concrete damage. A similar tendency was seen in binary-2 mixtures but to a lesser extent. The binary-2 AASCC cylinders exhibit surface erosion and the manifestation of aggregates after the exposure to 10% $MgSO_4$ and 10% $MgSO_4$ + 10% Na_2SO_4 sulfate solutions under drying and wetting cycles on top of the solution level.

Table 7. AASCC mixtures exposed to 10% sulfate solutions under drying and wetting cycles.

Sulfate Solution	Exposure IV				
	S-25	**B1-25**	**B2-25**	**T1-25**	**T2-25**
$MgSO_4$	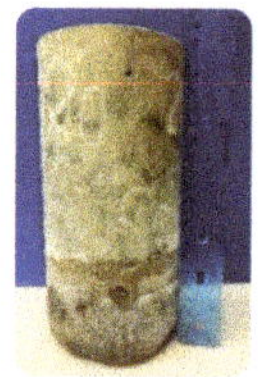				
Na_2SO_4		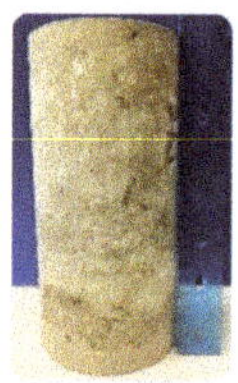	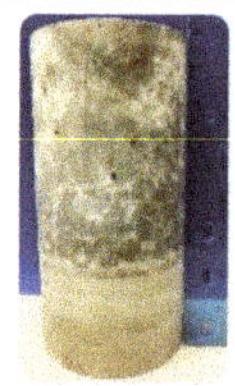	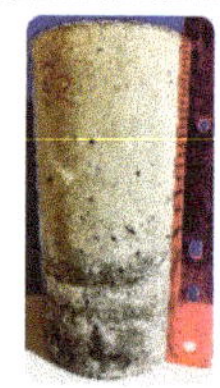	
Mixed		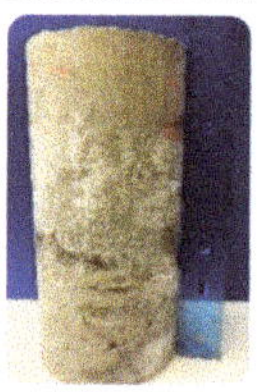			

Ternary-precursor mixtures. Ternary-1 and ternary-2 samples displayed some alterations in appearance, especially in the solution of 10% $MgSO_4$ under drying and wetting cycles. In the 10% Na_2SO_4 and 10% mixed sulfate solutions, samples were covered by a 1-mm-thick white cover, and a horizontal cracking in the ternary-2 specimens was found in the latter solution (Table 7). Ternary-1 samples deteriorated less than ternary-2 samples that reported the most significant mass and cross-section variations. Significant deterioration tends to occur when magnesium cation is present in the sulfate solution. For instance, ternary-2 samples in 10% $MgSO_4$ and 10% $MgSO_4$ +10% Na_2SO_4 lost ~5.4% and 5% of their weights, respectively.

4.4.2. Mass and Cross-Section Variations

Single-precursor mixtures. Figure 16 reports the time-dependent mass and cross-section changes made from 100% slag after 182 d of exposure to sulfate solutions partially and under 183 drying and wetting cycles. In all sulfate solutions, atypical surface scaling of the immersed portion of concrete cylinders was observed (Table 7), in contrast to the expected damage in part above the level of the solution. After 60 d in $MgSO_4$ solution, single-precursor cylinders lost ~8% of their mass, which dropped sharply to 2.4% at the end of the exposure time. Single-precursor samples exposed to 10% Na_2SO_4 solution lost 33% less mass than those exposed to 10% $MgSO_4$ solution, although the cross-section varied by roughly 2%. On the other hand, cylinders immersed in mixed sulfate solutions demonstrated an overall mass gain (0.8%) at 182 d, indicating that porosity decreased with immersion time. It should be noted that all single AASCC cylinders were intact above the solution level, indicating that the high sulfate concentration combined with drying and wetting cycles did not accelerate the damage mechanism.

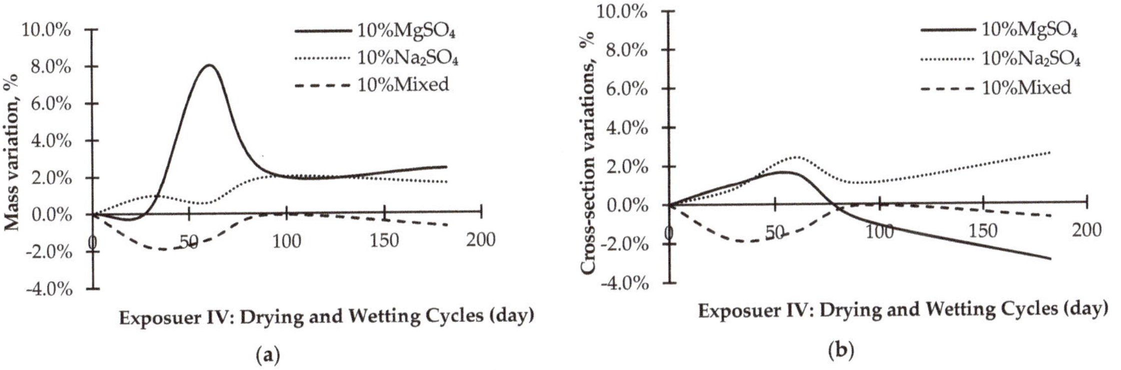

Figure 16. Single-precursor AASCC mixtures exposed to 10% sulfate solutions under drying and wetting cycles: (**a**) Time-dependent mass change; (**b**) Time-dependent cross-section change.

Binary-precursor mixtures. Figure 17 shows the mass and cross-section results of binary-precursor specimens immersed in various sulfate solutions and subjected to 183 drying–wetting cycles. For all binary samples exposed to different sulfate solutions, the mass loss and physical degradation arose due to spalling of the concrete surface above the solution level. This could be due to the nucleation and growth of sulfate crystals in the supersaturated solution, i.e., thenardite, which exerts stresses on the binary-precursor AASCC, causing damage and mass loss above the solution level and drying and wetting cycles. For example, the mass variations of binary-1 mixtures immersed in $MgSO_4$ solutions resulted in roughly 2% mass loss and cross-section variations compared with 1.4% mass loss and 2.2% cross-section variations in mixed sulfate solutions after 182 d. After 6 months of drying and wetting cycles, thenardite and ettringite were detected in binary-1 samples exposed to 10% Na_2SO_4 solution (Figure 18), even though a significant amount of gypsum was found in samples subjected to 10% $MgSO_4$ solution. This suggests that the drying and wetting cycles had a considerable effect on the rate of AASCC degradation.

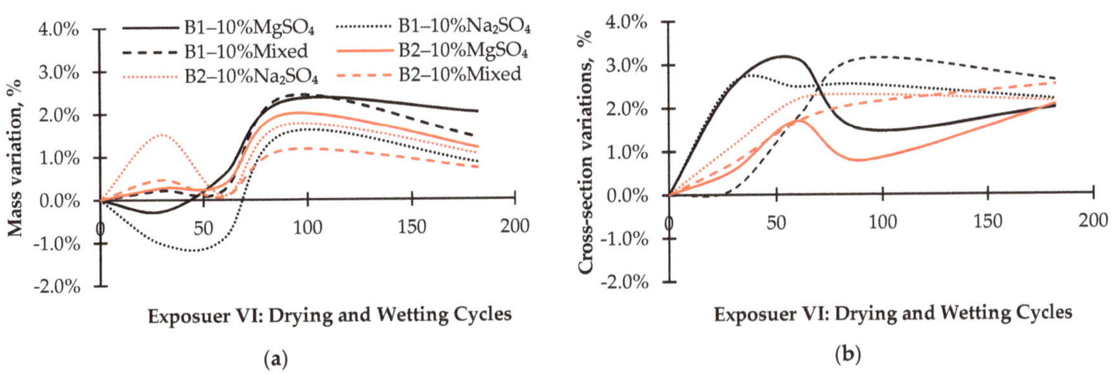

Figure 17. Time dependent (**a**) mass change and (**b**) cross section change of binary-precursor AASCC mixtures exposed to 10% sulfate solutions under drying and wetting cycles.

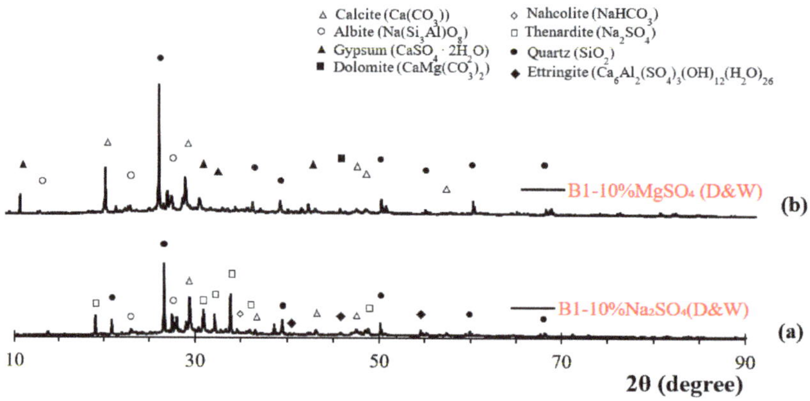

Figure 18. XRD spectra of binary-1 AASCC specimens exposed to sulfate attack under drying and wetting cycles: (**a**) In 10% Na_2SO_4 solution; (**b**) In 10% $MgSO_4$ solution.

Similarly, surface scaling (Table 7) began on the drying surfaces of the binary-2 cylinders immersed in both $MgSO_4$ and mixed sulfate solutions, i.e., 1.2% and 0.7% mass loss, respectively. Noteworthy is the fact that the crystalline phase of $MgSO_4$ (epsomite) would be stable close to room ambient temperature based on the thermodynamic data, but in dry conditions, epsomite would tend to dehydrate and produce the monohydrate kieserite phase; this process would develop a high crystallization pressure. However, after 60 d, binary-1 samples exposed to Na_2SO_4 showed a 1% mass gain, which reversed to a 1% mass loss after the testing time is over. This can be attributed to the direct precipitation of thenardite (above ~32.4 °C) and the quick crystallization of mirabilite following thenardite dissolution, which causes a higher degradation level in porous mixtures mirabilite crystallization alone [40].

Ternary-precursor mixtures. Figure 19a,b illustrates the ternary-precursor mixture mass fluctuations and their related specimen's cross-section variations after six months of physical sulfate exposure. After one month of exposure, i.e., 30 cycles of drying and wetting, both ternary-precursor cylinders lost around 5% of their mass, while the exposure to all sulfate solutions was most probably due to the crystallization pressure on pores walls. These values continued to increase till the end of the testing period. All ternary-1 samples with a higher CaO content had similar cross-section variations ranging from 1.4 to 1.9% when exposed to all sulfate solutions. However, cross-section results of low-calcium ternary-2 samples revealed higher variability in $MgSO_4$ (2%), Na_2SO_4 + $MgSO_4$ (1.2%), and Na_2SO_4 (0.3%) sulfate solutions. The dissolution and crystallization of soluble salts in porous materials like concrete due to the variations in temperature and RH can cause phase changes between thenardite (Na_2SO_4) and mirabilite ($Na_2SO_4 \cdot 10H_2O$) as well as kieserite ($MgSO_4$) and epsomite ($MgSO_4 \cdot 7H_2O$).

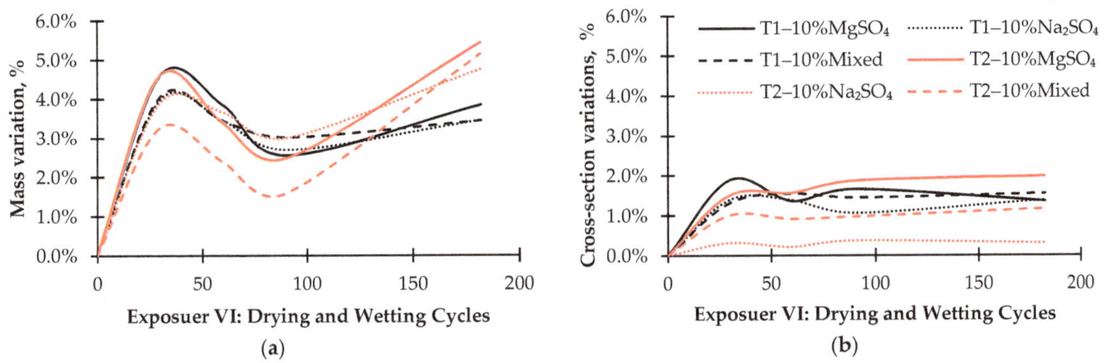

Figure 19. Time dependent (**a**) mass change and (**b**) cross-section change of ternary-precursor AASCC mixtures exposed to 10% sulfate solutions under drying and wetting cycles.

4.5. Thermogravimetry (TGA/DTG)

The TGA/DTG profiles for all AASCC mixtures are shown in Figure 20, with mass loss ranging from 18 to 24% during the period of observation. The differential thermograms (DTG) of various mixtures exhibited peaks at various temperatures. Differential thermograms (DTG) of single-, binary-, and ternary-precursor mixtures showed distinct peaks at different temperatures, designated as A–C. The first two peaks (A and B) were observed at temperatures ranging from 70 °C to 120 °C for all AASCC mixtures exposed to various sulfate solutions due to the elimination of free water in the pores and the dehydration of the C-A (N)-S-H, ettringite, or gypsum gel products, these phases were detected by XRD in all specimens. Weak endothermic peaks in the range of 375–485 °C were detected in all samples designated to the dehydration of hydrotalcite at the spot labeled with C [41]. The decomposition of carbonate minerals, notable calcite, was explained with the weight loss observed in all samples within a temperature range of 675–735 °C (spot D). The existence of various structural forms of carbonates, particularly weakly crystalline or amorphous forms, in the matrix may be indicated by the elevated decomposition temperature, which ranges between 820–930 °C (spot E).

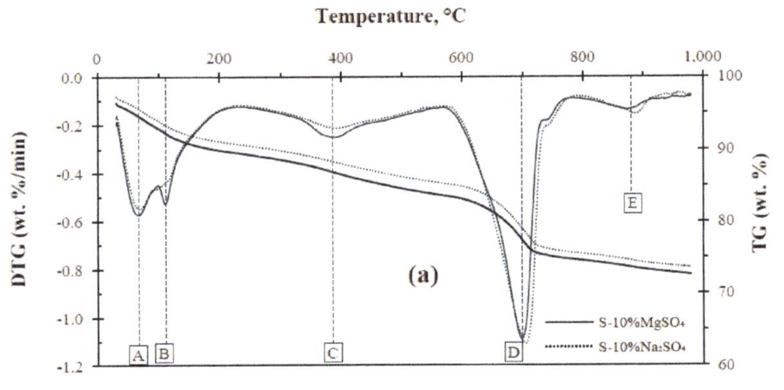

Figure 20. *Cont.*

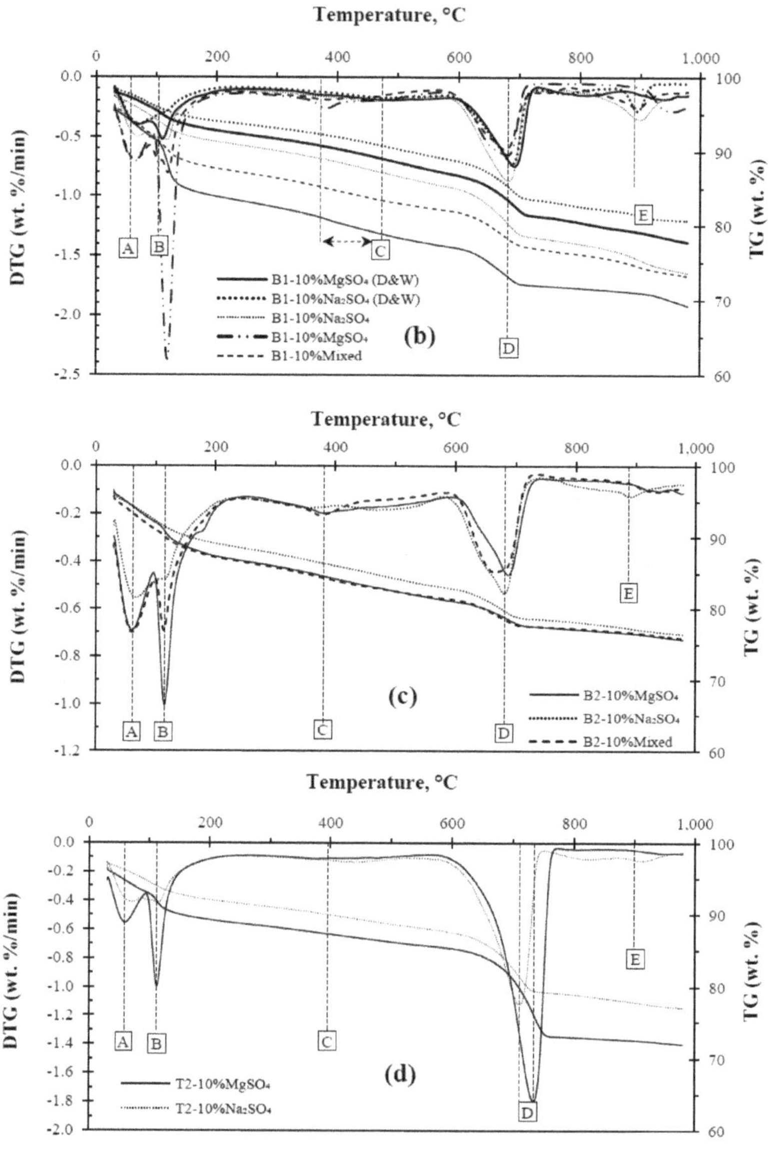

Figure 20. TGA/DTG data for AASCC exposed to different sulfate solutions (**a**) single, (**b**) binary-1, (**c**) binary-2, and (**d**) ternary-2 mixtures.

4.6. Proposed Deterioration Mechanisms

The suggested mechanism for magnesium sulfate attack, sodium sulfate, or a mixture of both sulfate solutions is shown in Figure 21. Step 1 entails exposing AASCC specimens to different sulfate solutions: (a) Na_2SO_4 and (b) $MgSO_4$ with a pH range of 6–8. After one month of exposure to sulfate solutions, the pH of the solution changes to 11–12. Therefore, sulfate solutions were renewed, and the pH of the solution was kept low during Step 2 to maintain a consistent rate of attack; thus, the attack was expected to proceed. In Step 3, expansive gypsum and ettringite were anticipated to form in the surface regions of the AASCC cylinders near the sulfate solutions. However, the core beneath tries to resist the sulfate attack or decalcify the strength-giving products, resulting in a complete breakdown and disintegration of the specimen.

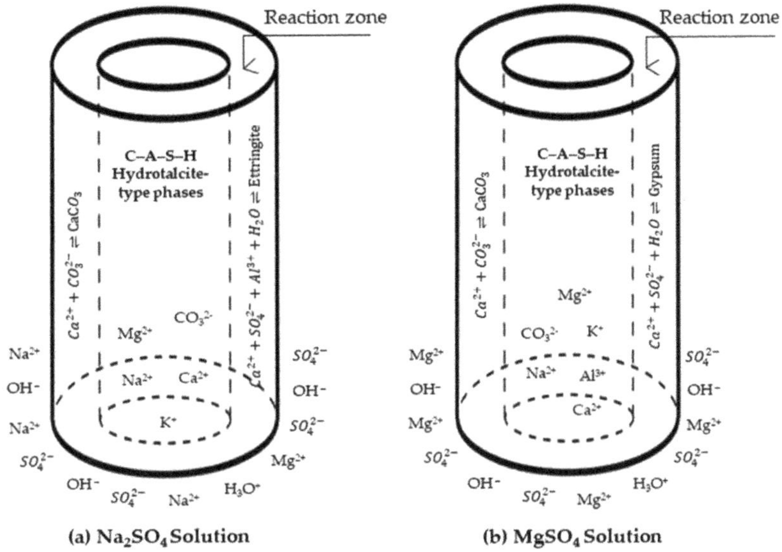

Figure 21. Proposed deterioration mechanism of AASCC cylinders in (**a**) Na_2SO_4 and (**b**) $MgSO_4$.

The reaction mechanisms between the AASCC matrix and different sulfate solutions differ depending on the type of sulfate and hydration products in AASCC, as shown in Figure 21. In Na_2SO_4 solution, ettringite formation, alkali leaching, slight decalcification, and dealumination from C-A-S-H phases (Figure 21a) are assumed to occur in AASCC cylinders, as noticed from the analytical results. Nevertheless, it is believed that the presence of carbonate ions from the sodium carbonate activator inhibits ettringite formation because of the presence and competition mechanisms of carbonate ions and sulfate ions.

Experience with magnesium sulfate has frequently revealed that regular concrete can suffer from the formation of a layer of brucite $Mg(OH)_2$ in the pore solution due to reactions between Mg^{2+} and OH^- ions. This can disrupt further attacks since brucite precipitation is known to be rapid at pH ~10.5 [42]. A direct decalcification of C-S-H may occur due to prolonged pH monitoring, which may later react with Mg^{2+} and SO_4^{2-} to form M-S-H. The potential degradation mechanisms of AASCC cylinders subjected to $MgSO_4$ sulfate solutions are shown in Figure 21b. The anticipated attack by magnesium sulfate prompted the development of a "gypsum-layer". Gypsum precipitates may form because of the reaction between available Ca^{2+} in the pore solution and SO_4^{2-}, as well as carbonate-bearing phases because of the presence of carbonates in the pore solution.

Cross-section variations of AASCC specimens in all sulfate solutions were followed by a two-stage procedure. During Stage 1, the primary stage, the cross-section of the specimens gradually increased up to 60 d of exposure. A very modest increase followed this gradual change in the cross-section results (Stage 2), where the level of change was nearly constant. Instead, cross-section variations fluctuated at a continuously increasing rate until the end of the testing phase, considering the multi-damaging effect of a monitored pH environment, high sulfate concentration, and 183 cycles of drying and wetting.

5. Conclusions

Considering the synergistic effect of various parameters in various sulfate solutions, this paper provided a thorough understanding of the resistance of different AASCC mixture designs, activated with 1:1 two dry-powder activators, i.e., anhydrous sodium metasilicate

(Na_2SiO_3) and sodium carbonate (Na_2CO_3). The findings of this investigation allow for the following inferences:

1. The consequences of continuous immersion in 10% sodium, 10% magnesium, and 10% mixed sulfate solutions indicate that, except for binary-1 mixtures with SF replacing slag, all the tested AASCC mixtures were resilient to sulfate for up to 6 months. This could be due to these materials' impermeability when different precursor blend ratios and compositions are used.
2. In all AASCC cylinders, the main product of degradation in 10% $MgSO_4$ and 10% $MgSO_4$ + 10% Na_2SO_4 sulfate solutions is gypsum, whereas ettringite was produced in 10% Na_2SO_4 sulfate solutions.
3. The ion chromatography analysis of the sulfate solutions revealed that, during sample exposure, there was severe leaching of Na, Ca, and Si from the structure of AASCC blends, particularly in $MgSO_4$ and mixed solutions.
4. The exposure of singles, binary-2, and ternary-precursor mixtures resulted in pastes rich in C-(N)-A-S-H gels, other carbonate products, and mesopores rather than micropores, which delayed or inhibited the crystal growth of expansive products.
5. To increase the hostility of the exposure environment, high sulfate concentrations, pH control, fractional immersion, and wetting–drying cycles were used; however, the synergistic action of all parameters increased the sulfate crystallization.
6. Compared to the 10% continuous partial immersion (Exposure I), the continuous partial immersion with an uncontrolled pH condition (Exposure II), the 5% continuous partial immersion (Exposure III), and the 10% continuous partial immersion with wetting–drying cycles (Exposure IV). Exposure IV and I had nearly identical damage mechanisms, with the latter being the most severe.

Author Contributions: Conceptualization, D.K. and A.M.S.; methodology, D.K. and A.M.S.; validation, A.M.S. and A.e.M.S.; formal analysis, D.K.; investigation, D.K.; resources, A.M.S.; data curation, D.K.; writing—original draft preparation, D.K.; writing—review and editing, A.M.S. and A.e.M.S.; visualization, D.K.; supervision, A.M.S.; project administration, A.M.S.; funding acquisition, A.M.S. All authors have read and agreed to the published version of the manuscript.

Funding: This research was funded by the Natural Sciences and Engineering Research Council of Canada, grant number RGPIN-2018-05094 (Holder: Ahmed M. Soliman).

Institutional Review Board Statement: Not applicable.

Informed Consent Statement: Not applicable.

Data Availability Statement: The data that support the findings of this study are available from the corresponding author, A.M.S., upon a reasonable request.

Acknowledgments: Authors would like to acknowledge Lafarge Canada for donating the slag and Sika for donating the used admixtures.

Conflicts of Interest: The authors declare no conflict of interest.

References

1. *ASTM C1012*; Test Method for Length Change of Hydraulic-Cement Mortars Exposed to a Sulfate Solution. ASTM International: West Conshohocken, PA, USA, 2018.
2. Taylor, H.F. *Cement Chemistry*; Thomas Telford: London, UK, 1997; Volume 2.
3. Santhanam, M.; Cohen, M.D.; Olek, J. Mechanism of Sulfate Attack: A Fresh Look: Part 1: Summary of Experimental Results. *Cem. Concr. Res.* **2002**, *32*, 915–921. [CrossRef]
4. Ismail, I.; Bernal, S.A.; Provis, J.L.; Hamdan, S.; van Deventer, J.S. Microstructural Changes in Alkali Activated Fly Ash/Slag Geopolymers with Sulfate Exposure. *Mater. Struct.* **2013**, *46*, 361–373. [CrossRef]
5. Yusuf, M.O. Performance of Slag Blended Alkaline Activated Palm Oil Fuel Ash Mortar in Sulfate Environments. *Constr. Build. Mater.* **2015**, *98*, 417–424. [CrossRef]
6. Valencia Saavedra, W.G.; Angulo, D.E.; Mejía de Gutiérrez, R. Fly Ash Slag Geopolymer Concrete: Resistance to Sodium and Magnesium Sulfate Attack. *J. Mater. Civ. Eng.* **2016**, *28*, 04016148. [CrossRef]

7. Kanaan, D.; Safhi, A.e.M.; Suleiman, A.R.; Soliman, A.M. Fresh, Hardened, and Microstructural Properties of Ambient Cured One-Part Alkali-Activated Self-Consolidating Concrete. *Sustainability* **2023**, *15*, 2451. [CrossRef]
8. Rebel, B.; Detwiler, R.J.; Gebler, S.H.; Hooton, R.D. The Right Sulfate Test Makes a Difference. *Concr. Int.* **2005**, *27*, 49–52.
9. EFNARC. *The European Guidelines for Self-Compacting Concrete: Specification, Production and Use*; International Bureau for Precast Concrete (BIBM): Brussels, Belgium, 2005.
10. Kanaan, D.; Soliman, A.M.; Suleiman, A.R. Zero-Cement Concrete Resistance to External Sulfate Attack: A Critical Review and Future Needs. *Sustainability* **2022**, *14*, 2078. [CrossRef]
11. Kanaan, D.M.; Soliman, A.M. Fresh and Mechanical Properties of One-Part Alkali-Activated Self-Consolidating Concrete. In Proceedings of the Canadian Society of Civil Engineering Annual Conference, Virtual, 26–29 May 2021; Springer Nature: Singapore, 2023; pp. 17–30.
12. Boyd, A.J.; Mindess, S. The Use of Tension Testing to Investigate the Effect of W/C Ratio and Cement Type on the Resistance of Concrete to Sulfate Attack. *Cem. Concr. Res.* **2004**, *34*, 373–377. [CrossRef]
13. Marchand, J.; Skalny, J.P. Materials Science of Concrete: Sulfate Attack Mechanisms. In *Seminar on Sulfate Attack Mechanisms*; American Ceramic Society: Westerville, OH, USA, 1998; p. 1999.
14. Mehta, P. Sulfate Attack on Concrete—A Critical Review. *Mater. Sci. Concr. IIIpp.* **1992**, *105*.
15. Kalousek, G.L.; Porter, L.C.; Benton, E.J. Concrete for Long-Time Service in Sulfate Environment. *Cem. Concr. Res.* **1972**, *2*, 79–89. [CrossRef]
16. de Almeida, I.R. Resistance of High Strength Concrete to Sulfate Attack: Soaking and Drying Test. *Spec. Publ.* **1991**, *126*, 1073–1092.
17. Nehdi, M.; Hayek, M. Behavior of Blended Cement Mortars Exposed to Sulfate Solutions Cycling in Relative Humidity. *Cem. Concr. Res.* **2005**, *35*, 731–742. [CrossRef]
18. Sahmaran, M.; Erdem, T.K.; Yaman, I.O. Sulfate Resistance of Plain and Blended Cements Exposed to Wetting–Drying and Heating–Cooling Environments. *Constr. Build. Mater.* **2007**, *21*, 1771–1778. [CrossRef]
19. Haynes, H.; O'Neill, R.; Neff, M.; Mehta, P.K. Salt Weathering Distress on Concrete Exposed to Sodium Sulfate Environment. *ACI Mater. J.* **2008**, *105*, 35.
20. Bassuoni, M.T.; Rahman, M.M. Response of Concrete to Accelerated Physical Salt Attack Exposure. *Cem. Concr. Res.* **2016**, *79*, 395–408. [CrossRef]
21. Doehne, E.; Selwitz, C.; Carson, D. The Damage Mechanism of Sodium Sulfate in Porous Stone. In Proceedings of the SALTeXPERT Meeting, Prague, Czech Republic, 27 November–1 December 2002; pp. 27–146.
22. Steiger, M.; Linnow, K.; Ehrhardt, D.; Rohde, M. Decomposition Reactions of Magnesium Sulfate Hydrates and Phase Equilibria in the $MgSO_4$–H_2O and Na+–Mg2+–Cl—SO42—H_2O Systems with Implications for Mars. *Geochim. Et Cosmochim. Acta* **2011**, *75*, 3600–3626. [CrossRef]
23. Chatterji, S. Aspects of Generation of Destructive Crystal Growth Pressure. *J. Cryst. Growth* **2005**, *277*, 566–577. [CrossRef]
24. Malhotra, V.M.; Carette, G.G.; Bremner, T.W. Current Status of CANMET's Studies on the Durability of Concrete Containing Supplementary Cementing Materials in Marine Environment. *Spec. Publ.* **1988**, *109*, 31–72.
25. Komljenović, M.; Baščarević, Z.; Marjanović, N.; Nikolić, V. External Sulfate Attack on Alkali-Activated Slag. *Constr. Build. Mater.* **2013**, *49*, 31–39. [CrossRef]
26. Elyamany, H.E.; Abd Elmoaty, M.; Elshaboury, A.M. Magnesium Sulfate Resistance of Geopolymer Mortar. *Constr. Build. Mater.* **2018**, *184*, 111–127. [CrossRef]
27. Thaulow, N.; Sahu, S. Mechanism of Concrete Deterioration Due to Salt Crystallization. *Mater. Charact.* **2004**, *53*, 123–127. [CrossRef]
28. Bakharev, T. Durability of Geopolymer Materials in Sodium and Magnesium Sulfate Solutions. *Cem. Concr. Res.* **2005**, *35*, 1233–1246. [CrossRef]
29. Baščarević, Z.; Komljenović, M.; Miladinović, Z.; Nikolić, V.; Marjanović, N.; Petrović, R. Impact of Sodium Sulfate Solution on Mechanical Properties and Structure of Fly Ash Based Geopolymers. *Mater. Struct.* **2015**, *48*, 683–697. [CrossRef]
30. Sindhunata; Provis, J.L.; Lukey, G.C.; Xu, H.; van Deventer, J.S. Structural Evolution of Fly Ash Based Geopolymers in Alkaline Environments. *Ind. Eng. Chem. Res.* **2008**, *47*, 2991–2999. [CrossRef]
31. Temuujin, J.; Minjigmaa, A.; Lee, M.; Chen-Tan, N.; Van Riessen, A. Characterisation of Class F Fly Ash Geopolymer Pastes Immersed in Acid and Alkaline Solutions. *Cem. Concr. Compos.* **2011**, *33*, 1086–1091. [CrossRef]
32. Palomo, A.; Blanco-Varela, M.T.; Granizo, M.L.; Puertas, F.; Vazquez, T.; Grutzeck, M.W. Chemical Stability of Cementitious Materials Based on Metakaolin. *Cem. Concr. Res.* **1999**, *29*, 997–1004. [CrossRef]
33. Ye, H.; Chen, Z.; Huang, L. Mechanism of Sulfate Attack on Alkali-Activated Slag: The Role of Activator Composition. *Cem. Concr. Res.* **2019**, *125*, 105868. [CrossRef]
34. Whittaker, M.; Black, L. Current Knowledge of External Sulfate Attack. *Adv. Cem. Res.* **2015**, *27*, 532–545. [CrossRef]
35. Nehdi, M.L.; Suleiman, A.R.; Soliman, A.M. Investigation of Concrete Exposed to Dual Sulfate Attack. *Cem. Concr. Res.* **2014**, *64*, 42–53. [CrossRef]
36. Flatt, R.J. Salt Damage in Porous Materials: How High Supersaturations Are Generated. *J. Cryst. Growth* **2002**, *242*, 435–454. [CrossRef]

37. Price, W.A. Prediction Manual for Drainage Chemistry from Sulphidic Geologic Materials, 579 p, CANMET Mining and Mineral Sciences Laboratories, British Colombia, December, 2009. Available online: https://mend-nedem.org/wp-content/uploads/1.20.1_PredictionManual.pdf (accessed on 23 December 2022).
38. Wang, H.-Y. Durability of Self-Consolidating Lightweight Aggregate Concrete Using Dredged Silt. *Constr. Build. Mater.* **2009**, *23*, 2332–2337. [CrossRef]
39. Kamali, S.; Moranville, M.; Leclercq, S. Material and Environmental Parameter Effects on the Leaching of Cement Pastes: Experiments and Modelling. *Cem. Concr. Res.* **2008**, *38*, 575–585. [CrossRef]
40. Rodriguez-Navarro, C.; Linares-Fernandez, L.; Doehne, E.; Sebastian, E. Effects of Ferrocyanide Ions on NaCl Crystallization in Porous Stone. *J. Cryst. Growth* **2002**, *243*, 503–516. [CrossRef]
41. Zuo, Y.; Nedeljković, M.; Ye, G. Coupled Thermodynamic Modelling and Experimental Study of Sodium Hydroxide Activated Slag. *Constr. Build. Mater.* **2018**, *188*, 262–279. [CrossRef]
42. Pokrovsky, O.S.; Schott, J. Experimental Study of Brucite Dissolution and Precipitation in Aqueous Solutions: Surface Speciation and Chemical Affinity Control. *Geochim. Et Cosmochim. Acta* **2004**, *68*, 31–45. [CrossRef]

sustainability

MDPI

Article

How the Carbonation Treatment of Different Types of Recycled Aggregates Affects the Properties of Concrete

Miren Etxeberria * **and Silvia Castillo**

Department of Civil and Environmental Engineering, Universitat Politècnica de Catalunya·BarcelonaTECH, 6, Campus Nord, 08034 Barcelona, Spain
* Correspondence: miren.etxeberria@upc.edu; Tel.: +34-934011788

Abstract: In this research work, two types of recycled aggregates were used: (1) the recycled concrete aggregate, RCA, obtained by crushing the parent concrete produced using limestone cement and (2) the recycled concrete aggregate RCA-FA produced by crushing parent concrete made with FA blended cement. After the carbonation treatment process, the carbonated RCA-C and RCA-FA-C recycled aggregates were produced. The recycled concrete mixtures were prepared using the four types of recycled aggregates (RCA, RCA-FA, RCA-C, and RCA-FA-C) in 50% (by volume) to replace natural coarse aggregates. The physical and mechanical properties and durability (sorptivity, chloride ion penetration, and carbonation resistance) were determined and analysed. The obtained results were also compared with those of conventional concrete (CC). It was concluded that the physical and mechanical properties of recycled concrete improved when RCA-C were employed in concrete production. In contrast, the recycled concrete produced with RCA-FA-C was found to have the worst property values. According to durability properties, the concrete made with RCA-C and RCA-FA aggregates achieved the highest chloride resistance, similar to CC concrete. Nevertheless, the concrete produced with uncarbonated RCA acquired carbonation resistance equivalent to CC concrete.

Keywords: carbonation treatment; parent concrete with FA; recycled aggregate concrete; mechanical properties; MIP; durability; carbonation; sorptivity; chloride

Citation: Etxeberria, M.; Castillo, S. How the Carbonation Treatment of Different Types of Recycled Aggregates Affects the Properties of Concrete. *Sustainability* **2023**, *15*, 3169. https://doi.org/10.3390/su15043169

Academic Editors: Woubishet Zewdu Taffese and Sandra Barbosa Nunes

Received: 31 December 2022
Revised: 18 January 2023
Accepted: 7 February 2023
Published: 9 February 2023

1. Introduction

Demolition of concrete buildings generates a considerable amount of waste that finishes in landfills. For example, in 2018, approximately 806 million tons of construction and demolition waste were generated in the European Union, of which only 54.2% were recycled [1]. Consequently, with respect to concrete production, the use of recycled concrete aggregates (RCA) obtained by crushing concrete structures can help reduce the environmental impact caused by concrete waste [2,3].

The employment of recycled aggregate concrete (RAC) as a structural material has been widely analysed and validated in many applications [3–6]. In addition, it has been determined that the use of 50% of RCA as a substitute for coarse natural aggregates had no effect on the lowering of compressive strength [7,8]. However, the durability of RAC is usually lower than that of natural aggregate concrete (NAC) because durability is influenced by the connectivity of the porous network and water content [4,9–12].

According to chloride ion diffusion resistance, it was found that the RAC had lower resistance than the NAC when concretes were produced with ordinary Portland cement (OPC) [9,12–14].

The studies' conclusions tend to be inconsistent with the carbonation resistance of RAC. Several researchers [12,14] concluded that the RAC prepared with a high proportion of RCA had a lower carbonation resistance than the NAC. In contrast, several researchers [9] determined that the RAC mixes had similar or higher carbonation resistance than the NAC because of the old attached mortar. Leemann and Loser [15] determined that the impact on the carbonation coefficient of RAC, independent of the replacement levels of RCA and if they are already carbonated, is low (around 10%) at a given compressive strength. According to Zeng [8], using up to 50% of the RCA in place of natural coarse aggregates avoided a decrease in carbonation resistance. However, Pedro et al. [16] documented that when the quality of the RCA's attached mortar was lower than that of the new paste, the RAC carbonation coefficient increased when higher percentages of RCA were employed in place of natural aggregates.

Over the last decade, many studies have been conducted based on RCA quality improvement via the carbonation process [17,18]. Natural or accelerated carbonation using gas in contact with RCA is a possible way to improve the properties of RCA [19–21]. Indeed, hydrates such as portlandite or C–S–H, which are contained in the cement paste adhered to the natural aggregates, could react with CO_2. The reaction has slow kinetics in natural conditions. Still, it could be accelerated using a gas with a higher CO_2 content, such as flue gas from cement factories, and a higher temperature or gas pressure. Accelerated carbonation increases the density of the attached cement paste and decreases the water absorption of the RCAs thanks to the formation of $CaCO_3$ [19,22]. Portlandite carbonation usually reduces pore size and meso- and macropore volume due to $CaCO_3$ precipitation [23]. In addition, the capillary porosity decreases due to the clogging of the pores.

The performance of mortars and concrete made with carbonated RCA improved [18]. According to Liang et al. [17], concrete with carbonated RCA has better workability, mechanical properties and durability than untreated RCA. Xuan et al. [24] found that concrete achieved lower water absorption and permeability when RAC was prepared with carbonated RCAs as well as improved the extent of the durability of RAC. They found that concrete prepared with 100% carbonated RCAs reduced chloride ion permeability by 36.4% for concrete produced with uncarbonated RCA.

More recently, in the trend of CO_2 emission reduction, fly ash (FA) has been used more widely to reduce cement consumption in building materials. In addition, using FA can result in concrete with better durability [25]. However, due to the reduced portlandite content in supplementary cementitious material (SCM)-containing systems, carbonation will occur more rapidly in the main CO_2-binding phases. When using blast furnace slag and FA as SCM, the C-A-S-H phases can induce coarsening of the pore structure upon carbonation [23]. Consequently, it is believed that when the structures produced using FA are demolished, the carbonation treatment will not be satisfied with the improvement of the RCA properties due to the results of coarsening of the pores of the attached mortar, resulting in the increased permeability of RCA.

This study aimed to determine how the properties of concrete are affected by the carbonation treatment on different compositions of recycled aggregates. The recycled aggregates were obtained by crushing two types of parent concretes. Firstly, the recycled concrete aggregates named RCA were obtained by crushing the parent concrete produced using cement type CEM IIAL. Secondly, the recycled concrete aggregates named RCA-FA were produced by crushing parent concrete produced with FA blended cement (50% OPC and 50% FA). After the carbonation treatment process, the RCA-C (carbonated aggregates) and RCA-FA-C (carbonated FA aggregates) were produced, respectively. Finally, the recycled concrete mixtures were prepared using the four types of recycled aggregates (RCA, RCA-FA, RCA-C, and RCA-FA-C) in 50% (by volume) to replace natural coarse aggregates. The physical, mechanical, and durability (sorptivity, chloride ion penetration, and carbonation resistance) properties of the concrete produced using the four recycled aggregates were determined and analysed. The consequent results were also compared with those obtained by conventional concrete (CC).

2. Materials and Methods

2.1. Materials

2.1.1. Cement and Admixture

The cement CEM II A-L 42.5R (88% clinker, 10% limestone, and 2% set regulator, gypsum) was used in all the concrete production. The composition and physical properties (loss of ignition, LOI; Blaine specific surface, Blaine; and density) of the cement are given in Table 1. In addition, a superplasticiser (S) that is polymer-based on PAE compound was used for concrete production.

Table 1. Chemical composition of cement.

	SiO_2	Fe_2O_3	Al_2O_3	CaO	MgO	SO_3	Na_2O	K_2O	LOI	Blaine (cm^2/g)	Density (g/cm^3)
CEM II A-L	18.27	3.29	4.09	61.82	1.35	2.9	0.10	0.77	5.07	3900	3.16
FA	58.4	7.3	21.6	2.3	1.9	0.2	0.9	2.1	3.1	3400	2.16
CEM I 52.5R	19.4	3.4	4.2	63.5	1.4	3.0	0.12	0.53	3.7	4900	3.15

2.1.2. Natural Aggregates

Three fractions of crushed limestone aggregates (gravel 10–20 mm, gravel 4–10 mm, and sand 0–4 mm) were used as raw aggregates (NA) to produce conventional concrete (CC) (see Figure 1). The fineness modulus of sand was 2.9. Figure 2 shows their grading distribution. It was determined following EN 933-1 specifications and complied with EN 12620 specifications.

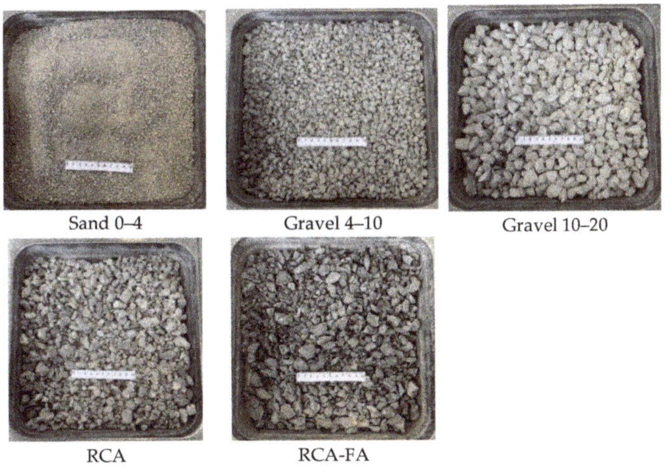

Figure 1. The raw and recycled aggregates (each line in the ruler is 1 cm).

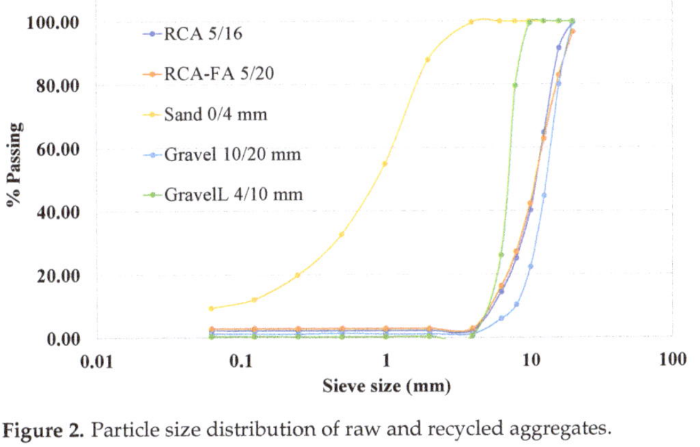

Figure 2. Particle size distribution of raw and recycled aggregates.

The physical properties of dry density (ρ_{rd}) and water absorption value (WA_{24}) were determined according to EN 1097-6 specifications (see Table 2).

Table 2. Physical properties of raw, recycled aggregates and carbonated recycled aggregates.

	Raw Aggregates, NA			Recycled Aggregates, RA		Carbonated RA			
						Process 1		Process 2	
	0/4 mm	4/10 mm	10/20 mm	RCA	RCA-FA	RCA-C	RCA-FA-C	RCA-C	RCA-FA-C
ρ_{rd} (kg/m^3)	2590	2640	2650	2310	2240	2360	2310	2370	2310
WA_{24} (%)	1.79	0.75	0.57	5.8	6.8	4.85	5.64	4.8	5.68

2.1.3. Recycled Aggregates

RCA and RCA-FA recycled aggregates (see Figure 1) were produced by crushing two different parent concretes. The RCA aggregate was produced by crushing the parent concrete made with CEM II A-L cement (see Table 1), and it had 30 MPa of compressive strength at 28 days. In addition, the RCA-FA aggregate was obtained by crushing concrete made using a blended cement: 50% of CEM I 52.5R and 50% of FA (equivalent to ASTM class F) (see Table 1 for their composition). The concrete had a compressive strength of 32 MPa after 28 days. Both parent concretes were produced with limestone aggregates, and they were approximately nine months old when they were crushed.

The RCA and RCA-FA aggregates were sieved to achieve a particle size distribution similar to that of coarse natural aggregates. Figure 2 and Table 2 describe the particle size distribution and the physical properties (ρ_{rd} and WA_{24}) of RCA and RCA-FA, respectively.

The recycled aggregates (RA) obtained a lower density and a higher absorption coefficient than NA due to the attached mortar presented in RA, which was found to provide more pores, causing a decrease in density and an increased absorption capacity.

Carbonated Recycled Aggregates

The carbonation treatment of the RCA and the RCA-FA aggregates was carried out in two processes (process 1 and 2). In process 1, the RCA and the RCA-FA were subjected to a carbonation process using an automatic carbonation control chamber with a concentration of 3% CO_2, 57% relative humidity (RH), and 20 °C for 24 days until they were fully carbonated. It was carried out following the conditions defined by UNE-EN 12390-12 specification. On the other hand, in process 2, the RCA and RCA-FA were submitted to a carbonation process in 20% CO_2, 57% RH, and 20 °C conditions for four days to achieve a complete carbonation. Process 2 was carried out following the GB T50082-2009 [26] standards. The physical properties (ρ_{rd} and WA_{24} were determined following EN 1097-6 [27] specifications) of the carbonated RA in both processes are described in Table 2.

Table 2 shows that the RCA-C and the RCA-FA-C achieved similar ρ_{rd} and WA_{24} values after they were submitted to carbonation processes 1 and 2. Consequently, due to the findings of carbonation process 2, using a reduced period (4 days), it was decided that all the RA volume required for concrete production would be carbonated under this process. As described by Liang et al. [17], an excessive concentration of CO_2 gas does not result in an evident increase in the carbonation percentage of RCA. However, the increased concentration of CO_2 gas also leads to the decalcification of C–S–H and even the complete disappearance of C–S–H under the condition of a 100% CO_2 concentration [28], resulting in an adverse effect on the properties of RCA.

According to Table 2, the ρ_{rd} increased and WA_{24} decreased when the recycled aggregates were submitted to carbonation process treatment. Table 2 shows that RCA and RCA-FA (uncarbonated) aggregates achieved a 5.8% and 6.8% absorption capacity, respectively. Once the aggregates were carbonated, the RCA-C and RCA-FA-C achieved absorption capacities of 4.8% and 5.7% with 17% and 16% reductions, respectively.

The Pore Structure of Recycled Aggregates

Porosity and pore structure were tested by mercury intrusion porosimetry (MIP) using a "Micromeritics Poresize 9320" mercury intrusion porosimeter following BS7591 Part 1. This test was performed on small RA samples weighing approximately 5.5 g. The samples were first soaked in acetone for four days to stop the hydration of the cement and then placed in a vacuum drier for 2 h to remove the remaining acetone. Before testing, the samples were dried in an oven at 50 °C for four days. The total porosities were 13.5% for RCA, 10.7% for RCA-C, 11.8% for RCA-FA, and 9.6% for RCA-FA-C.

Figure 3a shows the pore size distribution development of the RA before and after carbonation. As shown in Figure 3b, the pore size diameter can be divided into four pore size ranges following the Mindess [29] classification: >10 μm (air), 10–0.05 μm (macropores), 0.05–0.01 μm (mesopores), and <0.01 μm (micropores). Certain researchers [30,31] emphasised that concrete's mechanical and permeability properties depend on the mesopores and macropores. The RCA aggregate reduced the porosity from 13.5% to 10.7% upon carbonation. In addition, no increase in the pore volume of any size was appreciated. Xiao et al. [32] concluded that carbonated RAs (when OPC was the constituent of the attached mortar) decreased pore volume and pore amount. Although RCA-FA aggregates suffered a porosity reduction from 11.8% to 9.6%, after the carbonation process, the pore size distribution of carbonated RCA-FA-C aggregates was also found to have changed with respect to that of RCA-FA. An increase in mesopores and macropores from 0.05 to 0.2 μm was appreciated in the RCA-FA-C aggregate. In addition, the RCA-C and RCA-FA-C aggregates achieved very similar pore distributions.

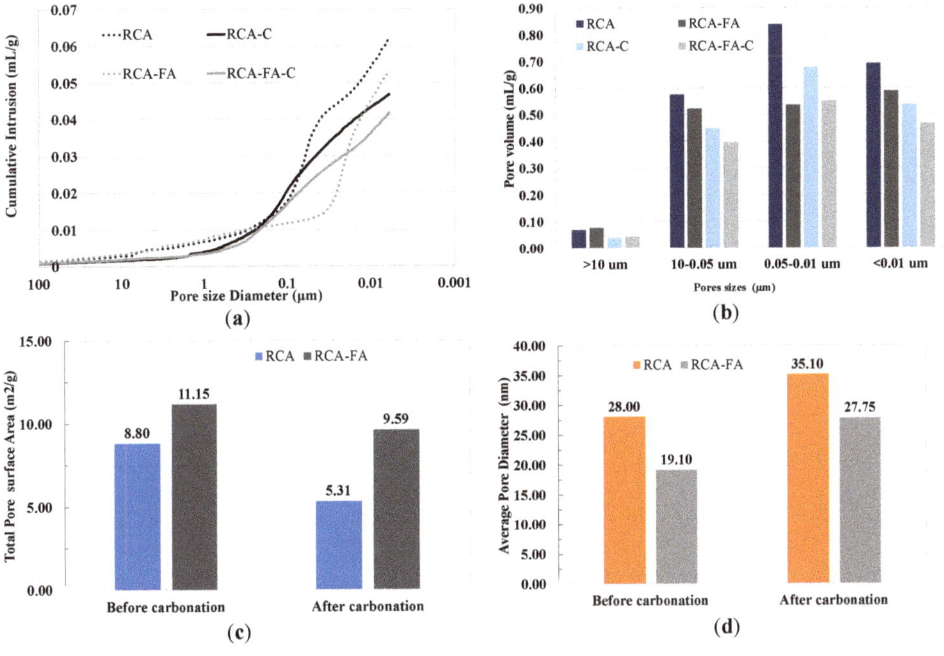

Figure 3. (**a**) Development of pore size distribution, (**b**) distribution of pore diameters, (**c**) total pore surface area before and after the carbonation process, and (**d**) average pore diameter before and after the carbonation process.

Figure 3c,d show the pore structure parameters of RA before and after carbonation. The pore structure of the RA improved when carbonated. The total pore surface area decreased by approximately 30% in the case of RCA and 14% in the case of RCA-FA.

Liang et al. [33] reached a similar conclusion regarding the pore surface area of RA (when the OPC was a composition of the attached mortar) after carbonation, which decreased by approximately 36%.

2.2. Concrete Production

All concrete mixes were prepared and manufactured in the laboratory. A total of five types of concrete mixes were prepared (see Table 3). Firstly, the CC concrete was produced with 100% raw fine and coarse aggregates. In addition, four recycled concretes (RAC50, RAC50-C, RAC50-FA, and RAC50-FA-C) were produced, replacing 50% (by volume) of both raw coarse aggregate fractions by the RCA, the RCA-C (carbonated), the RCA-FA, and the RCA-FA-C (carbonated) aggregates, respectively.

Table 3. Mix proportions of produced concretes (kg of material/m^3). Superplasticizer (S) is given in % with respect to cement weight.

Concrete Reference	CEM	Raw Aggregates			RA	Total Water	S (%)	SLUMP (mm)	DENSITY (kg/m^3)
		0–4 mm	4–10 mm	10–20 mm					
CC	350	900	275.2	644.6	-	192.5	0.5	170	2420
RAC50	350	900	137.6	322.3	399.6	211.1	0.5	150	2310
RAC50-C	350	900	137.6	322.3	411.8	208.9	0.5	170	2340
RAC50-FA	350	900	137.6	322.3	389.2	217	0.5	170	2310
RAC50-FA-C	350	900	137.6	322.3	401.3	211.8	0.6	190	2360

The concrete mix proportions were defined according to their maximum volumetric compaction. This mix proportion for the CC was 50% fine aggregates and 50% coarse aggregates. The distribution of coarse aggregate was 30% coarse 4–10 mm and 70% coarse 10–20 mm. The recycled concretes were produced using 50% of RA in substitution of coarse NA aggregates. All concrete production was subjected to a constant effective water–cement ratio (effect w/c) of 0.50 and a cement content of 350 kg/m^3. They were defined following the Spanish structural code [34] in order to guarantee the compressive strength of 44 MPa in cubic specimens and durability properties for reinforced concrete to be exposed to XC4 and XS 1 environment conditions. (A characteristic compressive strength of 30 MPa was required in cylindrical specimens.) Following Neville's [35] definition of the effective amount of water in the mixture occupying the space outside the aggregate particles, the effect w/c ratio was kept constant in all concretes. This was in order to achieve the same conditions with respect to the hydration of the cement paste caused by the high absorption of RA. As a result of the high absorption capacity of RA, all four types of RA were used with a humidity of up to 70–80%, thus reducing their water absorption value [36]. However, the most important aspect was that the aggregates employed were moist to reduce their absorption capacity [37]. Due to the moderate initial moisture content, the recycled aggregate absorbed some amount of free water, which lowered the initial w/c ratio in the interfacial transition zone (ITZ) and thus improved the interfacial bond between the aggregates and cement [38]. Table 3 describes the mix proportions used for concrete production. The weight of the aggregates is expressed as dry weight. In addition, the total amount of water was considered, including the effective water, the water absorbed by aggregates, and the humidity of RA at concrete production. Finally, 0.5 to 0.6% of superplasticisers were used relative to cement weight of all concretes to achieve the desired workability.

The consistency (slump test) of the concretes was determined in the fresh state following the standard UNE-EN 12350-2:2020 [39]. All the concretes achieved a slump value between 150–170 mm employing 0.5% of superplasticiser. The use of 0.6% of S in the RAC50-FA-C concrete increased the slump value to 190 mm. Although the total water content of the RAC was higher than that of the CC concrete, it did not affect the workability of concretes, as the RA were used with high humidity.

The fresh density was determined following UNE-EN 12350-6 [40] specification. Table 3 shows that the use of RA, carbonated or not, did not affect the properties of the concrete in the fresh state. Although the density of RAC was reduced by 100 kg/m^3 compared to CC concrete, all the concretes reached adequate density for structural concrete [35].

All the concrete mixtures were produced in a vertical axle mixer. The materials were always added manually in the same order. First was the aggregates (from coarser to finer); after they were mixed for 30 s, the cement was added. While the solid components were mixed for 1 min, water was added, followed by a superplasticiser. The complete mixture was then mixed for 1 min more. Once the concrete was produced, concrete specimens were produced after the slump test, and fresh density was determined. The concrete specimens were produced and cured according to the UNE-EN 12390-2: 2020 [39] standard and were manually compacted using a metal bar, then covered with a plastic sheet, and left to cure in the air for 24 h. The concrete specimens were kept in moulds for 24 h. After demoulding, they were stored in a humidity room with an RH of $\geq$95% and a constant temperature of 20 $\pm$ 2 °C until one hour before testing. All test elements were subjected to the same conditions before testing.

Three types of specimens were produced. Three cubic samples of 100 $\times$ 100 $\times$ 100 mm for each test were used to determine the physical properties, compressive strength, and sorptivity. Two cylinders with a 100 mm diameter and 200 mm height were employed to determine the properties of splitting tensile strength, modulus of elasticity, and chloride ion penetrability. In addition, two prismatic specimens of 100 $\times$ 100 $\times$ 400 mm were used in each concrete to determine carbonation resistance.

2.3. *Test Procedure*

2.3.1. Physical Properties

The density, absorption and voids were measured following the ASTM C 642-21 "Standard Test Method for Density, Absorption, and Voids in Hardened Concrete" [41] at 28 days. Three cube specimens of 100 $\times$ 100 $\times$ 100 mm were used for each type of concrete produced.

2.3.2. Mechanical Properties

The compressive strength, splitting tensile strength, and modulus of elasticity (E) were determined for each concrete. The compressive strength of concrete was determined using a 3000 kN loading capacity machine. The compressive strength was determined at 7, 28, and 56 days following UNE-EN 12390-3 [42] specifications. Three 100 $\times$ 100 $\times$ 100 mm cubic specimens were used for each age. The elastic modulus (E) and tensile strength were tested at 28 days following UNE 12390-13 [43] and UNE-EN 12390-6 [44] specifications, respectively. Two specimens (cylinders of 100 mm in diameter and 200 mm in height) were used in each test for each type of concrete produced.

2.3.3. Durability

The sorptivity value, the chloride ion penetration test, and the accelerated carbonation resistance of each concrete were determined. The capillary water absorption of concrete was assessed at 28 days using two of the 100 $\times$ 100 $\times$ 100 mm cubic specimens according to ISO 15148:2002(E) [45]. To determine the amount of capillary water absorption content, the bottom face of the specimens was submerged in water 5 mm. (The lateral surfaces were impregnated with an impermeable resin.) The cumulative water absorbed was recorded at different time intervals up to 48 h by weighing the specimen after removing the surface water using a dampened cloth. Sorptivity is the slope of the regression curve of the quantity of water absorbed by a unit surface area versus the square root of the elapsed time from the initial instant to 120 min. The capillary water absorption results are the average of the three measurements.

The chloride penetrability of concrete was measured according to ASTM C1202-22e1 [46] using a 100∅ mm concrete disc of 50 mm thickness cut from the 100∅/200 mm concrete sample. According to Ghanem et al. [47], this test could represent the true long-term chloride penetration when the concrete was produced with the cement employed in this study. The resistance of concrete to chloride ion penetration is described by the total charge (in Coulombs) passed during a test period of 6 h. In this study, the chloride ion penetrability test was performed on the concrete specimens at 28 and 56 days of age, and each result was the average of the two measurements taken.

The accelerated carbonation coefficients were determined according to the UNE-EN 12390-12 [48] specification. Two prismatic specimens of 100 × 100 × 400 mm were used for each produced concrete. The specimens subjected to an accelerated carbonation process were removed from the humidity chamber after 28 days and preconditioned in the laboratory for 14 days. Pre-conditioning was performed at a CO_2 concentration of 400 ppm, 21 ± 1°C, and 50–55% RH before placing them in the CO_2 chamber. After pre-conditioning for 14 days, the samples were placed in a chamber at 3% CO_2 and 57% RH at 20 °C. The carbonation depth of each sample was determined after 0, 14, 28, 56, and 91 days of exposure to the chamber environment. To determine the carbonation depth, a phenolphthalein indicator with a solution of 1 g phenolphthalein in 70 g ethanol and 30 g water was applied to the fractured concrete surface, following UNE-EN 14630 [49] specification.

3. Results

3.1. Physical Properties

Table 4 shows the physical properties obtained from the produced concretes. In addition, the dispersion of the results is described between brackets. All the concretes produced with RA had a higher absorption capacity than the CC concrete due to the high porosity of the adhered mortar. As a result, the RAC50 concrete achieved a 6.5% higher absorption capacity than that of the RAC50-C concrete, in which the carbonated RCA-C aggregates were employed. However, the absorption capacity of the RCA50-FA was 8% lower than that of the RCA50-FA-C concrete, in which the carbonated RCA-FA-C recycled aggregates were used. Thus, the CC concrete achieved the lowest absorption capacity, followed by RAC50-FA, RAC50-C, RAC50-FA-C, and RAC50 concrete. The RAC50 concrete achieved the highest percentage of accessible voids. It was noted that the standard deviation of all the properties was low.

Table 4. Physical properties of concretes (Standard deviation in brackets).

Concrete Reference	Water Absorption (%)	Dry Density (Kg/m^3)	Accessible Voids (%)	Sorptivity (mm/min$^{0.5}$)
CC	4.86 (0.13)	2290 (20)	11.13 (0.23)	0.035
RAC50	5.75 (0.14)	2240 (10)	12.85 (0.23)	0.039
RAC50-C	5.38 (0.14)	2240 (10)	12.06 (0.27)	0.036
RAC50-FA	5.2 (0.07)	2240 (10)	11.66 (0.13)	0.032
RAC50-FA-C	5.62 (0.05)	2230 (10)	12.53 (0.09)	0.048

3.2. Mechanical Properties

Table 5 shows the mechanical properties obtained for each type of concrete. In addition, the standard deviation value of the results is given in parenthesis. Regarding compressive strength, the RAC50-FA-C concrete reached the lowest strength at any age: 12% lower than the CC concrete at seven days. At 28 days, RAC50-C reached the strength of the CC concrete, and tests on the other three recycled concretes determined a 5% lower value than CC concrete. In addition, all the concretes achieved the compressive strength required by Spanish structural code [34]. At 56 days, RA concrete reached a similar strength: 7–9% lower than CC concrete. The standard deviation value obtained by all the concretes' properties was acceptable.

Table 5. Mechanical properties of the produced concretes.

Test	Time (Days)	CC	RAC50	RAC50-C	RAC50-FA	RAC50-FA-C
Compressive strength (MPa)	7	44.63 (0.31)	42.23 (0.19)	42.76 (0.52)	42.26 (0.33)	39.19 (2.76)
	28	45.92 (0.21)	44.10 (1.83)	45.71(1.48)	45.09 (1.22)	43.82 (2.22)
	56	54.0 (1.52)	50.29 (0.38)	50.65(1.30)	49.90 (1.22)	49.07 (1.76)
Splitting tensile (MPa)	28	3.19 (0.08)	2.89 (0.25)	3.36 (0.16)	3.14 (0.17)	3.02 (0.44)
E (GPa)	28	34.1 (0.87)	30.9 (0.35)	31.9 (0.33)	31.9 (0.71)	30.6 (0.19)

Regarding splitting tensile strength, the RAC50-C concrete achieved the highest strength: 18% higher than the RAC50. In contrast, while the RAC50-FA concrete achieved a similar strength to the CC concrete, the RAC50-FA-C concrete achieved a 5% lower strength value. However, all the recycled concrete achieved an adequate strength value for structural concrete [35]. Kou et al. [50] determined that the splitting tensile strength at 28 days and 90 days of concrete produced with carbonated recycled aggregates (when the recycled aggregates' attached mortar was composed with OPC) increased by 6% and 12%, respectively, compared to concrete produced with uncarbonated RA. In addition, the tensile strength with carbonated RA at 90 days was even higher than that of the reference concrete. Moreover, Liang et al. [17] determined that the effect of CO_2 treatment of RA on the increase in splitting tensile strength was more significant than that of compressive strength. Table 5 also shows that the standard deviation of the obtained values was low for all the produced concretes. Figure 4 describes the relationship between compressive strength and splitting tensile strength of each concrete at 28 days. The findings described indicate that a higher compressive strength value also resulted in an increase the splitting strength value. It is clear that the use of RCA-C aggregates improved the mechanical properties of concrete and, more considerably, the splitting tensile strength. However, the concrete produced with carbonated RCA-FA-C aggregates, the RAC50-FA-C, achieved lower values than those of the RAC50-FA concrete.

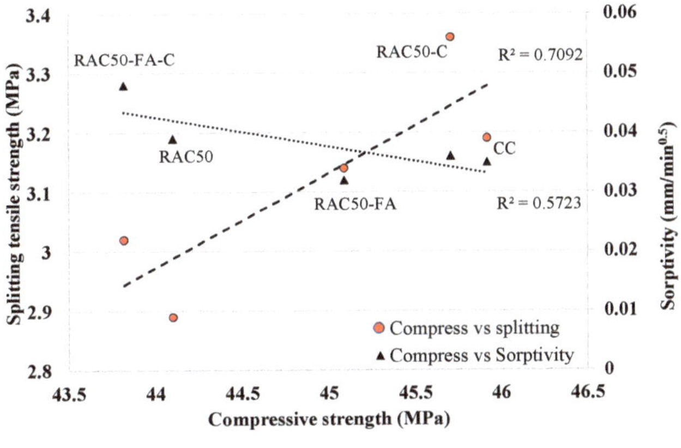

Figure 4. Compressive strength vs. splitting tensile strength and sorptivity.

Consistent with the modulus of elasticity test, all the concrete produced using the RA suffered a reduction in the values obtained for the CC concrete. However, while the RAC50-C and the RAC50-FA concrete mixtures values decreased by 6%, the values of the RAC50 and the RAC50-FA-C decreased by 10% compared to that of CC concrete. Table 5 also described that the standard deviation obtained in all the values was low.

Consequently, it can be concluded that the RAC50-C concrete achieved better mechanical properties than the RAC50. Furthermore, other researchers also found that in the concrete produced with carbonated RCA (when OPC is a component of attached mortar), the mechanical properties of concrete improved compared to those of the recycled concrete produced with uncarbonated recycled aggregates [17,24,50]. In contrast, the RAC50-FA-C concrete achieved lower mechanical properties than the RAC50-FA. When the composition of the attached mortar of RA included FA, the carbonation process worsened the properties of the RCA-FA-C with respect to those of the RCA-FA aggregates and, consequently, the mechanical properties of the RAC50-FA-C concrete with respect to the RAC50-FA, too.

3.3. Durability

3.3.1. Capillary Water Absorption—Sorptivity

Figure 5 shows the capillary absorption capacity of all the produced concretes. All the obtained data are described with error bars. While the RAC50-FA concrete proved to achieve the lowest absorption capacity value, the RAC50-FA-C achieved the highest. In addition, RAC50-FA reached a lower capillary absorption capacity than the CC concrete. Moreover, RAC50-C concrete achieved lower capillary absorption capacity than RAC50 concrete. Similarly, Table 4 shows that the RAC50-FA concrete achieved the lowest sorptivity value of 0.032 mm/min$^{0.5}$, followed by the CC concrete with 0.035 mm/min$^{0.5}$ and the RAC50-C with 0.036 mm/min$^{0.5}$. The concretes RAC50 and RAC50-FA-C achieved the highest values, with 0.039 and 0.048 mm/min$^{0.5}$, respectively. Similar to previously analysed properties, the RAC50-FA-C concrete produced using RCA-FA-C aggregates (carbonated aggregates with FA in attached mortar) achieved 50% higher sorptivity than the concrete produced with the RCA-FA aggregate.

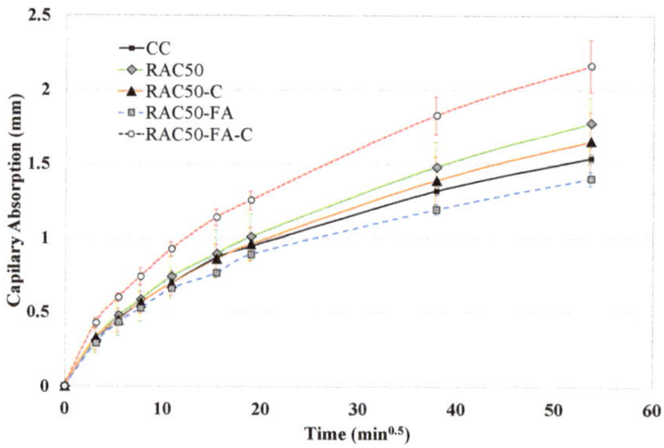

Figure 5. Capillary absorption of all the concrete.

Figure 4 shows the relation between compressive strength and sorptivity at 28 days of age. While the RAC50-FA-C concrete achieved 2% lower compressive strength than that of the RAC50-FA concrete, the sorptivity of the RAC50-FA-C was 50% higher than that of the RCA50-FA. In contrast, the RAC50-C concrete improved compressive strength and sorptivity values compared to those of the RAC50. While the RAC50-C concrete achieved a 4% higher compressive strength than the RAC50, the sorptivity of the RAC50-C was reduced by 8% with respect to that of the RAC50.

According to several researchers [51], concrete with a sorptivity value lower than 0.10 mm/min$^{0.5}$ is considered durable. However, other researchers [52] proposed lowering that value to 0.05 mm/min$^{0.5}$ for safety reasons. Pedro and Brito [53] found that the quality of concrete is low if the sorptivity value is higher than 0.2 mm/min$^{0.5}$, medium if the coefficient is between 0.2 mm/min$^{0.5}$ and 0.1 mm/min$^{0.5}$, and good if the coefficient is lower than 0.1 mm/min$^{0.5}$. Based on those criteria, all the concretes produced with 50% RA achieved a durable category according to the sorptivity property.

3.3.2. Chloride Ion Penetration

Table 6 shows the chloride ion penetration resistance value of the produced concretes at 28 days and 56 days of curing. According to the results of the ASTMC1202 test, every concrete had a high chloride ion penetrability, as the obtained electrical conductance values were higher than 4000 coulombs, which is the maximum admitted value to categorise as moderate resistance to the penetration of chloride ion concretes. In addition, resistance was reduced when RA was employed. However, the RAC50-FA concrete obtained only a 4% higher value than the CC concrete. Therefore, the use of RAC-FA aggregates (uncarbonated and FA in attached mortar) improved the property of recycled aggregate concrete. As known, the resistance to chloride ion penetration is higher when FA is used [54,55]. Consequently, the presence of FA in RA's attached mortar improved the concrete's property.

Table 6. Penetration of chloride ion resistance coulombs.

Concrete Reference	28 Days (Coulombs)	56 Days (Coulombs)	Increase of Resistance
			(%)
CC	6731 (49.2)	5180 (821.1)	23
RAC50	8799 (582.2)	6377 (810.2)	27
RAC50-C	7407 (411.0)	6166 (319.7)	17
RAC50-FA	7006 (352.5)	5412 (300.1)	23
RAC50-FA-C	9296 (1088.2)	7221 (270.2)	22

However, the RAC50-FA-C concrete achieved the lowest value to chloride penetration resistance. The carbonation process of the RCA-FA-C aggregate caused the coarsening of the pore structure of the C-A-S-H phase [23] presented in FA attached mortar. Table 6 shows that the RAC50-FA-C and RAC50-FA concretes achieved 38% and 4% higher values than the CC concrete, respectively. In contrast, the RAC50-C concrete reached higher chloride resistance than the RAC50 concrete. The RAC50-C and RAC50 achieved 10% and 30% higher values, respectively, than the CC concrete. At 56 days, the chloride ion penetration resistance of all the concretes increased compared to the values obtained at 28 days. The RAC50 achieved the highest increase with 27% and the RAC50-C the lowest with 16%. However, at 56 days, the CC concrete achieved the lowest chloride penetration resistance, followed firstly by the RAC50-FA, then the RAC50-C, the RAC50, and lastly the RAC50-FA-C concrete with the lowest resistance. Moreover, the RAC50-FA and the RAC50-FA-C achieved 4% and 39% higher values, respectively, than that of the CC concrete. However, it must be mentioned that the standard deviation of each of the properties was relatively high, requiring as a consequence more than two specimens to determine the average value.

3.3.3. Carbonation Resistance

Table 7 depicts the accelerated carbonation coefficient (k_{acc}) obtained in the testing of all the concretes. The standard deviation of each average carbonation depth value is also described. The k_{acc} was calculated assuming a steady-state condition (i.e., a constant carbonation coefficient) defined by Fick's first law of diffusion, as shown in Equation (1).

$$X_c(t) = k_{acc} \cdot (t)^{0.5} \tag{1}$$

where X_c is the carbonation depth (mm) determined experimentally, k_{acc} is the carbonation coefficient, and t is the time (day).

Table 7. Carbonation depth and carbonation coefficient of the various concretes.

Concrete Reference	Carbonation Depth (mm) at Exposure Time					Carbonation Coefficient	
	0 Days	14 Days	28 Days	56 Days	91 Days	k_{acc} (mm/day$^{0.5}$)	$K_{natTHEO}$ (mm/year$^{0.5}$)
CC	0 (0.00)	3.5 (0.29)	5.2 (0.77)	7.0 (0.82)	8.8 (0.27)	0.926	2.105
RAC50	0 (0.00)	4.4 (0.18)	5.4 (0.53)	7.0 (0.00)	9.5 (0.35)	0.958	2.180
RAC50-C	0 (0.00)	5.1 (0.62)	7.4 (0.44)	9.9 (0.32)	12.9 (1.18)	1.338	3.043
RAC50-FA	0 (0.00)	4.5 (0.53)	6.9 (0.29)	9.9 (0.74)	11.8 (1.41)	1.269	2.886
RAC50-FA-C	0 (0.00)	5.0 (0.06)	8.7 (1.36)	10.0 (0.18)	13.9 (1.97)	1.432	3.257

After 91 days of CO_2 exposure, the CC concrete reached the lowest carbonation depth, with an 8.8 mm value, followed by the RAC50 concrete, which achieved a carbonation depth of 9.5 mm. As several researchers have determined, the concrete produced with 50% of uncarbonated RCA, i.e., the RAC50 concrete, achieved a similar carbonation rate to that of the CC concrete [7,8,15]. In contrast, the RAC50-C concrete (produced with carbonated RCA-C) achieved 12.9 mm of carbonation, reaching an accelerated carbonation rate, kacc, of 1.35 mm/day$^{0.5}$. The RAC50-C concrete achieved a 35.5% higher carbonation rate than that of the RAC50 concrete. In addition, the recycled concrete produced with RCA-FA (obtained from FA parent concrete), i.e., the RAC50-FA and RAC50-FA-C concretes, achieved a carbonation rate of 1.27 and 1.44 mm/day$^{0.5}$, respectively. As expected, the concretes produced with the RCA-FA aggregates (composed with FA) achieved a higher carbonation rate than those produced with the RCA aggregates. In addition, the concrete made with uncarbonated RCA-FA aggregates achieved a lower carbonation rate than the carbonated aggregates. The CO_2- treated recycled aggregates worsened the carbonation resistance of recycled concrete.

The total porosity of the carbonated (nominated as CO_2) and uncarbonated area (No) of the five produced concretes mixtures after being submitted to the 91 days of accelerated carbonation attack were measured using a porosimetry mercury intrusion (MIP) tester (see Table 8). The porosity of the RAC50-FA concrete increased from 9.36 to 12.67%, probably due to the coarsening pore diameter of the attached mortar of recycled aggregates, which was composed of FA. In contrast, the total porosity of the RAC50-FA-C decreased after the carbonation process (RA were already carbonated) was concluded. Consequently, the new cement paste was only carbonated. All the concrete mixtures were produced using CEM IIAL cement, which did not have SCM; therefore, the carbonation of the new cement paste caused the decrease in pore size and the total volume of meso- and macropores due to $CaCO_3$ precipitation [23]. Thus, the porosity and average pore diameter of the CC, the RAC50, and the RAC50-C concretes also decreased after the carbonation process. Figure 6 shows that the RAC50-FA concrete was the only concrete that increased its cumulative volume due to the coarsening of the macro- and mesopores of the attached mortar composed of FA.

Table 8. Total porosity and average pore diameter of produced concrete submitted to the carbonation process.

	CC		RAC50		RAC50-C		RAC50-FA		RAC50-FA-C	
	No	CO_2	No	CO_2	No	CO_2	No	CO_2	No	CO_2
Porosity (%)	11.12 (1.6)	8.95 (0.98)	13.18 (1.3)	11.64 (2.01)	12.05 (1.1)	11.84 (1.45)	9.36 (0.8)	12.67 (1.6)	14.93 (2.2)	12.92 (1.9)
Average pore D (nm)	35.61 (1.1)	32.74 (0.7)	46.29 (2.3)	32.57 (1.5)	38.48 (1.8)	34.25 (1.4)	47.67 (1.9)	34.45 (1.7)	61.92 (2.6)	35.35 (0.7)

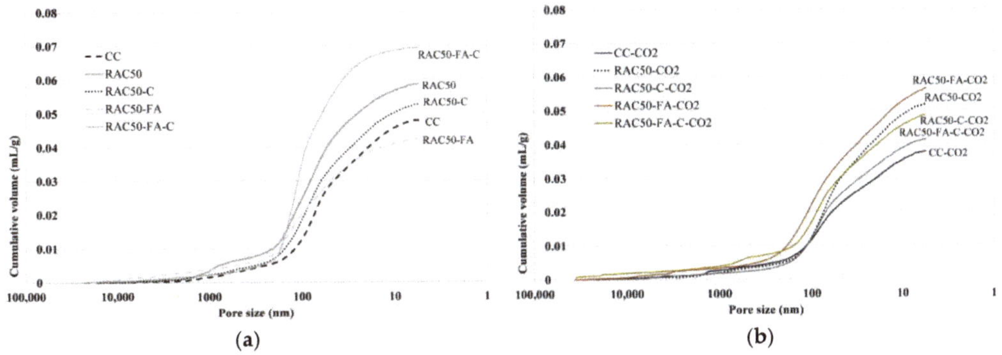

Figure 6. Development of pore size distribution of (**a**) uncarbonated area (No) and (**b**) carbonated area (CO_2) of concrete after exposure to the accelerated carbonation attack.

The k_{acc} can be used to estimate the theoretical natural carbonation coefficient ($k_{natTHEO}$)) [56,57] process of each produced concrete using Equation (2).

$$k_{acc}/k_{natTHEO} = (\varnothing_{acc})^{0.5}/(\varnothing_{natTHEO})^{0.5} \quad (2)$$

where $\varnothing_{acc}$ and $\varnothing_{natTHEO}$ are the CO_2 concentrations in the accelerated carbonation (3%) and natural carbonation processes (425 ppm, in Barcelona), respectively. Table 7 describes the $k_{natTHEO}$ of each concrete.

According to the Spanish concrete structural code (SC) [34], a concrete structure exposed to XC3 and XC4 environments (corrosion by carbonation) with a useful life of 50 and 100 years has limited the minimum cover depth. Therefore, in XC3 conditions, a minimum cover of 20 mm and 30 mm would be required. In addition, a minimum cover depth of 25 and 35 mm, respectively, would be required for concrete exposed to the XC4 environment conditions. Table 9 describes the carbonation depth achieved by each concrete according to the $k_{natTHEO}$ rate after a lifespan of 50 and 100 years.

Table 9. Carbonation depth after a lifespan of 50 and 100 years.

	CC	RAC50	RAC50-C	RAC50-FA	RAC50-FA-C	XC3	XC4
	Carbonation Depth (mm)					Min. Cover (mm)	
50 years ($k_{natTHEO}$)	14.9	15.5	21.6	20.5	23.1	20	25
100 years ($k_{natTHEO}$)	21.1	21.9	30.5	28.9	32.6	30	35

The obtained results from this study concluded that all the concretes achieved a lower carbonation depth value than the minimum cover depth value required for the defined useful lifespan expected in XC4 environment conditions. However, for exposure in XC3 environment conditions, the CC and RCA50 concretes would only be adequate for a lifespan of 50 years, and the CC, the RCA50, and the RCA50-FA would be adequate for exposure in XC4 environmental conditions. Thus, none of the recycled concrete produced with CO_2-treated recycled aggregates would achieve the governmental minimum requirements.

3.3.4. Chloride Ion Penetration vs. Carbonation Resistance

Figure 7 describes the chloride ion penetration (in coulombs) at 28 days (Table 7) and carbonation resistance (K_{acc}, Table 8) of recycled aggregate concretes with respect to the CC concrete. Figure 7 shows that the treatment and composition of recycled aggregates influenced the concrete's chloride and carbonation resistance in a different form.

While the RAC50-C concrete (concrete produced with carbonated RCA-C) and RAC50-FA concrete (concrete produced with uncarbonated RCA-FA) achieved the highest chloride resistance and similar values to the CC concrete, the RAC50 concrete (concrete produced with uncarbonated RCA) achieved carbonation resistance equivalent to that of the CC concrete.

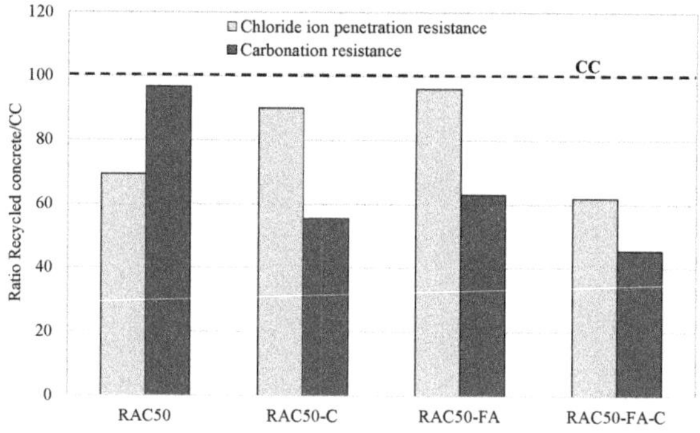

Figure 7. Influence of recycled aggregate concrete in Chloride ion penetration and carbonation resistance with respect to CC concrete.

4. Conclusions

The following conclusions can be deduced from the results obtained by testing how the properties of concrete are affected by the carbonation treatment on different compositions of recycled aggregates:

- The CO_2 treatment increased the density and reduced the absorption capacity of recycled aggregates. However, when the attached mortar of the recycled aggregates was composed with FA (RCA-FA), after carbonation, the pore size distribution was transformed, increasing in the mesopores and macropores. Consequently, the RCA-FA-C (carbonated) aggregate achieved higher permeability than the RCA-FA;
- Therefore, according to the properties of recycled aggregates, the carbonate treatment was effective when the attached mortar of the recycled aggregates was composed of OPC. In contrast, the RCA-FA-C (carbonated) aggregates achieved higher permeability than those of the RCA-FA aggregates;
- According to physical and mechanical properties, the following is true:
    - All the recycled concretes achieved adequate properties as structural concrete. They reached the compressive strength of the CC concrete at 28 days and a lower modulus elasticity than that of the CC concrete;
    - The recycled concrete produced with 50% of the RCA-C and RCA-FA aggregates achieved the most efficient properties of all the recycled concretes:
        - They only achieved a 7% higher absorption capacity than that of the CC concrete;
        - The concrete produced with RCA-C achieved the highest strength;
        - The recycled concrete produced with RCA-C and RCA-FA reached similar or higher splitting tensile strength to that of the CC concrete;
        - The recycled concrete produced with RCA-C and RCA-FA suffered a lower decrease of 6% in modulus elasticity compared to the CC concrete.

According to durability, the following is true:

- All the recycled concrete produced with 50% of RA achieved the durable category defined with a sorptivity value lower than 0.05 mm/min$^{0.5}$. The concrete made with 50% of the RCA-FA achieved the lowest sorptivity value of 0.032 mm/min$^{0.5}$, followed by the CC concrete with 0.035 mm/min$^{0.5}$ and the recycled concrete produced with 50% of the RCA-C with 0.036 mm/min$^{0.5}$;

- The recycled concrete achieved higher chloride ion penetration than that of the CC concrete. However, the concrete produced with the RCA-FA aggregates achieved the best value with only 4% higher penetration than the CC concrete, followed by the concrete produced with the RCA-C aggregates. In contrast, the concrete produced with the RCA-FA-C aggregates achieved the highest penetration: 38% higher than that of the CC concrete;
- The CC concrete achieved the lowest carbonation rate, followed by the concretes produced with 50% RCA and 50% RCA-FA. The CO_2-treated recycled aggregates worsened the carbonation resistance of the recycled concrete, with this worsening condition increasing when the RCA-FA-C aggregates were used;
- It was found that the CC and RAC50 concretes were the only concretes adequate for both XC3 and XC4 environment expositions.

It can be concluded that carbonation treatment was inefficient when the recycled aggregates were obtained by crushing a parent concrete composed of FA blended cement. The concretes made with the RCA-FA-C aggregates obtained the worst physical, mechanical, and durability property values. In contrast, the concrete produced with 50% of the RCA-C aggregates obtained the best property values except for carbonation resistance, which was higher in concrete produced with the RCA aggregate. In addition, the concrete properties made with the RCA-FA aggregates achieved better sorptivity and chloride ion penetration resistance than those produced with the RCA-C aggregates.

Author Contributions: Conceptualization, M.E.; methodology, M.E.; validation, M.E. and S.C.; investigation, M.E. and S.C.; resources, M.E. writing—original draft preparation, M.E.; writing—review and editing, M.E.; project administration, M.E.; funding acquisition, M.E. All authors have read and agreed to the published version of the manuscript.

Funding: This research received no external funding.

Institutional Review Board Statement: Not applicable.

Informed Consent Statement: Not applicable.

Data Availability Statement: Not applicable.

Acknowledgments: The first author also wants to thank the staff of the Laboratory of Technology of Structures and Materials "Lluis Agulló" of the UPC for their support.

Conflicts of Interest: The authors declare no conflict of interest.

References

1. European Statistics-Eurostat. *Energy, Transport and Environment Statistics*, 2020th ed.; European Statistics-Eurostat: Luxembourg, 2020.
2. de Carvalho, J.M.F.; Fontes, W.C.; de Azevedo, C.F.; Brigolini, G.J.; Schmidt, W.; Peixoto, R.A.F. Enhancing the eco-efficiency of concrete using engineered recycled mineral admixtures and recycled aggregates. *J. Clean. Prod.* **2020**, *257*, 120530. [CrossRef]
3. Kaplan, G.; Gulcan, A.; Cagdas, B.; Bayraktar, O.Y. The impact of recycled coarse aggregates obtained from waste concretes on lightweight pervious concrete properties. *Environ. Sci. Pollut. Res.* **2021**, *28*, 17369–17394. [CrossRef]
4. Le, H.B.; Bui, Q.B. Recycled aggregate concretes—A state-of-the-art from the microstructure to the structural performance. *Constr. Build. Mater.* **2020**, *257*, 119522. [CrossRef]
5. Silva, S.; Evangelista, L.; de Brito, J. Durability and shrinkage performance of concrete made with coarse multi-recycled concrete aggregates. *Constr. Build. Mater.* **2021**, *272*, 121645. [CrossRef]
6. Pedro, D.; De Brito, J.; Evangelista, L. Influence of the use of recycled concrete aggregates from different sources on structural concrete. *Constr. Build. Mater.* **2014**, *71*, 141–151. [CrossRef]
7. Etxeberria, M. Evaluation of eco-efficient concretes produced with fly ash and uncarbonated recycled aggregates. *Materials* **2021**, *14*, 7499. [CrossRef]
8. Zeng, X. Progress in the research of carbonation resistance of RAC. *Constr. Build. Mater.* **2020**, *230*, 116976. [CrossRef]
9. Thomas, C.; Setién, J.; Polanco, J.A.; Alaejos, P.; De Juan, M.S. Durability of recycled aggregate concrete. *Constr. Build. Mater.* **2013**, *40*, 1054–1065. [CrossRef]
10. Kwan, W.H.; Ramli, M.; Kam, K.J.; Sulieman, M.Z. Influence of the amount of recycled coarse aggregate in concrete design and durability properties. *Constr. Build. Mater.* **2011**, *26*, 565–573. [CrossRef]

11. Berredjem, L.; Arabi, N.; Molez, L. Mechanical and durability properties of concrete based on recycled coarse and fine aggregates produced from demolished concrete. *Constr. Build. Mater.* **2020**, *246*, 118421. [CrossRef]
12. Guo, H.; Shi, C.; Guan, X.; Zhu, J.; Ding, Y.; Ling, T.C.; Zhang, H.; Wang, Y. Durability of recycled aggregate concrete—A review. *Cem. Concr. Compos.* **2018**, *89*, 251–259. [CrossRef]
13. Otsuki, N.; Miyazato, S.; Yodsudjai, W. Influence of recycled aggregate on interfacial transition zone, strength, chloride penetration and carbonation of concrete. *J. Mater. Civ. Eng.* **2003**, *15*, 443–451. [CrossRef]
14. Adessina, A.; Fraj, A.B.; Barthélémy, J.F.; Chateau, C.; Garnier, D. Experimental and micromechanical investigation on the mechanical and durability properties of recycled aggregates concrete. *Cem. Concr. Res.* **2019**, *126*, 105900. [CrossRef]
15. Leemann, A.; Loser, R. Carbonation resistance of recycled aggregate concrete. *Constr. Build. Mater.* **2019**, *204*, 335–341. [CrossRef]
16. Pedro, D.; de Brito, J.; Evangelista, L. Performance of concrete made with aggregates recycled from precasting industry waste: Influence of the crushing process. *Mater. Struct. Constr.* **2015**, *48*, 3965–3978. [CrossRef]
17. Liang, C.; Pan, B.; Ma, Z.; He, Z.; Duan, Z. Utilization of CO2 curing to enhance the properties of recycled aggregate and prepared concrete: A review. *Cem. Concr. Compos.* **2020**, *105*, 103446. [CrossRef]
18. Lu, B.; Shi, C.; Cao, Z.; Guo, M.; Zheng, J. Effect of carbonated coarse recycled concrete aggregate on the properties and microstructure of recycled concrete. *J. Clean. Prod.* **2019**, *233*, 421–428. [CrossRef]
19. Zhang, J.; Shi, C.; Li, Y.; Pan, X.; Poon, C.-S.; Xie, Z. Performance Enhancement of Recycled Concrete Aggregates through Carbonation. *J. Mater. Civ. Eng.* **2015**, *27*, 04015029. [CrossRef]
20. Pu, Y.; Li, L.; Wang, Q.; Shi, X.; Fu, L.; Zhang, G.; Luan, C.; Abomohra, A.E.F. Accelerated carbonation treatment of recycled concrete aggregates using flue gas: A comparative study towards performance improvement. *J. CO_2 Util.* **2021**, *43*, 101362. [CrossRef]
21. Sereng, M.; Djerbi, A.; Metalssi, O.O.; Dangla, P.; Torrenti, J.M. Improvement of recycled aggregates properties by means of CO_2 uptake. *Appl. Sci.* **2021**, *11*, 6571. [CrossRef]
22. Li, Y.; Zhang, S.; Wang, R.; Zhao, Y.; Men, C. Effects of carbonation treatment on the crushing characteristics of recycled coarse aggregates. *Constr. Build. Mater.* **2019**, *201*, 408–420. [CrossRef]
23. von Greve-Dierfeld, S.; Lothenbach, B.; Vollpracht, A.; Wu, B.; Huet, B.; Andrade, C.; Medina, C.; Thiel, C.; Gruyaert, E.; Vanoutrive, H.; et al. Understanding the carbonation of concrete with supplementary cementitious materials: A critical review by RILEM TC 281-CCC. *Mater. Struct. Constr.* **2020**, *53*, 136. [CrossRef]
24. Xuan, D.; Zhan, B.; Poon, C.S. Durability of recycled aggregate concrete prepared with carbonated recycled concrete aggregates. *Cem. Concr. Compos.* **2017**, *84*, 214–221. [CrossRef]
25. Ding, Z.; Quy, N.X.; Kim, J.; Hama, Y. Evaluations of frost and scaling resistance of fly ash concrete in terms of changes in water absorption and pore structure under the accelerated carbonation conditions. *Constr. Build. Mater.* **2022**, *345*, 128273. [CrossRef]
26. *GB/T50082-2009*; Standard for test method of long term performance and durability of ordinary concrete. China Architecture and Building Press: Beijing, China, 2009.
27. *UNE-EN 1097-6*; Tests for mechanical and physical properties of aggregates. Determination of particle density and water absorption—Ensayos para determinas las propiedds mecánicas y físicas de los áridos. (Parte 6: Determinación de la densidad de partículas y la absorción de agua). AENOR: Madrid, Spain, 2014.
28. Hyvert, N.; Sellier, A.; Duprat, F.; Rougeau, P.; Francisco, P. Dependency of C-S-H carbonation rate on CO_2 pressure to explain transition from accelerated tests to natural carbonation. *Cem. Concr. Res.* **2010**, *40*, 1582–1589. [CrossRef]
29. Mindess, S.; Young, J.F. *Concrete*; Prentice-Hall, Inc.: Englewood Cliffs, NJ, USA, 1981.
30. Igarashi, S.I.; Watanabe, A.; Kawamura, M. Evaluation of capillary pore size characteristics in high-strength concrete at early ages. *Cem. Concr. Res.* **2005**, *35*, 513–519. [CrossRef]
31. Jenings, H.M. Design of high strength cement based materials: Part 2 Microstructure. *Mater. Sci. Technol.* **1988**, *4*, 285–290. [CrossRef]
32. Xiao, J.; Zhang, H.; Tang, Y.; Deng, Q.; Wang, D.; Poon, C.S. Fully utilizing carbonated recycled aggregates in concrete: Strength, drying shrinkage and carbon emissions analysis. *J. Clean. Prod.* **2022**, *377*, 134520. [CrossRef]
33. Liang, C.; Ma, H.; Pan, Y.; Ma, Z.; Duan, Z.; He, Z. Chloride permeability and the caused steel corrosion in the concrete with carbonated recycled aggregate. *Constr. Build. Mater.* **2019**, *218*, 506–518. [CrossRef]
34. del de Hormigón, C. *Código estructural. Spanish Structural Code*; Boletón Oficial del Estado: Madrid, Spain, 2015; pp. 61561–61567.
35. Neville, A. *Properties of Concrete*, 5th ed.; Longman: London, UK, 2011.
36. Etxeberria, M.; Vázquez, E.; Marí, A.; Barra, M. Influence of amount of recycled coarse aggregates and production process on properties of recycled aggregate concrete. *Cem. Concr. Res.* **2007**, *37*, 735–742. [CrossRef]
37. Poon, C.S.; Shui, Z.H.; Lam, L. Effect of microstructure of ITZ on compressive strength of concrete prepared with recycled aggregates. *Constr. Build. Mater.* **2004**, *18*, 461–468. [CrossRef]
38. Etxeberria, M.; Vazquez, E.; Mari, A. Microstructure analysis of hardened recycledaggregate concrete. *Mag. Concr. Res.* **2006**, *58*, 683–690. [CrossRef]
39. *UNE-EN 12350-2*; Testing hardened concrete. Making and curing specimens for strength tests. Ensayos de hormigón endurecido. Parte 2. Fabricación y curado de probetas para ensayos de resistencia. AENOR: Madrid, Spain, 2020.
40. *UNE-EN 12350-6*; Testing fresh concrete. Density. Ensayos de hormigón fresco. Parte 6. Densidad. AENOR: Madrid, Spain, 2014.

41. *ASTM C642-21*; Standard Test Method for Density, Absorption, and Voids in Hardened Concrete. ASTM: West Conshohocken, PA, USA, 2021.
42. *UNE-EN 12350-3*; Testing hardened concrete. Compressive strength of test specimens. Ensayos de hormigón endurecido. Parte 3: Determinación de la resistencia a compresión de probetas. AENOR: Madrid, Spain, 2020.
43. *UNE-EN 12350-13*; Testing hardened concrete. Determination of secant modulus of elasticity in compressionEnsayos de hormigón endurecido. Parte 13: Determinación del módulo secante de elasticidad en compresión. AENOR: Madrid, Spain, 2022.
44. *UNE-EN 12350-6*; Testing hardened concrete. Tensile splitting strength of test specimens. Ensayos de hormigón endurecido. Parte 6: Resistencia a tracción indirecta de probetas. AENOR: Madrid, Spain, 2010.
45. *ISO 15148:2002(E)*; Hygrothermal performance of building materials and products—Determination of water absorption coefficient by partial immersion. ISO: Geneva, Switzerland, 2002.
46. *ASTM C1202*; Standard Test Method for Electrical Indication of Concrete's Ability to Resist Chloride Ion Penetration. ASTM: West Conshohocken, PA, USA, 2019.
47. Ghanem, H.; Trad, A.; Dandachy, M.; ElKordi, A. Effect of Wet-Mat Curing Time on Chloride Permeability of Concrete Bridge Decks. In *Advances and Challenges in Structural Engineering: Proceedings of the 2nd GeoMEast International Congress and Exhibition on Sustainable Civil Infrastructures, Egypt 2018–The Official International Congress of the Soil-Structure Interaction Group in Egypt (SSIGE)*; Springer International Publishing: Berlin/Heidelberg, Germany, 2019; pp. 194–208.
48. *UNE-EN 12390-12*; Testing hardened concrete. Determination of the carbonation resistance of concrete. Accelerated carbonation method. Ensayos de hormigón endurecido. Parte 12: Determinación de la resistencia a la carbonatación del hormigón. Método de carbonatación acelerada. AENOR: Madrid, Spain, 2020.
49. *UNE-EN 14630*; Products and systems for the protection and repair of concrete structures—Test methods—Determination of carbonation depth in hardened concrete by the phenolphthalein method; Productos y sistemas para la protección y reparación de estructuras de hormigón. Métodos de ensayo. Determinación de la profundidad de carbonatación en un hormigón endurecido por el método de la fenolftaleína. AENOR: Madrid, Spain, 2007.
50. Kou, S.C.; Zhan, B.J.; Poon, C.S. Use of a CO_2 curing step to improve the properties of concrete prepared with recycled aggregates. *Cem. Concr. Compos.* **2014**, *45*, 22–28. [CrossRef]
51. Alexander, M.G.; Ballim, Y.; Stanish, K. A framework for use of durability indexes in performance-based design and specifications for reinforced concrete structures. *Mater. Struct. Constr.* **2008**, *41*, 921–936. [CrossRef]
52. Menéndez, G.; Bonavetti, V.L.; Irassar, E.F. Ternary blend cements concrete. Part II: Transport mechanism. *Mater. Construcción* **2007**, *57*, 31–43. [CrossRef]
53. Pedro, D.; de Brito, J.; Evangelista, L. Structural concrete with simultaneous incorporation of fine and coarse recycled concrete aggregates: Mechanical, durability and long-term properties. *Constr. Build. Mater.* **2017**, *154*, 294–309. [CrossRef]
54. Lam, L.; Wong, Y.L.; Poon, C.S. Effect of fly ash and silica fume on compressive and fracture behaviors of concrete. *Cem. Concr. Res.* **1998**, *28*, 271–283. [CrossRef]
55. Dinakar, P.; Reddy, M.K.; Sharma, M. Behaviour of self compacting concrete using Portland pozzolana cement with different levels of fly ash. *Mater. Des.* **2013**, *46*, 609–616. [CrossRef]
56. Parrott, L. *A Review of Carbonation in Reinforced Concrete*; BCA, British Cement Association Crowthorne: Berkshire, UK, 1987.
57. Van Den Heede, P.; De Belie, N. A service life based global warming potential for high-volume fly ash concrete exposed to carbonation. *Constr. Build. Mater.* **2014**, *55*, 183–193. [CrossRef]

Table 2 shows the chemical Portland cement composition, determined thorugh X-ray fluorescence (XRF-7000 from Shimadzu, Kyoto, Japan). Local available natural basaltic coarse aggregates (supplied by Mineração Grandes Lagos, Itapura, Brazil) were used to produce PC. Three narrow-sized aggregates (A, B, and C) were combined in this study.

Table 2. Chemical Portland cement composition determined by X-ray fluorescence (wt.%).

SiO_2	Al_2O_3	Fe_2O_3	CaO	Na_2O	K_2O	SO_3	MgO	Others	LOI
19.9	4.4	4.6	62.8	0.1	1.0	1.6	1.1	0.2	4.3

2.2. Experimental Procedure

2.2.1. Aggregates Characterization and Packing Procedure

The physical characterization of each basaltic aggregate (named A, B and C) was performed following the instructions of ASTM C127 [30] and ASTM C29 [31]. To determine both the unit weight and % of voids, the same compaction procedure used during the production of PC samples was followed: compaction was carried out on a vibration table in two layers (10 s per layer); simultaneously, 10 blows (approx. 1 per second) were made with a steel rod to accommodate aggregates in the cylindrical mold (100 mm × 200 mm). Finally, five more seconds were spent on the vibration table. The procedure was repeated three times. This modification in relation to ASTM C29 [31] in the compaction procedure was performed to guarantee the same conditions to accommodate aggregate particles while producing PC.

The packing procedure, which applied the maximum density methodology proposed by O'Reilly [27], was performed by combining two near narrow-sized aggregates (A and B or B and C). The maximum density methodology is based on the refinement of pores and, consequently, the increase in unit weight from optimizing the packing of the employed aggregates (see Figure 1).

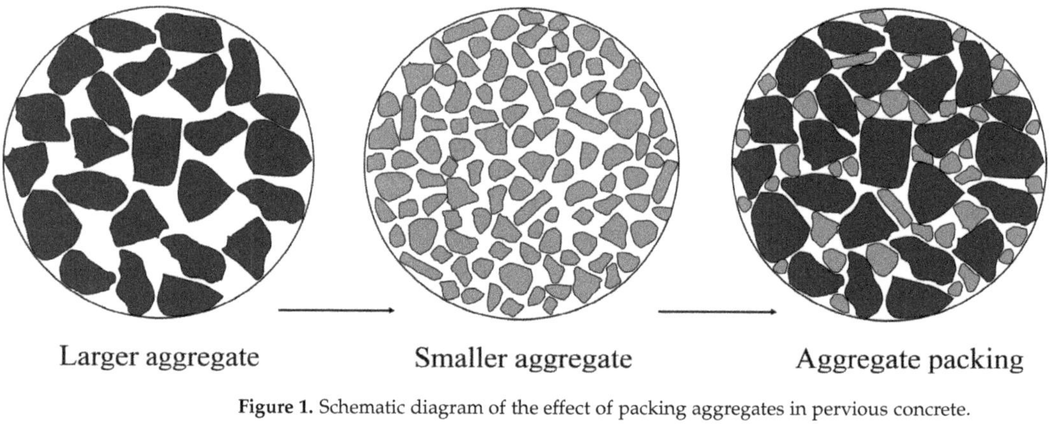

Figure 1. Schematic diagram of the effect of packing aggregates in pervious concrete.

According to O'Reilly, the unit weight (kg/m^3) of the largest aggregate should be first determined. Then, a small amount of the larger aggregate (usually around 10% per weight) is replaced with a smaller aggregate, and the unit weight is measured again. The replacement percentage should be incrementally increased to observe its effect on unit weight. The unit weight was obtained using the compaction procedure described above and calculated according to ASTM C29 [31]. By graphing the replacement percentage versus unit weight, a maximum value can be determined to indicate the optimal arrangement of aggregates.

This procedure was run by combining narrow-sized aggregates A–B and B–C. The A–B–C combination was performed using the previous results obtained via the above-mentioned combinations (A–B and B–C).

2.2.2. Pervious Concrete Production

A cement:aggregate mass ratio of 1:5 and an effective water/cement ratio of 0.26 were set for all the PC mixtures. Six PC mixtures were assessed, as previously mentioned, with three selected different narrow-sized aggregates (named A, B, and C), and three combinations of packed aggregates (named AB, BC, and ABC) designed using the maximum density methodology for aggregates packing. The nomenclature adopted for the different PC is related to the employed aggregate: PC-Z, where "Z" is the used aggregate size (A, B, C, AB, BC, or ABC). For all the mixtures, aggregates were utilized by considering the saturated surface dry (SSD) condition. The PC mix proportions are presented in Table 3.

Table 3. Mix proportions of the prepared PC.

Mixture ID	Aggregate (kg/m^3)			Cement (kg/m^3)	Water (kg/m^3)	Water/Cement	Cement/Aggregate
	A	B	C				
PC-A	1581.25	-	-	316.25	82.23	0.26	1:5
PC-B	-	1530.79	-	306.16	79.60		
PC-C	-	-	1523.96	304.79	79.25		
PC-AB	841.78	841.78	-	336.71	87.54		
PC-BC	-	633.10	949.65	316.55	82.30		
PC-ABC	333.73	333.73	1001.20	333.73	86.77		

The PC mixtures were prepared in a mechanical drum mixer following these steps: first, the aggregate was added to the mixer along with 50% of the water, and the mixture was stirred for 30 s. Next, 100% Portland cement was added, and the mixture was stirred for another 30-s period. Finally, the remaining water was added, and the mixture was stirred for 60 s. Cylindrical specimens (100 × 200 mm) were then molded for compressive strength tests (five specimens), and also for density, porosity, infiltration rate and pore characterization by an image analysis (three specimens). Specimens were molded in two layers by following the same compaction process to measure the unit weight for aggregates (see Section 2.2.1).

To conduct the drop-weight impact resistance test, hexagonal samples were molded with a side length of 145 mm and a height of 60 mm (5 samples for each mixture). The molding process of the hexagonal specimens involved the following steps: first, the mold was filled to the top and placed on a vibrating table for 15 s, while 20 blows were applied simultaneously with a steel rod to accommodate concrete. Next, 5 s on the vibrating table were applied to finish compaction.

All the specimens were demolded after 24 h and transferred to a curing room with a relative humidity of approximately 95% for 28 days until the age test. After this period, physical, hydraulic and mechanical tests were run.

2.2.3. Porosity and Density of Pervious Concrete

The PC samples' porosity and density were tested in accordance with ISO 17785-2 [32]. Three samples from each mixture were tested. Samples were dried in an oven until their mass was constant. The samples' dry mass (kg), denoted as "m_d", was then recorded. Next, the samples were immersed in water at a temperature of 20 °C ± 2 °C and left for 30 min. To eliminate entrapped air in voids, samples were tapped 10 times around the specimen's circumference with a rubber mallet. Finally, the immersed mass (kg) of samples, denoted as "m_s", was recorded.

Porosity was calculated using Equation (1), where "ρ_w" is the density of water at the water bath temperature and "V_d" is the specimen's volume. Density (kg/m^3) can be calculated using Equation (2).

$$\text{Porosity}\ (\%) = \left\{1 - \frac{(m_d - m_s)/\rho_w}{V_d}\right\} \times 100\% \tag{1}$$

$$\text{Density}\ (\%) = \frac{m_d}{V_d} \tag{2}$$

2.2.4. Infiltration Rate

The infiltration rate test was conducted in the laboratory following the recommendations proposed by ISO 17785-1 [33] (see Figure 2). Three samples of each mix proportion were analyzed, and infiltration rate "k" was determined via Equation (3).

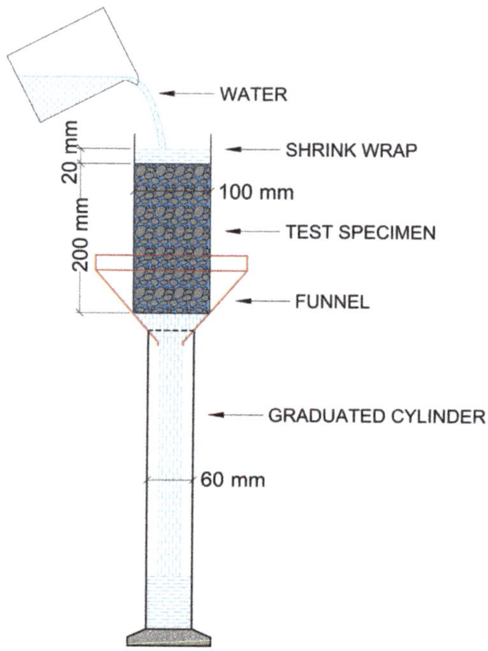

Figure 2. Schematic representation of the infiltration rate test.

Specimens were wrapped along their length with three layers of plastic film, and a 50 mm lip was left above the top surface. To prevent water percolation on the sides and to adjust shrink wrap to sample surfaces, shrink wrap was heated with a heat gun. Specimens were then placed in a funnel to leave the bottom face free.

Infiltration rate "k" was calculated based on the time "t" required for the percolation of 2000 mL of water "V", poured while maintaining a head of water of approximately 20 mm at the top of the specimen with a known cross-sectional area "A".

$$\text{k} = \frac{V}{A \times t} \tag{3}$$

2.2.5. Compressive Strength

Compressive strength was evaluated by testing five samples for each mix proportion using a universal testing machine (EMIC) with a maximum capacity of 200 tons at a load rate of 0.45 MPa/s, as described in NBR 5739 [34].

2.2.6. Drop-Weight Impact Resistance Test

The drop-weight impact resistance test is quantified by the number of blows required to cause a certain level of damage. This test provides an estimate of the energy absorbed by a sample at various load levels. To perform the test, a hexagonal sample is supported by a bed of sand, and a steel ball (63 mm diameter, 1.06 kg mass) is dropped onto the sample from a height of 1.0 m (see Figure 3). The test is repeated several times until the sample fractures. The energy of each impact is calculated using Equation (4), which takes into account the height of the drop, the steel ball's mass and the gravitational constant. This test was used to evaluate the drop-weight impact resistance of the samples in this study. According to the literature [35,36], the energy absorbed by each sample during the drop-weight impact resistance test can be calculated using Equation (4), where "IE" is the impact energy (J), "h" is the drop height (m), "m" is the steel ball's mass (kg) and "g" is the gravity constant (9.81 m/s^2).

$$\text{IE} = h \times m \times g \tag{4}$$

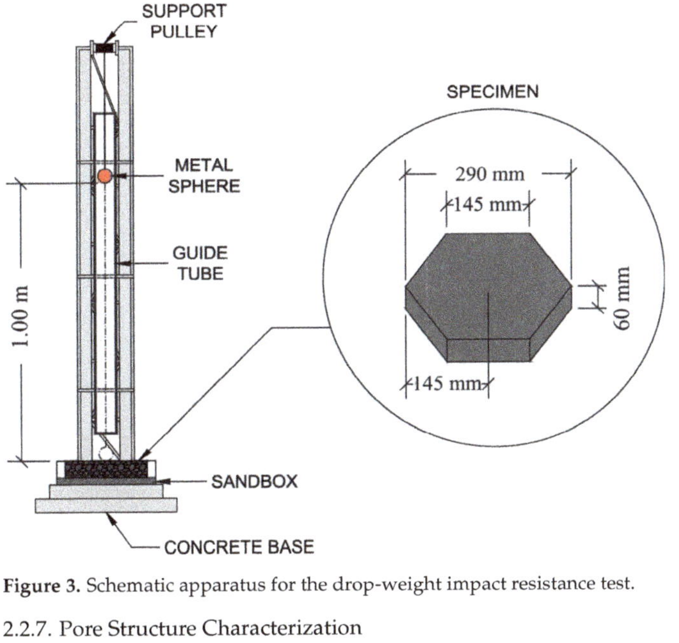

Figure 3. Schematic apparatus for the drop-weight impact resistance test.

2.2.7. Pore Structure Characterization

A digital image analysis is a powerful tool for obtaining detailed information about materials' internal pore structure. The procedure described in Figure 4 involves half-sectioning specimens and painting each PC face to improve contrast. High-quality two-dimensional optical images are captured and then analyzed with the ImageJ software to determine parameters such as total porosity, total number of pores, average pore diameter and average pore area. This approach provides a more detailed understanding of the pore structure than traditional methods such as method ISO 17785-2 [32].

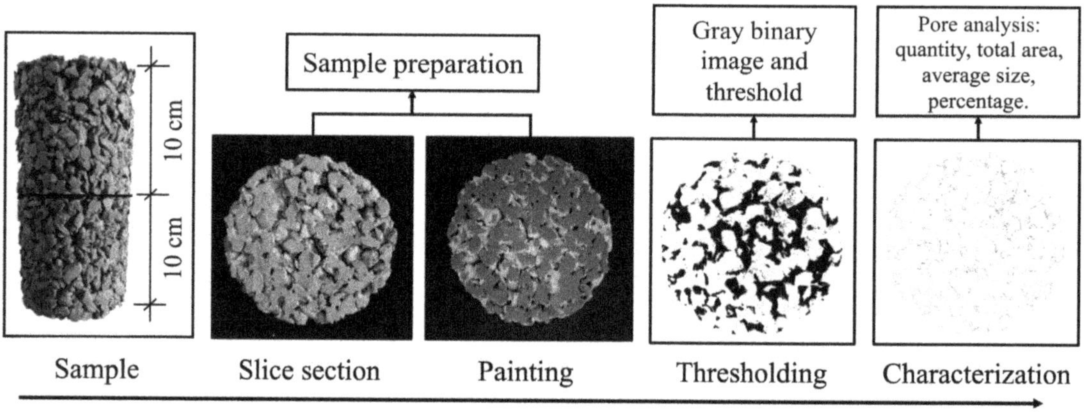

Figure 4. Schematic pore analysis of specimens using software ImageJ.

2.2.8. Correlations among Pervious Concrete Properties

Multiple regression is a statistical method used to explore the relation between a dependent variable and two independent variables or more by determining the coefficients that best fit the data [37]. In this study, a multiple regression analysis with a 95% confidence level was performed to investigate the relation between the compressive strength and impact resistance of PC and its physical properties, including density, porosity and aggregate size.

The model is expressed by Equation (5), where y represents the predicted values (compressive strength or impact resistance), x_1, x_2, x_3, and x_n denote the independent variables (density, porosity, aggregate size), and a_1, a_2, a_3 refer to the regression coefficients. The model's accuracy can be assessed by calculating the coefficient of determination (R-squared value), which measures the proportion of total variation in the dependent variable that can be explained by the independent variables. A higher R-squared value indicates the model's better data fit.

$$y(x_1, x_2, x_3, x_n) = \beta_0 + a_1 \times x_1 + a_2 \times x_2 + a_3 \times x_3 + \ldots + a_n \times x_n \tag{5}$$

3. Results and Discussions

3.1. Aggregates Characterization and Packing Results

Three different narrow-sized aggregates, as well as their combination, were utilized in the study. Figure 5a,b depicts the morphological characteristics and the single-sized particle distributions, respectively. Table 4 outlines their physical properties. According to previous studies [38,39], a coefficient of uniformity below 2.0 is desirable for achieving higher porosity in PC. Of the tested aggregates, "B" and "C", respectively, demonstrated coefficients of uniformity of 1.5 and 1.4, which indicate a narrow particles range. However, aggregate "A" displayed a coefficient of 2.4, which suggests a slightly wider particle size range.

The results obtained from the maximum density methodology for the evaluated aggregates are presented in Figure 5c. The maximum unit weight was achieved when 50% of aggregate "A" was substituted for "B", which resulted in a 10.7% increase in unit weight compared to employing aggregate "B" alone. This outcome can be explained by the smaller particle size and higher coefficient of uniformity exhibited by aggregate "A", which can significantly broaden particle size distribution when combined with "B".

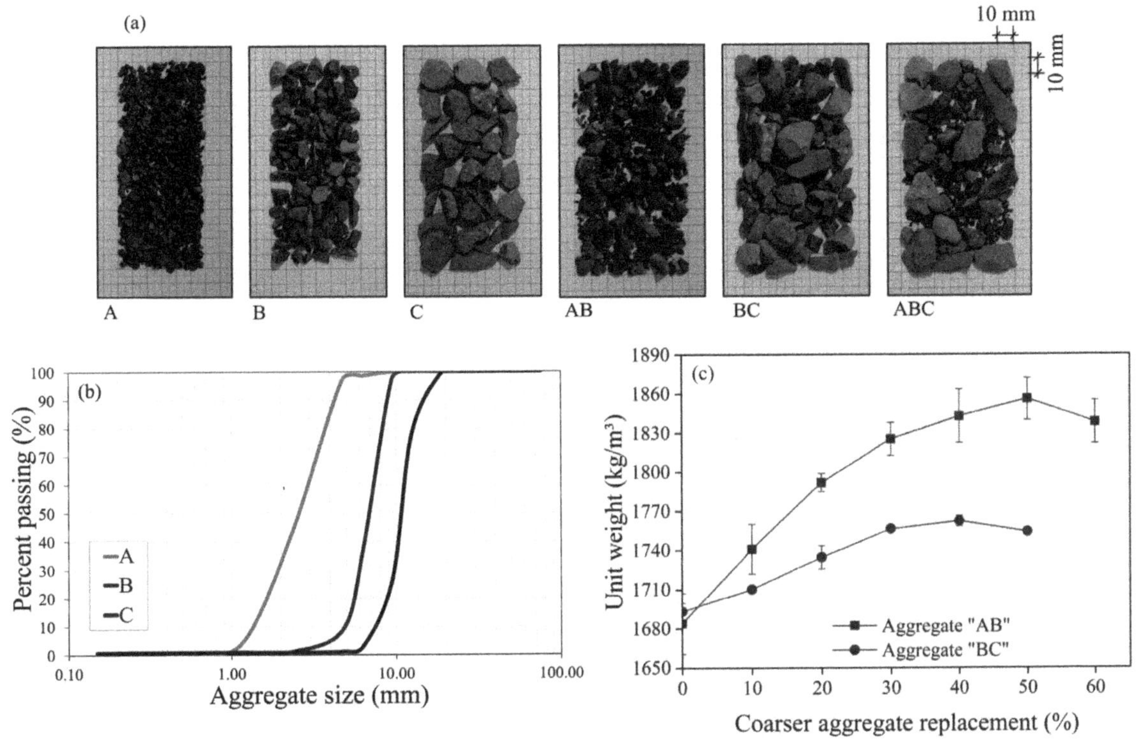

Figure 5. (**a**) Morphological aspect of the narrow-sized and combined aggregates; (**b**) Gradation curves of basaltic aggregates; (**c**) Unit weight for combined aggregates AB and BC at different replacement percentages.

Table 4. Physical properties of basaltic aggregates.

Properties	A	B	C	AB	BC	ABC	Standard
Water absorption (%)	4.45	2.31	1.97	3.38	2.11	2.53	ASTM C127 [30]
Specific gravity	2.97	3.04	3.03	3.01	3.03	3.02	ASTM C127 [30]
Specific gravity OD *	2.62	2.84	2.86	2.73	2.85	2.81	ASTM C127 [30]
Specific gravity SSD **	2.74	2.91	2.92	2.83	2.92	2.88	ASTM C127 [30]
Unit weight (g/cm^3)	1.75	1.68	1.69	1.86	1.76	1.84	ASTM C29 [31]
% Voids	40.96	44.63	44.11	38.22	41.72	39.09	ASTM C29 [31]

* OD—oven-dry; ** SSD—saturated-surface-dry.

With the packed aggregate "BC", optimal replacement was determined as 40% of aggregate "B". This resulted in a density of 1762.2 kg/m^3, which corresponds to a 4.1% increase in unit weight compared to using aggregate "C" alone. To produce the packed aggregate "ABC", 60% of aggregate "C", 20% of "B", and 20% of "A" were mixed in line with the results previously obtained for "BC" and "AB". This combination gave a density of 1835.8 kg/m^3, which represents an 8.9% increase in the unit weight compared to using aggregate "C" alone. Table 4 presents the physical parameters for the packed aggregates.

3.2. *Pervious Concrete Properties*

3.2.1. Porosity and Density

Figure 6a,b displays the porosity and density of all the produced PCs. An increase in porosity and a decrease in density took place in the PCs containing only a narrow-sized aggregate (A, B, or C) when aggregate size was increased. In contrast, PC-A, which was prepared with aggregate "A", had lower porosity and higher density, likely because its unit weight and coefficient of uniformity values were, respectively, higher. Additionally, previous research has indicated that smaller coarse aggregates generally lead to denser and less porous PCs [21].

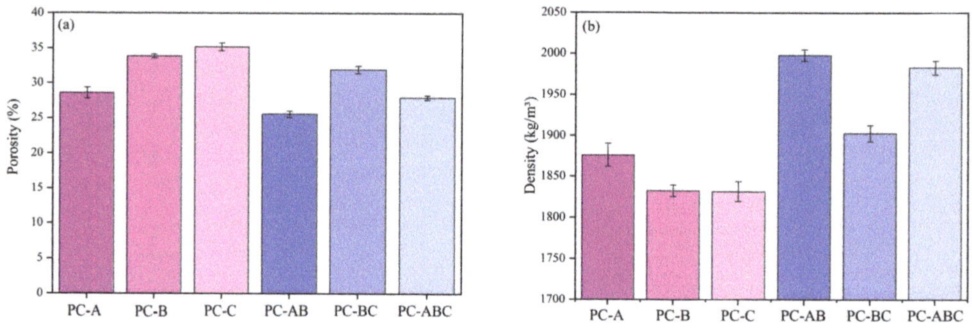

Figure 6. (**a**) Porosity of PC mixtures; (**b**) density of PC mixtures.

Despite having different size range distributions, PC-B and PC-C obtained similar porosity and density values (roughly 34% and 1830 kg/m^3, respectively). This result can be attributed to the presence of the comparable voids content and unit weight of aggregates, which are the predominant factors that influence PCs' porosity and density [40].

The maximum density methodology was applied to combine aggregates, and resulted in PC's decreased total porosity and increased density. The incorporation of aggregate A with aggregate B (PC-AB) reduced total porosity by 24.3% and increased density by 9.0% compared to the mixture created solely with aggregate B. Similarly, the combination of aggregate B with C (PC-BC) reduced porosity by 9.2% and increased density by 3.9% compared to the concretes created only with aggregate C. The inclusion of aggregate A (PC-ABC) further enhanced the particles packing of aggregate C, reduced porosity by 20.5% and increased density by 8.3% compared to PC-C. These findings are consistent with the literature, which reports porosity ranging from 25.6% to 31.9% and density falling within the 1903.2 kg/m^3 to 1998.0 kg/m^3 range [7,12,35]. Wu [25] obtained similar results, with a reduction in porosity from 13.9% to 35.7% by combining basalt aggregates (2–5 mm) with steel slag aggregates (1–2 mm). Figure 7 presents the correlation between density and porosity, where a good correlation was obtained (R^2 = 0.78).

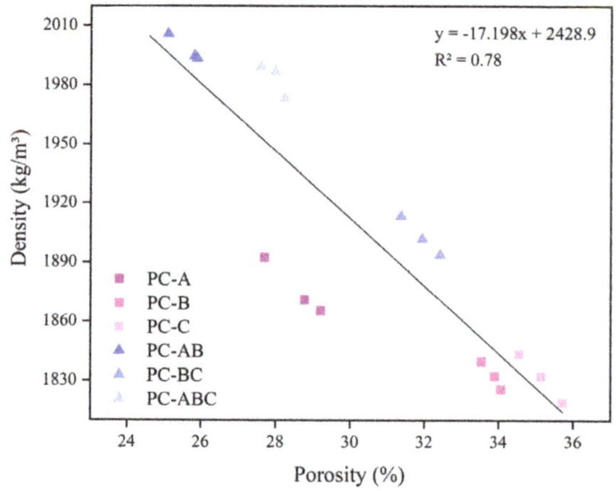

Figure 7. Relation between PCs' porosity and density.

3.2.2. Infiltration Rate

The results of the hydraulic conductivity tests for the infiltration rate are presented in Figure 8. Standard NBR 16416 [41] states that the minimum infiltration rate for PC should be 0.1 cm/s if used for paving. The infiltration rate for the PC created with narrow-sized aggregates (PC-A, PC-B, and PC-C) ranged from 0.46 to 2.62 cm/s. Coarser narrow-sized aggregates led to higher infiltration rates due to the presence of larger interconnected pores [9,42]. Conversely, smaller coarse aggregates tend to decrease PC's drainage capacity by creating smaller pores, which impede water flow [20,43]. Yu et al. [43] found that the number of seepage flow paths lowered and became thinner in a PC created with smaller aggregates, which led to reduced hydraulic efficiency.

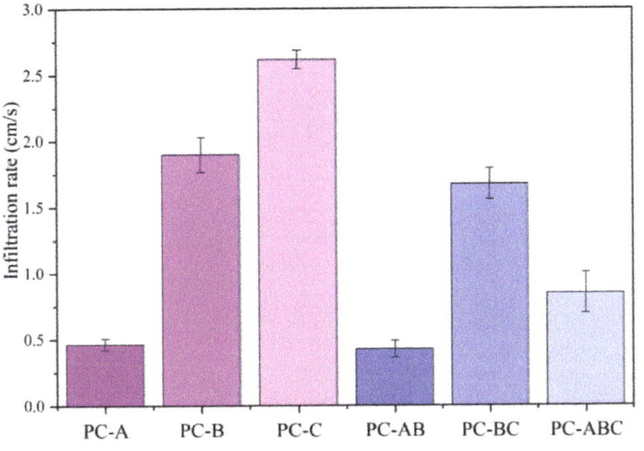

Figure 8. Infiltration rate for PC mixtures.

As expected, the narrow-sized aggregates combination resulted in a lower infiltration rate. PC-AB had the lowest infiltration rate (0.43 cm/s) of all the assessed PCs, which is a reduction of 67.6% compared to PC-B. The reduction of PC-BC was 36.3% compared to PC-C. It is interesting to note that the PCs created with two packed aggregates had a similar infiltration rate to those prepared with smaller sized aggregates. Only PC-ABC had a higher infiltration rate compared to the PC created with the smallest aggregate (PC-A). The decrease observed in the infiltration rate was similar to that reported by other authors, who combined smaller sized aggregates (10–94%) [2,26,28]. In all cases, however, the observed reduction in the infiltration rate did not block the seepage flow. Samples met the minimum requirements set out by the standard and the data found in the literature [14,44].

3.2.3. Compressive Strength and Drop-Weight Impact Resistance

The results of the 28-day compressive strength tests of the PC made with different aggregate sizes are presented in Figure 9a. Of the mixtures prepared with narrow-sized aggregates, PC-A exhibited the highest compressive strength. In contrast, PC-B and PC-C mixtures obtained a reduction of 23.8% and 32.6%, respectively, compared to PC-A. The heavier unit weight and the lower aggregate A porosity likely contributed to better compressive strength. It is noteworthy that using a coarser aggregate led to lower compressive strength, as reported by previous studies [9,15]. This trend could be attributed to smaller aggregates' larger surface area, which creates more binder bonds in the internal structure, thus improving compressive strength [20].

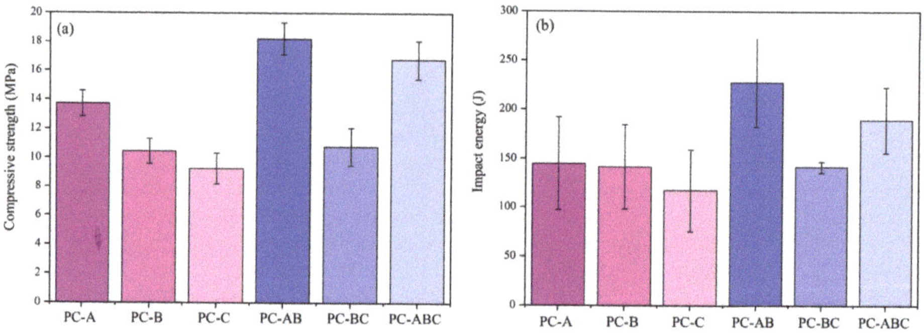

Figure 9. (**a**) 28-day compressive strength; (**b**) 28-day impact energy for PC mixtures.

Employing combined aggregates in PC resulted in a significantly improved compressive strength compared to the PCs made solely with narrow-sized coarse aggregates. Specifically, a 74.4% increase was observed for the PC-AB mix compared to PC-B. A slight increase of 16.4% was also recorded for the PC-BC mix compared to PC-C. When the smallest particle size aggregate was combined in the PC-ABC mix, the increase in compressive strength was 81.2%, and the resulting compressive strength was 16.76 MPa, which came very close to that of the PC-AB mix. Increased compressive strength can be attributed to greater aggregates packing, which densifies concrete. Although the literature reports that combinations of aggregates can either reduce or increase compressive strength depending on the amount of each aggregate to be incorporated, the increase herein observed is consistent with other reported values that range from 14.3 to 18.4 MPa (yielding increments up to 150%) [2,24,25]. It is important to note that no harmful reduction in the infiltration rate took place, which suggests that enhanced PC could be produced by combining narrow-sized aggregates with the maximum density methodology.

Figure 9b illustrates the influence of aggregate size on PC's impact energy for hexagonal-shaped samples. For narrow-sized aggregates, an increase in aggregate size is associated with a slight decrease in impact energy: PC-A, PC-B and PC-C required 144.7 J, 141.2 J, and 117.1 J, respectively. However, the combination of two narrow-sized aggregates significantly increased impact resistance. The PC-AB mix required 227.3 J for rupture, which represents a 61.0% increase compared to PC-B. Similarly the PC-BC mix required 141.2 J, a 20.6% increase compared to the PC-C mix. The incorporation of the finer narrow-sized aggregate "A" resulted in a more significant increment in impact resistance with energy rising to 189.4 J in the PC-ABC mixture, which is the equivalent to a 61.8% increase compared to the PC-C mix. This may be due to the presence of finer aggregates, which promote enhanced bridging connections and support greater impact energy. Although there is a gap in the literature about applying impact resistance tests in PC, the results of this study came close to those obtained by Mastali and Dalvad [45] in self-compacting concrete reinforced with 1.25% recycled carbon fiber-reinforced polymer, with maximum absorbed energy to the first and ultimate crack impact resistance of 223 J and 293 J, respectively.

The combination of aggregates in PC leads to a significant increase in both compressive strength and impact resistance. The smallest aggregate's packing ability is more efficient in improving mechanical properties by being more effective in filling pores [25]. Therefore, the packaging methodology can be a useful tool for optimizing aggregate mixtures to obtain more resistant PC mixes without having to resort to trial-and-error testing.

3.2.4. Pore Structure Characterization by Digital Image Analyses

In addition to total porosity, analyzing the pore structure can provide valuable insights into the behavior of the materials used in mixtures and how they influence PC properties [17,46]. In this study, two-dimensional optical images were used to determine the total porosity, average diameter and pore area for the various evaluated mix proportions. The digitized and thresholded images of samples appear in Figure 10.

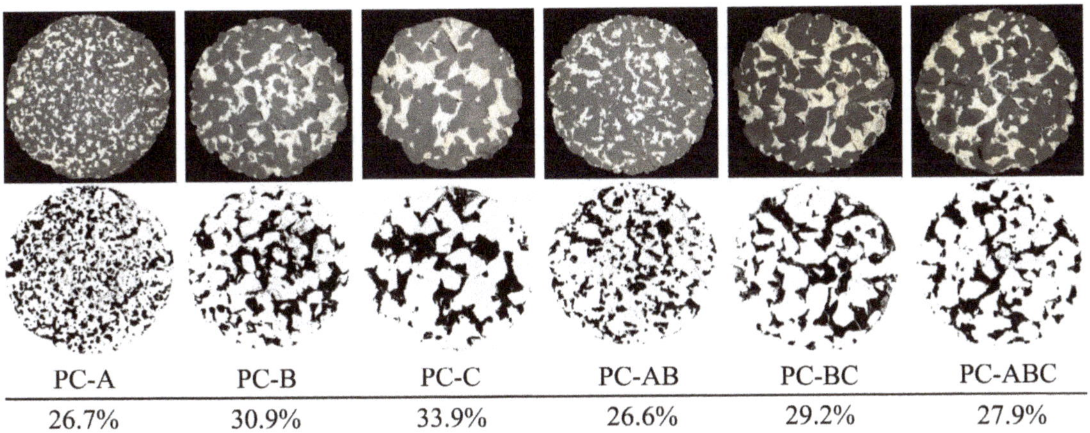

Figure 10. Scanned and binary (threshold) images of different PC sample surfaces and their respective total porosity.

The total porosity obtained through the digital image analysis can be compared to the value calculated according to ISO 17785-2 [32]. Both results are presented in Figure 11, and a similar trend and a linear fit with an R^2 of 0.85 were obtained. These findings are consistent with previous research by Chandrappa and Biligiri [11], who reported similar results.

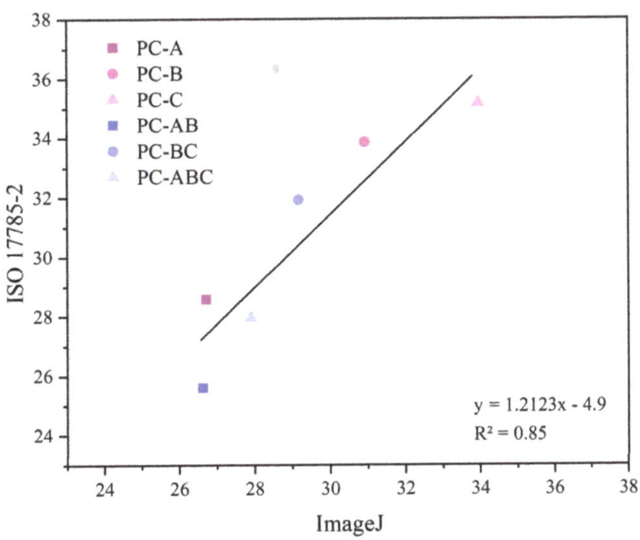

Figure 11. Comparative analysis of porosity using digital image analyses (ImageJ) and ISO 17785-2.

Table 5 presents the pore structure analysis results obtained with image analysis software. Employing coarser aggregates in PC resulted in a small number of total pores in a slice section compared to the finer aggregate, which ranged from 61 to 263 pores. This trend is consistent with the findings of Sumanasooriya and Neithalath [47] for recycled aggregates. Furthermore, the average pore diameter and pore area also reduced when finer aggregates were used, and PC-A obtained values of 2.52 mm and 7.39 mm^2, respectively, which represent 61.7% and 18.4% of the values for PC-C. The smaller pore diameters in PC-A and their lesser connectivity were responsible for the lower infiltration rate compared to PC-C [20,47].

Table 5. No. of pores, average pore diameter and average pore area for all the mixtures.

Mixture ID	Number of Pores	Average Pore Diameter (mm)	Average Pore Area (mm^2)
PC-A	263	2.52	7.39
PC-B	89	3.45	25.79
PC-C	61	4.08	40.22
PC-AB	144	3.03	13.45
PC-BC	61	4.09	33.56
PC-ABC	77	3.61	26.56

The packing methodology had a notable impact on the PC pore structure. PC-AB showed an increased number of pores compared to PC-B, which ranged from 89 to 144. However, the average pore diameter and pore area decreased from 3.45 to 3.03 mm and from 25.79 to 13.45 mm^2, respectively. These results represent a reduction of 12.2% and 47.8% in the average pore diameter and average pore area, respectively.

When comparing PC-BC to PC-C, no difference in the total number of pores and average pore diameter was seen. However, the average pore area showed a reduction of 16.6% due to the fine aggregates filling the voids of coarser aggregates during the packing process. This reduction in pore area is supported by the lower total porosity obtained for PC-BC vs. PC-C using ISO 17785-2 (see Figure 11) [32].

PC-ABC had a larger total number of pores than PC-C (77 vs. 61). However, the reductions in the average pore diameter and average pore area were 11.5% and 34.0%, respectively, which indicate pore structure refinement. This result explains the lower infiltration rate and the increase in the mechanical properties of the PCs with a combination of aggregates. The packing process provided by the finer aggregate combined with the larger particles reduced pores, as herein observed and by other authors [20,43].

3.3. Correlations among Pervious Concrete Properties

The relation among PCs' density, porosity, infiltration rate and mechanical properties is generally linear or exponential [16,18,22]. Figure 12 illustrates the relation among compressive strength, impact energy and infiltration rate concerning the density and porosity of the PC mixtures for which different sized aggregates were used.

The compressive strength showed a linear increase with rising density and a decrease with increasing porosity ($R^2 = 0.66$ and $R^2 = 0.93$, respectively). The impact energy for rupture exhibited similar behavior when related to density ($R^2 = 0.82$) and porosity ($R^2 = 0.78$). Conversely for the infiltration rate, an increase in density and a decrease in porosity gave a lower infiltration rate ($R^2 = 0.53$ and $R^2 = 0.89$, respectively). These findings agree with the literature [9,15,19,48], and suggest a close relation between the infiltration rate and porosity.

A multiple regression model (using 18 experimental data) was conducted using a statistical program (Statgraphics Centurion) to establish the correlation among compressive strength, impact energy, density, porosity and aggregate size. With the linear regression model (Equation (5)), this analysis identified the parameters (density, porosity, aggregate size) that most significantly affected the assessed PC mixtures' compressive strength and impact energy.

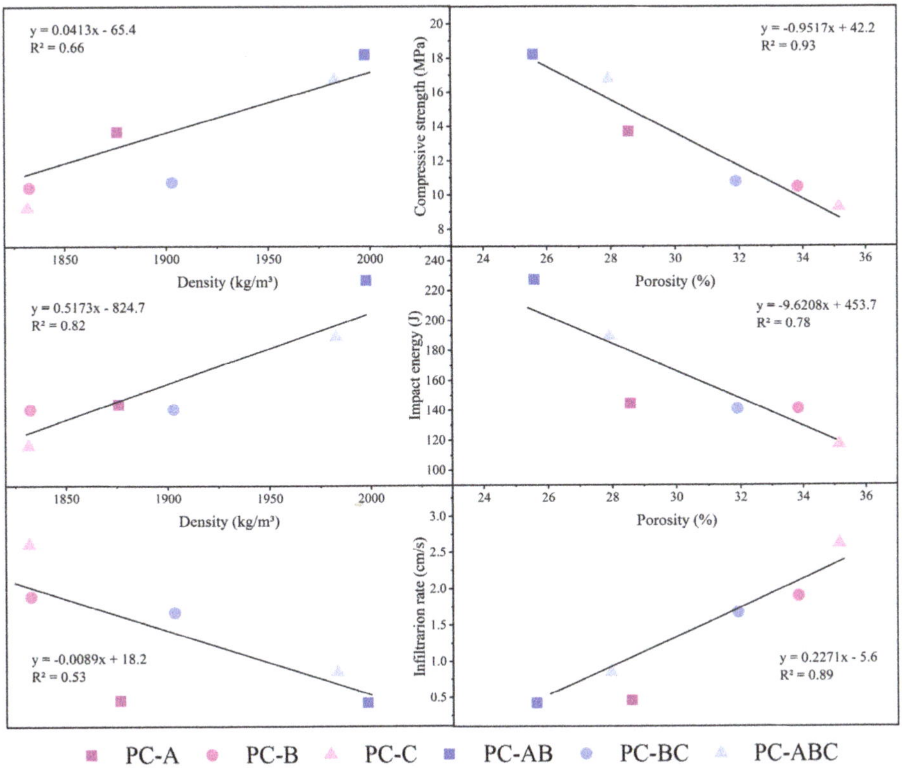

Figure 12. Relation of compressive strength, impact energy and infiltration rate with density and porosity for pervious concrete.

Table 6 summarizes the coefficients obtained for a_1 (density), a_2 (porosity), a_3 (aggregate size) and β_0 (interception), the *p*-values significance, and an analysis of variance (ANOVA) at the 95% confidence level. According to the ANOVA, there was a statistically significant relation between the assessed parameters (p-value < 0.05). The multiple regression model for compressive strength showed that density was the most crucial parameter to affect PC's compressive strength (*p*-value 0.014). However, aggregate size was also a variable with a strong influence in the regression model (*p*-value 0.056). For impact energy, all the variables strongly impacted the regression model with *p*-values below 0.05. Although all the variables were influential, density and aggregate size were the prominent influencers. For both compressive strength and impact resistance, density was the main influencing variable, followed by aggregate size.

The predicted compressive strength and impact resistance values were compared to the obtained experimental data. The results are shown in Figure 13. Both models exhibited a strong linear correlation between the predicted and experimental values, with $R^2 = 0.95$ for compressive strength and $R^2 = 0.86$ for impact strength, which indicate a good fit.

The conducted tests and the statistical analysis confirmed that aggregate size significantly impacted PC samples' mechanical properties. Optimizing the aggregate combination is, thus, an effective way to enhance PC's mechanical properties. The proposed methodology provides a simple approach for combining aggregates, which reduces both time consumption and consumed resources. Therefore, its application can be an efficient and effective means to obtain desired mechanical PC properties.

Table 6. Summary of the multiple regression model results that considers the dependence of compressive strength and impact energy vs. density, porosity and aggregate size.

	Coefficients		*p*-Value	
	Compressive Strength	Impact Energy	Compressive Strength	Impact Energy
ANOVA	-	-	0.000	0.000
R^2	0.95	0.86	-	-
Intercept (β_0)	−118.831	−4299.440	0.058	0.003
Density (a_1)	0.071	2.095	0.014	0.001
Porosity (a_2)	0.043	19.981	0.910	0.022
Aggregate size (a_3)	−0.987	−38.692	0.056	0.001

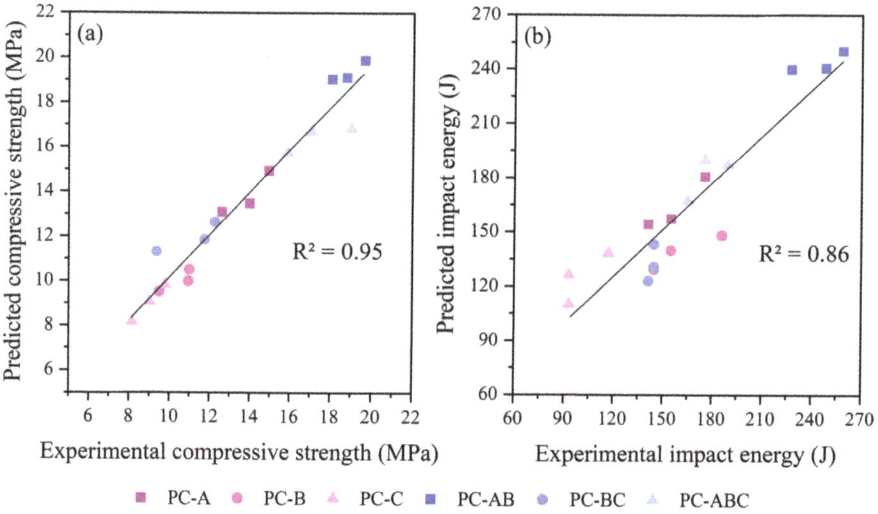

Figure 13. Prediction results of the multiple regression model versus the experimental data of pervious concrete: (**a**) compressive strength; (**b**) impact energy.

4. Conclusions

By following the maximum density methodology, this study aimed to investigate how packing aggregates could improve PC's properties by making it more efficient and sustainable. The experimental results led to the following conclusions:

- Aggregate size directly impacts PC's properties. Packing smaller aggregates makes concrete denser by reducing its porosity by up to 24.3%.
- Aggregate size and porosity both affect the infiltration rate, with smaller aggregates leading to lower infiltration rates due to narrower seepage flow paths and more efficient pore filling when packed with larger aggregates. However, even the lowest infiltration rate (0.43 cm/s) meets the recommendations that appear in the literature.
- Employing smaller aggregates tends to increase mechanical PC properties. Both compressive strength and impact resistance improve with smaller packed aggregates. The statistical analysis shows that density and aggregate size are the primary parameters that influence mechanical properties.
- Using smaller aggregates increases the total number of pores, but reduces both the average pore diameter and the average pore area, as observed with the image analysis.

The proposed methodology can be beneficial for designing denser PC by efficiently combining aggregates. This can be useful in certain applications, such as pavement design, which requires improved mechanical properties, or in adsorptive processes for pollutant removal via slow drainage in PC. Moreover, this methodology simplifies the mix design process by reducing the need for trial-and-error methods to achieve efficient PC mixtures. As a result, fewer resources are consumed, which makes PC more sustainable and environment-friendly.

Author Contributions: Conceptualization—K.H.A., R.G.d.S., M.V.B., J.M. and J.L.A.; methodology—K.H.A., R.G.d.S., J.P. and M.M.T.; software—K.H.A., R.G.d.S., L.S. and M.M.T.; validation—K.H.A., R.G.d.S., L.S., M.V.B., J.M., J.P. and M.M.T.; formal analysis—K.H.A., R.G.d.S., L.S., J.P. and M.M.T.; investigation—K.H.A., R.G.d.S., L.S. and M.M.T.; resources—M.M.T. and J.L.A.; writing—original draft preparation—K.H.A., R.G.d.S., L.S., M.V.B., J.M., J.P. and M.M.T.; writing—review and editing—K.H.A., R.G.d.S., L.S., J.P., M.M.T. and J.L.A.; visualization—K.H.A., R.G.d.S., L.S., J.P., M.M.T. and J.L.A.; supervision, J.P., M.M.T. and J.L.A.; project administration—M.M.T. and J.L.A. All authors have read and agreed to the published version of the manuscript.

Funding: Coordenação de Aperfeiçoamento de Pessoal de Nível Superior—Brasil (CAPES—Finance Code 001 to CNPq—Conselho Nacional de Desenvolvimento Científico e Tecnológico—Brasil, and to European Union-Next generation for the grant "María Zambrano for attraction of international talent" of M.M. Tashima.

Data Availability Statement: Not applicable.

Acknowledgments: The authors would like to acknowledge Mineração Grandes Lagoas Ltd.a by supplying the basaltic aggregates. In the same way, authors would like to acknowledge Coordenação de Aperfeiçoamento de Pessoal de Nível Superior—Brasil (CAPES—Finance Code 001 and, to CNPq—Conselho Nacional de Desenvolvimento Científico e Tecnológico—Brasil. Finally, M.M. Tashima wishes to thank the Spanish Ministry of Universities and the Universitat Politècnica de València for grant "María Zambrano for attraction of international talent" funded by European Union-Next generation.

Conflicts of Interest: The authors declare no conflict of interest.

Notations

PC	pervious concrete
SSD	saturated-surface-dry
OD	oven-dry
md	mass of oven-dried samples (kg)
MS	mass of samples submerged in bath (kg)
ρw	density of water at the water bath temperature (kg/m^3)
Vd	specimen's volume (m^3)
k	infiltration rate (cm/s)
t	time required for the water percolation (s)
V	volume of water poured (cm^3)
A	specimen's cross-sectional area (cm^2)
IE	impact energy (J)
h	drop height (m)
m	steel ball's mass (kg)
g	gravity constant (9.81 m/s^2)

References

1. Gartland, L. *Heat Islands: Understanding and Mitigating Heat in Urban Areas*, 1st ed.; Routledge: London, UK, 2010; pp. 2–22.
2. Hung, V.V.; Seo, S.Y.; Kim, H.W.; Lee, G.C. Permeability and Strength of Pervious Concrete According to Aggregate Size and Blocking Material. *Sustainability* **2021**, *13*, 426. [CrossRef]
3. Sandoval, G.F.B.; Galobardes, I.; Campos, A.; Toralles, B.M. Assessing the Phenomenon of Clogging of Pervious Concrete (Pc): Experimental Test and Model Proposition. *J. Build. Eng.* **2020**, *29*, 101203. [CrossRef]
4. NRMCA. *Pervious concrete: Guideline. Concrete in Practice—What, Why, How?—CIP 38*; NRMCA: Silver Spring, MD, USA, 2004.
5. Ferguson, B.K. *Porous Pavements—Integrative Studies in Water Management and Land Development*, 1st ed.; France, R.L., Ed.; Taylor & Francis: Boca Raton, FL, USA, 2005; ISBN 0-8493-2670-2.
6. Li, V.C. High-Performance and Multifunctional Cement-Based Composite Material. *Engineering* **2019**, *5*, 250–260. [CrossRef]

7. Tennis, P.D.; Leming, M.L.; Akers, D.J. *Pervious Concrete Pavements*; Portland Cement Association: Skokie, IL, USA; National Ready Mixed Concrete Association: Silver Spring, MD, USA, 2004.
8. Wang, J.; Meng, Q.; Zhang, L.; Zhang, Y.; He, B.J.; Zheng, S.; Santamouris, M. Impacts of the Water Absorption Capability on the Evaporative Cooling Effect of Pervious Paving Materials. *Build. Environ.* **2019**, *151*, 187–197. [CrossRef]
9. Xie, X.; Zhang, T.; Wang, C.; Yang, Y.; Bogush, A.; Khayrulina, E.; Huang, Z.; Wei, J.; Yu, Q. Mixture Proportion Design of Pervious Concrete Based on the Relationships between Fundamental Properties and Skeleton Structures. *Cem. Concr. Compos.* **2020**, *113*, 103693. [CrossRef]
10. American Concrete Institute. *ACI 522R-10 Report on Pervious Concrete*; American Concrete Institute: Farmington Hills, MI, USA, 2011.
11. Chandrappa, A.K.; Biligiri, K.P. Effect of Pore Structure on Fatigue of Pervious Concrete. Road Mater. *Pavement Des.* **2018**, *14*, 1525–1547. [CrossRef]
12. Pereira da Costa, F.B.; Haselbach, L.M.; da Silva Filho, L.C.P. Pervious Concrete for Desired Porosity: Influence of w/c Ratio and a Rheology-Modifying Admixture. *Constr. Build. Mater.* **2020**, *268*, 121084. [CrossRef]
13. Yap, S.P.; Chen, P.Z.C.; Goh, Y.; Ibrahim, H.A.; Mo, K.H.; Yuen, C.W. Characterization of Pervious Concrete with Blended Natural Aggregate and Recycled Concrete Aggregates. *J. Clean. Prod.* **2018**, *181*, 155–165. [CrossRef]
14. Debnath, B.; Sarkar, P.P. Permeability Prediction and Pore Structure Feature of Pervious Concrete Using Brick as Aggregate. *Constr. Build. Mater.* **2019**, *213*, 643–651. [CrossRef]
15. Ćosić, K.; Korat, L.; Ducman, V.; Netinger, I. Influence of Aggregate Type and Size on Properties of Pervious Concrete. *Constr. Build. Mater.* **2015**, *78*, 69–76. [CrossRef]
16. Shen, P.; Zheng, H.; Liu, S.; Lu, J.X.; Poon, C.S. Development of High-Strength Pervious Concrete Incorporated with High Percentages of Waste Glass. *Cem. Concr. Compos.* **2020**, *114*, 103790. [CrossRef]
17. Lori, A.R.; Hassani, A.; Sedghi, R. Investigating the Mechanical and Hydraulic Characteristics of Pervious Concrete Containing Copper Slag as Coarse Aggregate. *Constr. Build. Mater.* **2019**, *197*, 130–142. [CrossRef]
18. Deo, O.; Neithalath, N. Compressive Response of Pervious Concretes Proportioned for Desired Porosities. *Constr. Build. Mater.* **2011**, *25*, 4181–4189. [CrossRef]
19. Yu, F.; Sun, D.; Wang, J.; Hu, M. Influence of Aggregate Size on Compressive Strength of Pervious Concrete. *Constr. Build. Mater.* **2019**, *209*, 463–475. [CrossRef]
20. Liu, R.; Liu, H.; Sha, F.; Yang, H.; Zhang, Q.; Shi, S.; Zheng, Z. Investigation of the Porosity Distribution, Permeability, and Mechanical Performance of Pervious Concretes. *Processes* **2018**, *6*, 78. [CrossRef]
21. Ferić, K.; Kumar, V.S.; Romić, A.; Gotovac, H. Effect of Aggregate Size and Compaction on the Strength and Hydraulic Properties of Pervious Concrete. *Sustainability* **2023**, *15*, 1146. [CrossRef]
22. Yahia, A.; Kabagire, K.D. New Approach to Proportion Pervious Concrete. *Constr. Build. Mater.* **2014**, *62*, 38–46. [CrossRef]
23. Meddah, M.S.; Al-Jabri, K.; Hago, A.W.; Al-Hinai, A.S. Effect of Granular Fraction Combinations on Pervious Concrete Performance. *Mater. Today Proc.* **2017**, *4*, 9700–9704. [CrossRef]
24. Huang, J.; Luo, Z.; Khan, M.B.E. Impact of Aggregate Type and Size and Mineral Admixtures on the Properties of Pervious Concrete: An Experimental Investigation. *Constr. Build. Mater.* **2020**, *265*, 120759. [CrossRef]
25. Wu, F.; Yu, Q.; Brouwers, H.J.H. Mechanical, Absorptive and Freeze–Thaw Properties of Pervious Concrete Applying a Bimodal Aggregate Packing Model. *Constr. Build. Mater.* **2022**, *333*, 127445. [CrossRef]
26. Neithalath, N.; Sumanasooriya, M.S.; Deo, O. Characterizing Pore Volume, Sizes, and Connectivity in Pervious Concretes for Permeability Prediction. *Mater. Charact.* **2010**, *61*, 802–813. [CrossRef]
27. O'Reilly, D.V. *Método de Dosagem de Concreto de Elevado Desempenho*, 1st ed.; PINI: São Paulo, Brazil, 1998.
28. Associação Brasileira de Normas Técnicas. *NBR 16697—Portland Cement—Requirements*; Associação Brasileira de Normas Técnicas: Rio de Janeiro, Brazil, 2018.
29. *ASTM C150*; Standard Specification for Portland Cement. ASTM International: West Conshohocken, PA, USA, 2022.
30. *ASTM C127*; Standard Test Method for Density, Relative Density (Specific Gravity), and Absorption of Coarse Aggregate. ASTM International: West Conshohocken, PA, USA, 2015.
31. *ASTM C29/C29M-09*; Standard Test Method for Test Method for Bulk Density ("Unit Weight") and Voids in Aggregate. ASTM International: West Conshohocken, PA, USA, 2009.
32. *ISO 17785-2*; Testing Methods for Pervious Concrete—Part 2: Density and Void Content. International Organization for Standardization: Vernier, Geneva, 2018.
33. *ISO/DIS 17785-1*; International Organization for Standardization—Test Methods for Pervious Concrete—Part 1: Infiltration Rate. Chiswick: London, UK, 2014.
34. *NBR 5739*; Associação Brasileira de Normas Técnicas—Concrete—Compression Test of Cylindrical Specimens. Associação Brasileira de Normas Técnicas: Rio de Janeiro, RJ, Brazil, 2018.
35. Rossignolo, J.A. Concreto Leve de Alto Desempenho Modificado Com SB Para Pré-Fabricados Esbeltos—Dosagem, Produção, Propriedades e Microestrutura. Ph.D. Thesis, São Paulo University, São Paulo, Brazil, 2003.
36. Abid, S.R.; Abdul Hussein, M.L.; Ali, S.H.; Kazem, A.F. Suggested Modified Testing Techniques to the ACI 544-R Repeated Drop-Weight Impact Test. *Constr. Build. Mater.* **2020**, *244*, 118321. [CrossRef]

37. Rajakarunakaran, S.A.; Lourdu, A.R.; Muthusamy, S.; Panchal, H.; Jawad Alrubaie, A.; Musa Jaber, M.; Ali, M.H.; Tlili, I.; Maseleno, A.; Majdi, A.; et al. Prediction of Strength and Analysis in Self-Compacting Concrete Using Machine Learning Based Regression Techniques. *Adv. Eng. Softw.* **2022**, *173*, 103267. [CrossRef]
38. Zheng, M.; Chen, S.; Wang, B. Mix Design Method for Permeable Base of Porous Concrete: I. *J. Pav. Res. Tech.* **2012**, *5*, 102–107.
39. Silva, R.G.; Bortoletto, M.; Bigotto, S.A.M.; Akasaki, J.L.; Soriano, L. Effect of Wastes from Sugar Cane Industry on the Mechanical and Hydraulic Properties of Pervious Concrete. *Road Mater. Pavement Des.* **2022**, *23*, 1–18. [CrossRef]
40. Shen, P.; Lu, J.X.; Zheng, H.; Liu, S.; Sun Poon, C. Conceptual Design and Performance Evaluation of High Strength Pervious Concrete. *Constr. Build. Mater.* **2021**, *269*, 121342. [CrossRef]
41. *NBR 16416l*; Brasileira de Normas Técnicas Pavimentos Permeaveis de Concreto Requisitos e Procedimentos. Associação Brasileira de Normas Técnicas: Rio de Janeiro, RJ, Brazil, 2015.
42. Kant Sahdeo, S.; Ransinchung, G.D.; Rahul, K.L.; Debbarma, S. Effect of Mix Proportion on the Structural and Functional Properties of Pervious Concrete Paving Mixtures. *Constr. Build. Mater.* **2020**, *255*, 119260. [CrossRef]
43. Yu, F.; Sun, D.; Hu, M.; Wang, J. Study on the Pores Characteristics and Permeability Simulation of Pervious Concrete Based on 2D/3D CT Images. *Constr. Build. Mater.* **2019**, *200*, 687–702. [CrossRef]
44. Ibrahim, H.A.; Goh, Y.; Ng, Z.A.; Yap, S.P.; Mo, K.H.; Yuen, C.W.; Abutaha, F. Hydraulic and Strength Characteristics of Pervious Concrete Containing a High Volume of Construction and Demolition Waste as Aggregates. *Constr. Build. Mater.* **2020**, *253*, 119251. [CrossRef]
45. Mastali, M.; Dalvand, A. The Impact Resistance and Mechanical Properties of Self-Compacting Concrete Reinforced with Recycled CFRP Pieces. *Compos. B. Eng.* **2016**, *92*, 360–376. [CrossRef]
46. Lu, J.X.; Yan, X.; He, P.; Poon, C.S. Sustainable Design of Pervious Concrete Using Waste Glass and Recycled Concrete Aggregate. *J. Clean. Prod.* **2019**, *234*, 1102–1112. [CrossRef]
47. Sumanasooriya, M.S.; Neithalath, N. Pore Structure Features of Pervious Concretes Proportioned for Desired Porosities and Their Performance Prediction. *Cem. Concr. Compos.* **2011**, *33*, 778–787. [CrossRef]
48. Kia, A.; Wong, H.S.; Cheeseman, C.R. Clogging in Permeable Concrete: A Review. *J. Environ. Manag.* **2017**, *193*, 221–233. [CrossRef] [PubMed]

sustainability

MDPI

Article

Investigation of Physical and Mechanical Properties of Cement Mortar Incorporating Waste Cotton Fibres

Waiching Tang *, Ryan Monaghan and Umer Sajjad

School of Architecture and Built Environment, University of Newcastle, Callaghan, NSW 2308, Australia
* Correspondence: patrick.tang@newcastle.edu.au; Tel.: +61-249-217-246

Abstract: There is a lack of effective disposal methods for the increasing amount of textile waste that is being generated worldwide. This is creating environmental concerns and burdening waste management facilities. In this study, we propose that cotton fibres that have been recycled from textile waste could be used as fibre reinforcement in cement mortar. Seven mix designs were prepared, which were based on the quantity (0.4%, 0.8%, 1.6% and 2.0% by the weight of the cement) and length (20 mm, 30 mm and 40 mm) of the cotton fibres. The physical properties, including workability, compressive strength, flexural strength, density and water absorption, were investigated. The workability of the cement mortar was reduced with the addition of the cotton fibres. The flexural strength of the cement mortar with the added cotton fibres was improved by up to 9%, compared to the flexural strength of the control samples. The compressive strengths of the samples generally decreased with the increase in the fibre content and length. However, the C0.8 mix showed a comparable compressive strength to the control mix at all curing ages, with a slight decrease of 2.5% on day 56 of curing. The results were further clarified using SEM images. The improvement in the flexural properties showed that the cotton fibres could be implemented as fibre reinforcement in cementitious composites.

Keywords: cotton fibre; cement mortar; waste management; natural fibre; cement composites

check for updates

Citation: Tang, W.; Monaghan, R.; Sajjad, U. Investigation of Physical and Mechanical Properties of Cement Mortar Incorporating Waste Cotton Fibres. *Sustainability* **2023**, *15*, 8779. https://doi.org/10.3390/su15118779

Academic Editors: Woubishet Zewdu Taffese and Sandra Barbosa Nunes

Received: 27 April 2023
Revised: 25 May 2023
Accepted: 25 May 2023
Published: 29 May 2023

1. Introduction

Textile waste is an increasingly common issue for waste management facilities. There have been extensive investigations, globally, to find the most effective way to reduce, reuse and recycle textile waste materials; however, there is still an overwhelming amount of excess textile waste ending up in landfill. In Australia alone, on average, 27 kg of new clothing is acquired per person annually, of which 23 kg ends up in landfill [1,2]. In total, more than 800,000 tonnes of leather, rubber and textiles were discarded in 2018–2019, with a recycling rate of just 7% [3].

To reduce the environmental impact and relieve the pressure on waste management facilities [4], the cotton fibres extracted from textile waste could be recycled for use in construction applications. The current methods used for the fibre reinforcement of cementitious materials typically use synthetic fibres to improve their mechanical properties. However, the cost and environmental impacts associated with the manufacturing of synthetic fibres are unsustainable and comparatively worse than processing natural fibres such as cotton [5]. The use of waste/recycled material will lead to sustainable development in the construction industry [6,7]. Saleh et al. [8] used waste fibres to improve the properties of mortar and recommended the use of waste fibres for various construction applications.

Cement-based composites are widely utilised materials in civil engineering practices. Though, due to their brittle nature and poor resistance to crack formation, there have been significant studies exploring the potential use of natural fibres in cement composites to prevent these mechanical defects [9–12]. In addition, the fibre reinforcement could enhance the durability of cementitious composites. Jiang et al. [13] concluded that the inclusion of fibres improved the energy dissipation characteristics of cementitious composites. Due to the relatively high tensile strength and complex orientation of natural fibres, they have the ability to reinforce brittle cementitious composites [14]. Some recent studies have indicated that the implementation of natural fibre reinforcement in cement composites can significantly improve their physical, mechanical and microstructural properties [15]. Furthermore, Eskander and Saleh [16] used natural fibres to reinforce cement mortar and concluded that the addition of fibres showed resistance against radioactive waste and freeze–thaw cycles. However, there has been inconclusive research regarding the influence of fibre quantity and fibre length on specific mechanical properties.

The mechanical properties of individual cotton fibres depend on a fibre's dimensions, density and orientation. In general, increasing a fibre's length tends to decrease its tensile strength. Studies have shown that increasing the lengths of fibres can lead to a higher probability of kink banding and fibre entanglement, compared to shorter fibre lengths. This leads to a reduced mechanical response from individual fibres [10]. Kink banding is another phenomenon that can occur. It involves the dislocation of fibres on the surface when under axial loads [17]. The fibre entanglement can cause multiple fibres to interlock, forming complex fibrous structures that can alter the bonding between the cement matrix and the fibres [18].

Previous studies have compared the use of synthetic fibres with natural fibre reinforcement and have found comparable improvements in the flexural properties of natural-fibre-reinforced cement composites [12,19]. Pichardo et al. [12] reported a 40% and 7% improvement in the compressive and flexural strengths, respectively, of concrete by incorporating cotton fibres. Some studies have demonstrated the feasibility of using recycled cotton fibres as a renewable alternative to inorganic fibre reinforcement. Alomayri et al. [20] reported improvements of up to 50% in the flexural strength of composites with the addition of 2.1% of cotton fibres. Several studies [21–24] have been conducted on the use of various types of natural fibres that can be used; however, very few studies have focused on the cotton fibres extracted from textile waste. Therefore, further investigations are required to determine the effect of different percentages of cotton fibres to obtain an optimum quantity for improving the properties of mortar. In addition, there is a limited amount of literature available on the effect of the length of cotton fibres on the properties of mortar. The use of cotton fibres, unlike synthetic fibres, will reduce the carbon footprint of the construction industry.

In this research, to enhance textile recycling and to promote sustainability in the construction industry, cotton-derived fibres were incorporated in cement-based mortar. This was carried out by shredding recycled cotton sheets down to a fibrous state for use in the reinforcement of cement composites, and to assess the fresh, hardened and microstructural properties. Primarily, four mixes were prepared based on the addition of cotton fibres by the weight of the cement, in quantities ranging from 0.4%, 0.8%, 1.6% and 2.0%. The influence of the fibre content on the mechanical properties were then investigated. Moreover, to assess the influence of fibre length on the properties of mortar, cotton fibres with lengths of 20 mm, 30 mm and 40 mm were studied.

To determine the effects of fibre length and fibre quantity on the cement composites in this study, the physical and mechanical properties were assessed through workability, density, absorption, compressive strength and flexural strength tests. The microstructural properties were analysed through scanning electron microscopy (SEM) imaging, to assess the morphology of each mixture and to identify the interfacial bonding between the cotton fibres and the cement matrix.

2. Materials and Methods

2.1. Materials

The materials used to prepare mortar samples included cement, fine aggregates, cotton fibres and superplasticizer. In this research, Boral general-purpose cement, complying with Australian Standard AS 3972-2010 [25], was used. The fine aggregates, with particle sizes ranging from 0.075 mm to 4.75 mm and a bulk density of 2800 kg/m^3, were used.

The bed sheet and extracted cotton fibres are shown in Figure 1. To maintain a consistent water-to-cement ratio, MasterGlenium SKY 8100 [26] superplasticiser was added to all mixtures.

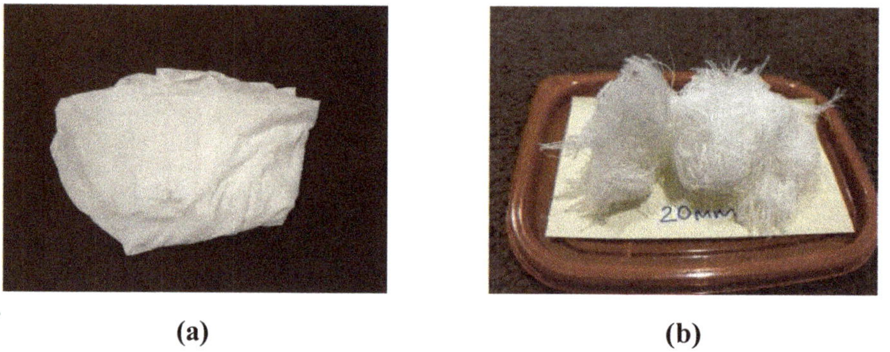

Figure 1. Raw materials: (**a**) bed sheet and (**b**) extracted cotton fibres.

The cotton fibres used in this research were extracted from recycled, white bed sheets, obtained from Lifeline Newcastle (Newcastle, Australia)—a textile recycling facility. The white sheet was utilized as it did not contain any dye material. The sheets were cleaned and then cut into strips. Each strip was then cut to the desired width and then further cut into squares, according to the required length. The individual fibres were collected from these square sheets to obtain the required amount of cotton fibres. The process for obtaining the required fibres is summarised in Figure 2.

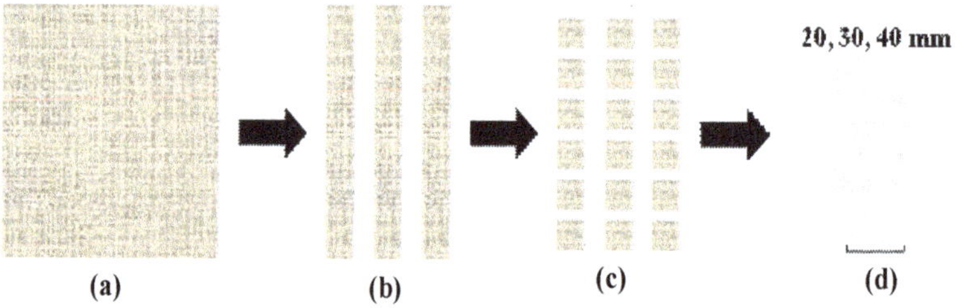

Figure 2. Recycled cotton fibre extraction process: (**a**) recycled bed sheet, (**b**) cut into strips, (**c**) cut into square of required lengths and (**d**) cotton fibres.

The SEM image of cotton fibres is shown in Figure 3, below. The diameter of cotton fibres ranged from 5 μm to 20 μm.

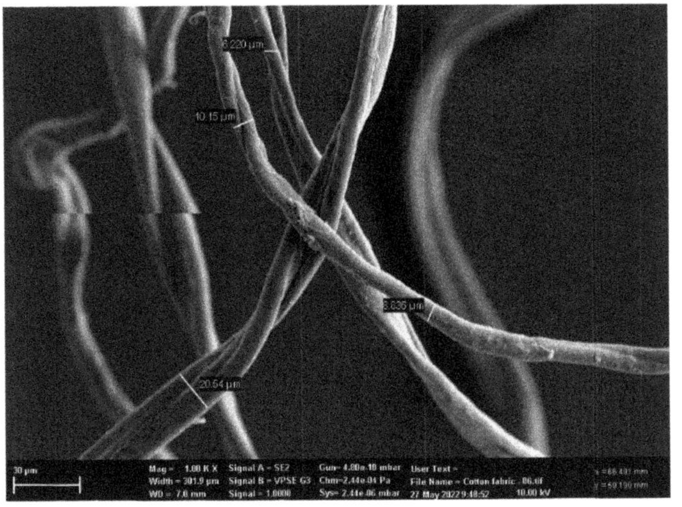

Figure 3. SEM image of cotton fibres.

2.2. Mix Design

The mix design for the preparation of mortar samples is presented in Table 1. The mix design was based on the trial mixes used to obtain high-strength mortar samples. The amounts of cement and fine aggregates were kept constant in all mixes. The quantity of water used in each mix was maintained to account for the absorption of water by cotton fibres and fine aggregates. Seven mixes were prepared, and the mixes were based on the quantity and length of the cotton fibres. The four mortar mixes were prepared with the addition of 0.4%, 0.8%, 1.6% and 2.0% of cotton fibres, by weight of cement in each mix. Cotton fibres of 30 mm in length were used for these four mixes. Two mixes were prepared with cotton fibres of 20 mm and 40 mm in length and the percentage of cotton fibres added to these two mixes were kept 0.8% of the weight of the cement. One control mix was prepared without the addition of cotton fibres to compare the effect of different amounts and lengths of cotton fibres on the properties of mortar. The percentage of cotton fibres added was based on previous studies [14,27].

Table 1. Designed mix proportions.

Mix Label	Fine Aggregates (kg)	Cement (kg)	Water (kg)	Cotton Fibre (wt%)	Cotton Fibre (kg)	Superplasticiser (L)
C0.0	1750	525	233	0.0	0.0	3.75
C0.4	1750	525	233	0.4	2.1	3.75
C0.8	1750	525	233	0.8	4.2	6.25
C1.6	1750	525	233	1.6	8.4	11.25
C2.0	1750	525	233	2.0	10.5	15.00
C0.8-S	1750	525	233	0.8	4.2	6.25
C0.8-L	1750	525	233	0.8	4.2	6.25

To identify each mix, a symbolic naming convention was adopted. For example, the letter 'C' at the start of each mixture indicates the presence of cotton fibres in each mix, and the amount of cotton fibres in that mix is then indicated by the decimal number following the letter 'C'. Where symbols 'S' and 'L' are present, this indicates that the smaller (20 mm) and larger (40 mm) cotton fibres were used. For example, C0.8-S indicates the mix contains

0.8% cotton fibre, relative to the cement weight, and that the smaller (20 mm) fibres were used. A water-to-cement ratio of 0.5 was maintained for all mixes, and superplasticiser was added to maintain comparable workability.

2.3. Preparation of Samples

All mixes were prepared in the engineering department's laboratory at the University of Newcastle and followed AS 2350.12-2006 [28] for the preparation of standard mortar and the moulding of samples for testing.

After all ingredients had been weighed and prepared, half of the fine aggregate was poured into the mortar mixer, followed by the entire cement bucket. The remaining fine aggregate was then added and the machine was turned on and mixed for 2 min. After 2 min, the dry materials were inspected to ensure adequate mixing. After inspection, the mortar mixer was then turned back on, and cotton fibres were added incrementally to reduce clumping phenomena between fibres. Fibres were mixed for 2 min, then the mixing machine was stopped for an additional inspection to observe the dispersion of cotton fibres, as shown in Figure 4.

Figure 4. Dry mixture with cotton fibres.

If fibres were well-dispersed in dry materials, the mortar mixer was then turned on and the water (with superplasticiser) was added slowly. The wet mixture was blended for an additional 2 min before stopping to check the homogeneity of the mix and dispersion of cotton fibres. The workability of the mix was determined using a flow table test, in accordance with ASTM C1437 [21]. After mixing, the fresh mortar was placed in different moulds for mechanical tests. The samples were removed from the moulds after 24 h and cured in the climate-controlled fog room.

2.4. Physical and Mechanical Testing

2.4.1. Water Absorption of Cotton Fibres

To estimate the 24 h water absorption of the cotton fibres, a test was conducted before the mortar mixing. To do this, 1 gram of cotton fibres was weighed in a precision, vacuum-chambered scale. The fibres were then submerged in water for 24 h. After 24 h, the fibres were removed from the water and placed on a drying rack to be air-dried for 1 h, ensuring the fibres were well dispersed.

After an hour, the fibres were transferred to a clean aluminium dish and placed in an oven at 50 °C, for 24 h. After 24 h, the fibres were re-weighed and the change in water content was measured using Equations (1)–(3), as follows:

$$24\text{ h } Absorption\ \% = \frac{M_{Dry\ final} - M_{Dry\ Initial}}{M_{Dry\ Initial}} \times 100 \quad (1)$$

$$M_{Dry\ Final} = Mass\ of\ Cotton\ Fibres\ after\ 24\text{ h } in\ Oven \quad (2)$$

$$M_{Dry\ Initial} = Mass\ of\ Cotton\ Fibres\ before\ being\ submerged\ in\ water \quad (3)$$

2.4.2. Workability and Density

The workability of the fresh mortar was assessed using the flow table test, in accordance with ASTM C1437 [29]. Two readings were taken and the average was reported. The densities of hardened mortar samples at 7, 28 and 56 days were determined in accordance with AS 1012.12.1-1998 (R2014) [30]. Three cube samples of each mix were used and the average was reported.

2.4.3. Mechanical Tests

For this research study, mechanical tests involved compressive and flexural strength. These tests were performed to observe the effect of cotton fibres on the mechanical properties of mortar. The compressive strength test was carried out following the method outlined in AS 1012.9:2014 [31]. The tests were performed on cube samples with dimensions of 70 mm, using a universal testing machine and keeping a loading rate of 0.5 $N/mm^2/s$. They were tested on days 7, 28 and 56 of curing.

The flexural strength tests of the cement mortar prisms were conducted using a three-point bending test in accordance with ASTM C348 [32]. The flexural tests were performed on the prisms of 40 × 40 × 160 mm, using a Geo-con 10 kN loading frame, at a displacement rate of 0.05 mm/min. Tests were carried out on day 28 of curing.

2.4.4. Water Absorption

The water absorption test involved the preparation of 3 samples on day 56 of curing. The samples followed AS1012.21-1999 (R2014) [33] for the water absorption of hardened concrete. The mass of each sample was recorded before being submerged in room-temperature water for 48 h. After 48 h, the samples were removed and re-weighed at surface-dry conditions (M1). The samples were then moved to a 100 °C oven for a further 24 h. After 24 h, the samples were removed and re-weighed (M2). The total absorption for each sample was calculated using Equation (4), as follows:

$$\text{absorption } (\%) = \frac{\text{M1} - \text{M2}}{\text{M2}} \times 100 \quad (4)$$

2.4.5. Scanning Electron Microscopy (SEM) Analysis

To observe the microstructure of each mix, a trimming of a sample from each mix was prepared for scanning electron microscopy (SEM) analysis. The SEM used to complete this analysis was a Zeiss Sigma VP, with low-vacuum conditions for very porous materials. To obtain clear images of the microstructure, backscattered detection was used at 15 kV acceleration voltage. To confirm the chemical composition and location of cotton fibres throughout each mix, energy dispersion X-ray spectroscopy (EDS) was also used.

3. Results and Discussion

3.1. Water Absorption of Cotton Fibres

A water absorption test on the cotton fibres indicated that their 24 h absorption was approximately 2 grams per 1 gram of cotton fibre. Therefore, the additional water added to each mix was two times the quantity of the cotton fibres. While there is insufficient literature to support the exact absorption rate of cotton fibres, this 24 h approximation should be sufficient for the low quantity of fibres added in this research study.

3.2. Workability

As the cotton fibre content increased, a steady decrease in the workability of the mixture was observed, and the added superplasticiser began to have a minimal effect on the flow diameter. The results are shown in Table 2. There were minimal improvements to the workability of the mix as the amounts of superplasticiser were increased. This was likely to be because the cotton fibres absorbed some of the superplasticiser while in the wet mixture; therefore, it was not well dispersed considering the cement particles [11]. Similar results were reported in [34]: the addition of fibres absorbed the superplasticizer and nullified its effect in cementitious composites. It is believed that the absorbing phenomenon is generalized for every type of superplasticizer and must be considered while working on the mix design of cement composites with cotton fibres. The dosage of superplasticizer used was similar to the dosage used for the optimum mix, C0.8, and the flow diameter of mixes with 20 mm and 40 mm fibres was 170 mm. It seems that the fibre length showed no significant effect on the workability of the mortar, compared to the flow diameter of mix C0.8. Similar results were reported by Banfill et al. [35]: increasing the fibre length and fibre content increased the workability of the cement mortar.

Table 2. Flow table workability measurements.

Mix	Average Flow Diameter (mm)	Superplasticiser Used (mL)
C0.0	170	15
C0.4	170	15
C0.8	160	25
C1.6	140	45
C2.0	130	60
C0.8-S	170	25
C0.8-L	170	25

3.3. Density

The density of the mortar samples in their hardened states on days 7, 28 and 56 of curing are demonstrated in Figure 5. The reported densities were the average of the three samples for each mix at the specific curing intervals. The density of the control mix increased on days 28 and 56, compared to the density on day 7. The densities of the mixes with cotton fibres showed irregular patterns; however, there was not much difference among the densities of the mixes with cotton fibres on days 28 and 56. The C0.8 sample had the highest density—2401 kg/m^3—on day 7 of curing, but this reduced by 8.45% and 10.58% on day 28 and 56, respectively. The slight change in the densities may have been because of the amount of superplasticizer used in each mix. The presence of superplasticizer affected the workability and compaction of the samples and had a relative effect on the densities of the mixes with cotton fibres. The presence of superplasticizer induced air entrainment in the mixes and caused slight variations in the densities of the mortar samples [36].

When we analysed the influence of the length of the cotton fibres on the density of the mortar mixes, as shown in Figure 5, we observed a slight difference among the densities of the mortar samples. The samples with 30 mm fibre lengths showed the highest densities on days 7, 28, and 56 of curing. The densities of all the mortar samples, with varying fibre lengths, showed a similar trend, with decreases in their densities on days 28 and 56 of curing. The density of the samples with 20 mm cotton fibres on day 28 of curing were exceptions to this trend; a slight increase in density was observed on these 28-day mortar samples, with 20 mm cotton fibres. Primarily, this was because of the lower rate of water loss compared to the other samples. These density results indicate that the presence of cotton fibres produces additional pores in the microstructure, causing a reduction in density. This aligns with the results in a previous study [37].

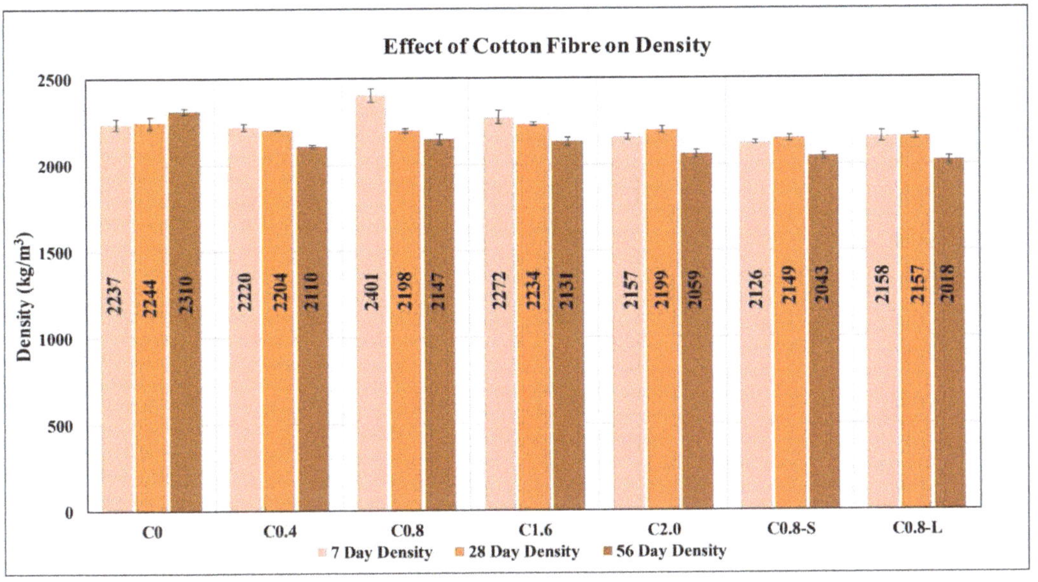

Figure 5. Density comparison at all curing ages for varying fibre contents.

3.4. Compressive Strength

The compressive strength tests were performed for all of the mixes on days 7, 28 and 56 of curing. Three samples per curing age were tested to determine the average compressive strengths. The failure pattern in the samples with and without cotton fibres are shown in Figure 6. The control samples without cotton fibres typically failed in a brittle and explosive manner, while the samples with cotton fibres showed ductile failure. The ductile failure pattern indicated the bridging effect provided by the cotton fibres.

Figure 6. Failure mechanism: (**a**) without cotton fibres and (**b**) with cotton fibres.

When comparing the compressive strengths at all of the curing ages, the incorporation of cotton fibres showed an overall reduction in compressive strength, as shown in Figure 7. The mixes containing lower cotton fibre quantities showed better compressive strength than the mixes with a higher cotton fibre content at all curing ages.

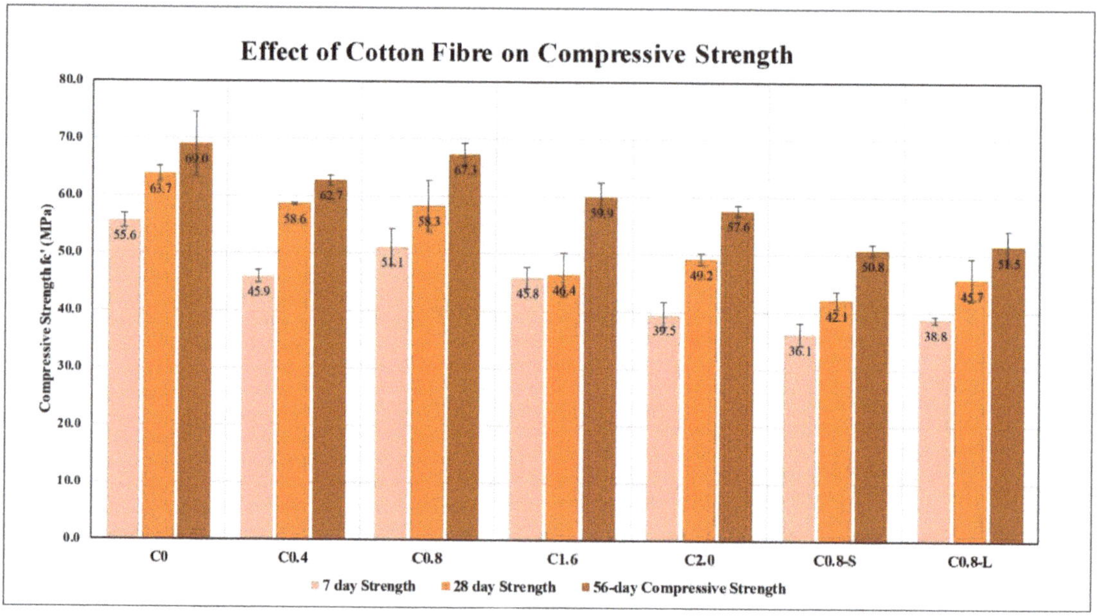

Figure 7. Effect of cotton fibre content on compressive strength.

The control mix demonstrated the highest compressive strengths: 55.6 MPa, 63.7 MPa and 69 MPa on days 7, 28 and 56, respectively. When comparing the 7-day compressive strength of the control mix with the compressive strength of the mixes containing cotton fibres, there was a noticeable decline, with mixes C0.4, C0.8, C1.6 and C2.0 reducing by 17.4%, 8.1%, 17.6% and 29%, respectively. A similar trend was observed on both days 28 and 56 of curing.

The C0.8 mix had the highest compressive strength of all mixes containing cotton fibres and was the most comparable to the control mix at all curing ages. The decrease in the compressive strength of mix C0.8 compared to the control mix were only 8.1%, 8.5% and 2.5% on days 7, 28 and 56 of curing, respectively.

When comparing the influence of the fibre content on the compressive strength of cement mortar, an optimal quantity (C0.8) was found. At this quantity, we observed improvements in compressive strength, and a decrease in the samples' compressive strengths were observed as the quantity of cotton fibres were increased [12,38]. The decrease in the compressive strength of the mixes containing cotton fibres was due to the complex structure of the cotton fibres, which occupied additional space in the matrix in a highly variable orientation. The orientational variability that was interfering with the interfacial bonding between the cement and aggregate led to a reduction in the compressive strength of the mortar mixes containing cotton fibres. Vailati et al. [38] reported that the orientation of fibres played an important role in the strength mechanism of mortar. In addition, the evaporation of the initially absorbed water by the cotton fibres led to a decrease in the compressive strength of the mixes containing cotton fibres. Kubica and Galman [39] reported that the retention of the initially absorbed water has a significant influence on the strength of the mortar. When observing the effects of the fibre length on the compressive strengths of the mixes, the 30 mm fibre length had the highest compressive strengths at all curing ages, as shown in Figure 7.

There was a significant decrease in the compressive strength of the mixes containing 20 mm and 40 mm fibres—29.3% and 24.2%, respectively—compared to the 7-day compressive strength of the mix containing 30 mm fibres. Similarly, the 28-day compressive strengths of the mixes with 20 mm and 40 mm fibre lengths were 27.7% and 21.6% lower, respectively, than the 28-day compressive strength of the mixes with 30 mm fibre lengths. Furthermore, the 56-day compressive strengths for the 20 mm and 40 mm fibre lengths were 24.5% and 23.5% lower, respectively, than the 56-day compressive strength of the mixes with 30 mm fibre lengths. As all the mixes were based on the C0.8 mix design, it can be concluded that fibre length does affect the compressive strength of mortar. The lower compressive strength for the samples containing 20 mm and 40 mm fibres could be attributed to the quantity and dimensions of the fibres in the matrix. As found in the literature, the dimensions of the fibre correlate to a higher probability of defects such as kink banding and fibre entanglement [10]. While fibre entanglement would be the reason for the reduced compressive strength in the samples containing 20 mm fibres, kink banding could be the reason in samples containing 40 mm cotton fibres. This is because an increased fibre length is more likely to develop kink banding in individual fibres, thus reducing the compressive strength of the mortar.

3.5. Flexural Strength

The flexural strength of each mix was determined using the three-point bending test, after 28 days of curing. All the mixes containing cotton fibres displayed a brittle failure mechanism, with some fibres visibly bridging between the failure surface. When comparing the cotton fibre content of all of the mixes to the control mix, there was a general decrease in flexural strength with the addition of cotton fibres, as shown in Figure 8. Mixes C0.4, C1.6 and C2.0 had an average flexural strength of 8.94 MPa, 9.1 Mpa and 8.95 Mpa, respectively, which were approximately 13.5% lower than the control mix. However, the C0.8 mix displayed an improved flexural strength of 11.35 MPa, which was a 9% improvement on the control mix.

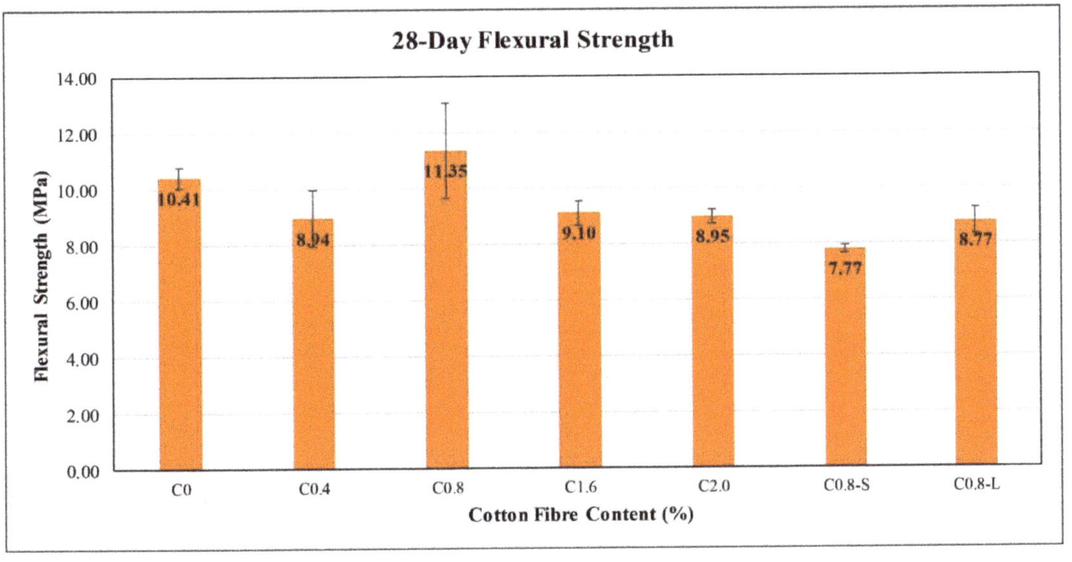

Figure 8. Effect of cotton fibre content on flexural strength.

When observing the influence of cotton fibre lengths on the flexural strength of the samples, the mix containing 30 mm fibres had the highest flexural strength, at 11.35 MPa, as shown in Figure 8. This improvement might be due to a combination of fibre-bridging in the microstructure and of a reduction in the crack propagation, as discussed above (see Figure 6). Furthermore, sufficient fibre bonding might occur in the interfacial transition zone (ITZ). A further discussion on microstructure can be found in Section 3.7, where the microstructure is discussed in detail, alongside microscopy images. Ozerkan et al. [40] also reported similar findings: the addition of fibres improved the flexural performance of the mortars but decreased their compressive strength. This could be attributed to the orientation of the cotton fibres in the matrix [41]. The orientation of the fibres in the cement composites played a vital role. The fibres that were oriented perpendicular to the crack may have helped in delaying or reducing the crack opening through a bridging effect [42].

The mix containing 20 mm and 40 mm fibres showed a reduction in flexural strength. The flexural strength of the mixes containing 20 mm and 40 mm fibres were 31.5% and 22.7% lower, respectively, than in the mix containing 30 mm fibres. This suggested that the geometry of the 20 mm fibres could be too small to cause the flexural improvements needed for the cotton fibres to bond effectively with the cement matrix. However, the 30 mm fibres were of sufficient dimensions to influence the microstructure and to develop a fibre-bridging effect. The reduced flexural strength of the mix containing 40 mm fibres could be a result of fibre entanglement throughout the sample, which could have led to a weaker bonding with the matrix.

3.6. Water Absorption

The water absorption test was performed for all mixes on day 56 of curing. Given the hydrophilic properties of cotton fibres, it was expected that the mixes with an increased fibre content would retain additional water. When comparing all of the mixes to the control mix, a steady increase in absorption could be observed, as shown in Figure 9. Mixes C0.4, C0.8, C1.6, C2.0, C0.8-S and C0.8-L showed an increase of 16.5%, 9.0%, 49.5%, 55.1%, 58.6% and 64.0%, respectively. The mix containing 40 mm fibres had the greatest increase in water absorption. By comparing the average 56-day densities to the 56-day water absorption of all the mixes, a strong correlation was observed between the absorption and density of all in them.

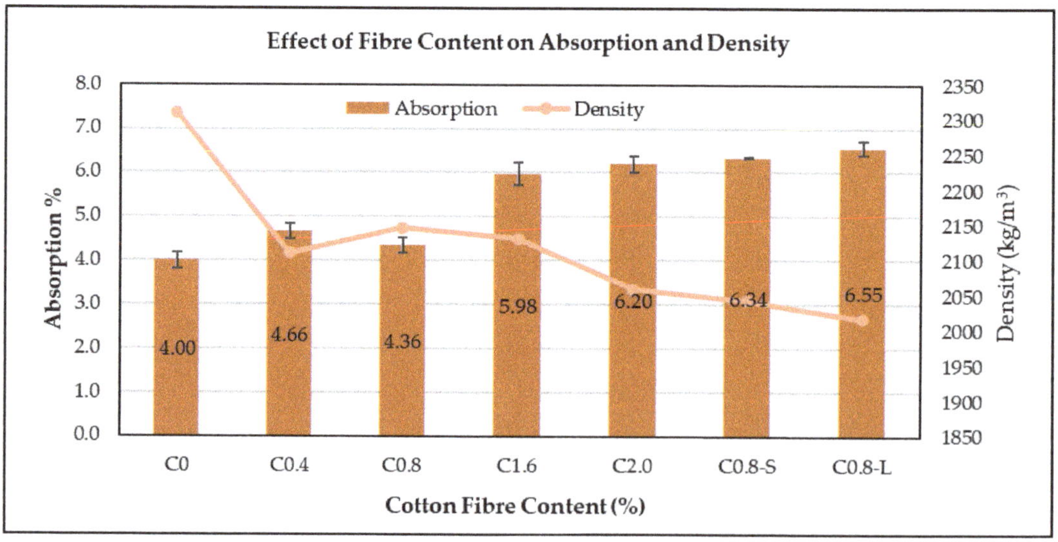

Figure 9. Effect of fibre content on 56-day water absorption and density.

As shown in Figure 9, the mixes with higher densities had a lower absorption rate. This suggests that mix C0.0 and mix C0.8 contained less voids than C0.4, C1.6, C2.0, C0.8-S or C0.8-L. When comparing the effects of fibre length, the mix containing 30 mm fibres had the highest density and lowest absorption, whereas the 40 mm mix had the highest absorption and lowest density. This further confirmed that the increased presence of fibres generated more pores in the microstructure. Similar results were reported by Ziada et al. [43]: the addition of fibres increased the water absorption of cementitious composites. In other studies, natural fibres have been pre-treated using various techniques to reduce the water absorption [44]. In this study, the cotton fibres were pre-soaked and dried for this reason. Although the synthetic fibres might have lower water absorption, the use of natural fibres enhances the sustainability of the building materials [45].

3.7. Scanning Electron Microscopy (SEM) Results

The microstructural formation of the mortar mixes with cotton fibres were observed through scanning electron microscopy (SEM). The SEM analysis was conducted on each of the mixes containing 30 mm cotton fibres. To validate the SEM findings, an additional energy dispersion X-ray spectroscopy (EDS) was used to determine the chemical composition corresponding to each SEM image. The morphologies of all of the samples are shown in Figure 10. It was confirmed that the microstructure was uniform in the cement regions of all of the mixes and that, overall, it was very dense, with some porous regions present.

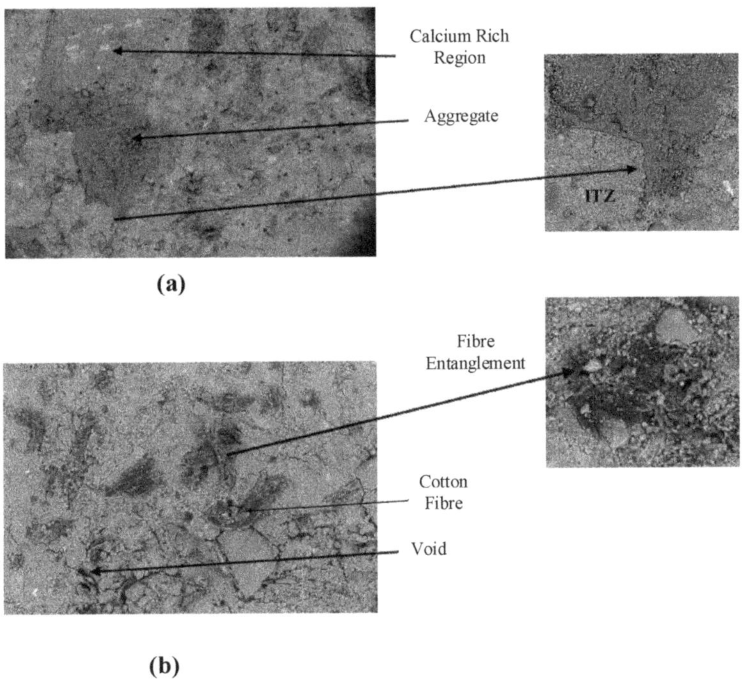

Figure 10. SEM images: (**a**) without fibres and (**b**) with cotton fibres (C1.6).

When comparing the SEM images for all the mixes, the C0.0 mix (Figure 10a) displayed very dense cement regions and minor cracking between the ITZ, in a mostly linear orientation. Some dark spots could be observed, but very minimal voids were present. Figure 10b shows a sample with cotton fibres, which showed higher porosity and significant dark spots throughout the matrix. In Figure 10b, two fibres appeared to intersect with one another, causing some fibre entanglement. The close-up image appears to capture an agglomeration of cotton fibres, causing fibre entanglement. These fibres did not bond effectively with the matrix, and there was significant micro-cracking spalling from the edges of the fibres.

When comparing these SEM images to the mechanical properties of each sample, Figure 10b confirmed that the increased cotton fibre content caused agglomerations of fibres throughout the matrix, increasing the probability of fibre entanglement and kink banding. When cracking propagates at sharp angles it indicates a poor interfacial bonding between the cement and the aggregate, or the presence of aggregates embedded beneath the surface of the matrix. When analysing the microstructural images for the C1.6 mix, we observed improvements in the cotton fibre and cement bonding, as shown in Figure 11.

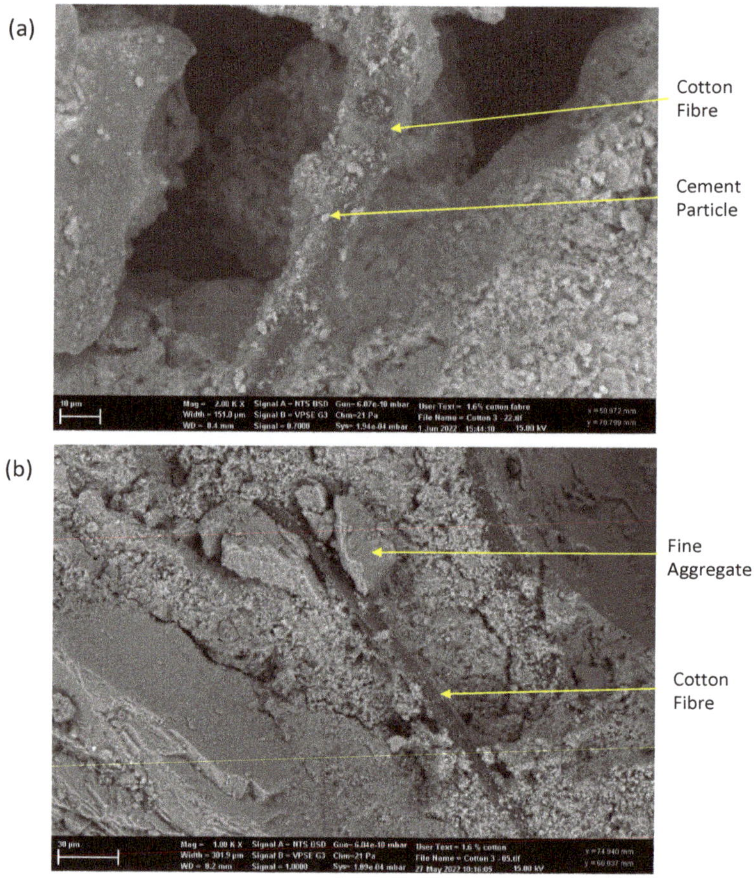

Figure 11. Fibre bridging in sample C1.6: (**a**) cement-coated fibre and (**b**) poorly-coated fibre.

Figure 11a shows a single, cement-coated cotton fibre bridging a void, with fine aggregate embedded within the void. The cotton fibre was well coated with cement particles, hence, the strong bonding between the void boundaries. Similarly, Figure 11b shows a cotton fibre bridging between two different cement regions. This fibre was poorly coated with cement particles and was most likely only fixed at the ends of the fibres. The flexural improvements in all the samples could have been because of the fibre bridging effect. However, due to the increased quantity of fibres in the matrix, kink banding and fibre entanglement was most likely the reason for the variability in the flexural strengths [10].

This suggested that the cotton fibre distribution was likely to be the most influential reason for the flexural improvements shown in mix C0.8, rather than the quantity of the fibres. The use of the proper type and quantity of superplasticizer also helps in the distribution of fibres. In addition, care must be taken while adding the fibres into the matrix to reduce the chances of the agglomeration of fibres in the cement matrix [14].

To confirm the chemical composition of the mixes, an EDS analysis was used to confirm the location of the cotton fibres in the SEM images and to investigate the dark spots throughout the cementitious regions. The EDS analysis was conducted at two locations to compare the atomic differences, and to confirm the darker regions in all of the SEM images where cotton fibres were present.

Provided that the cellulosic chemical composition of cellulose is $C_6H_{10}O_5$ [46], it was expected that a carbon to oxygen ratio of 1.2:1 would be observed in the spectroscopy results. As shown in Table 3, the quantitative analysis of both points showed significant traces of carbon and oxygen, with residual quantities of silicon and calcium.

Table 3. EDS chemical composition of points 1–6.

Point 1		Point 2		Point 3		Point 4		Point 5		Point 6	
Element	Atom (%)	Element	Atom (%)	Element	Atom (%)	Element	Atom (%)	Element	Atom (%)	Element	Atom (%)
O	44.68	C	61.45	C	62.69	C	60.68	O	44.80	O	64.78
C	43.48	O	33.26	O	33.33	O	36.03	C	40.53	C	13.15
Ca	7.19	Ca	2.36	Ca	1.76	Ca	1.75	Ca	8.25	Ca	12.33
Si	2.94	Si	1.88	Si	1.44	Si	1.02	Si	4.02	Si	5.97
Al	0.75	Al	0.48	Al	0.38	Al	0.26	Al	1.01	Al	1.79
Mg	0.30	Na	0.23	Na	0.13	Na	0.11	Na	0.040	Na	0.61
Na	0.22	K	0.14	Fe	0.11	K	0.08	Mg	0.33	Mg	0.44
Fe	0.19	Fe	0.13	K	0.11	Mg	0.06	Fe	0.28	Fe	0.41
K	0.17	Mg	0.07	Mg	0.06			K	0.24	K	0.34
S	0.08							S	0.14	S	0.17

When quantifying the atomic distribution for both point one and two, approximately 66% of the chemical composition was carbon and 30% was oxygen. Calcium and silicon only represented 1% of the chemical composition, and the trace amounts of aluminium, sodium and potassium were 0.45%, 0.24% and 0.31%, respectively. The EDS results at points one and two suggested that the carbon to oxygen content of 2:1 was not expected for a cellulosic fibre. Therefore, additional EDS locations were selected, as shown in Figure 12.

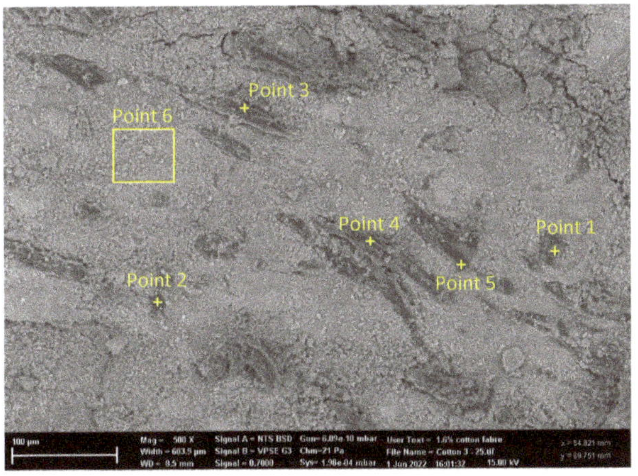

Figure 12. EDS Spectroscopy, points 1–6.

Points one and two were chosen to investigate the dark spots that recurred in all the mixes. Points 3, 4 and 5 were selected as these were expected to display cotton fibres. The area at point six was selected as a control point, to determine the chemical composition of the cementitious region.

As shown in Table 3, the EDS concluded that points one and five contained comparable chemical distributions, with carbon to oxygen ratios of 1.1:1, indicating the expected cotton fibre properties. Points two, three and four showed an increased carbon content, similar to the early cotton fibre findings, where the carbon to oxygen ratio was 1.6:1. The high content of carbon might be due to the presence of superplasticizer, as reported by Ma et al. [47], who suggested that the polycarboxylate-based superplasticizer caused significant concentrations of carbon in cement paste. The cement region at point six contained a typical cement–mortar chemical composition, where higher amounts of oxygen, calcium and silicon were present compared to points one to five [48].

After analysing the spectroscopy for the points shown in Figure 12, we concluded that the superplasticiser was most likely being absorbed by the cotton fibres, which increases the carbon content and leads to a carbon to oxygen ratio of 1.6:1.

When comparing the SEM analysis to the mechanical properties of each sample, we concluded that the inclusion of cotton fibres introduced additional voids in the matrix, therefore, reducing its density. This supports the results obtained through absorption tests, where the samples with higher quantities of fibres increased the voids, and where the water was likely to have been retained. This caused a higher absorption than in the samples with less fibres.

When justifying why the C0.8 mix displayed the best mechanical properties compared to other samples, it was likely that the dispersion of the fibres was more consistent throughout the sample, allowing cement particles to bond more effectively with the fibres. When the fibres were more evenly distributed in the cement matrix, there were less observable void areas and better interfacial bonding between the cement, aggregate and fibres. The microstructural analysis concluded that the dispersion of fibres was more influential on the mechanical properties than the quantity and length of the fibres. In general, the efficiency of the cotton fibres in cementitious composites greatly depends upon the dispersion, orientation, quantity and absorption capacity of the fibres. These factors can be minimized by implementing proper techniques and using a suitable mix design. The impact of the dispersion of cotton fibres on the durability properties of cement mortar could be considered in further investigations.

4. Conclusions

This research investigated the properties of cement mortar that incorporates waste cotton fibres. A total of seven mixes were prepared to determine the effects of the quantity and length of cotton fibres on the physical properties of cement mortar. The percentages of the cotton fibres that were added were 0.4%, 0.8%, 1.6% and 2.0% of the weight of the cement. The lengths of cotton fibres that were used were 20 mm, 30 mm and 40 mm. The addition of the cotton fibres reduced the workability and density of the cement mortar. The flexural strength of the cement mortar containing cotton fibres was enhanced by up to 9%, compared to the flexural strength of the control samples. The compressive strength of the mortar samples generally decreased with the increase in the fibre content and length. However, the C0.8 mix showed comparable compressive strengths to the control mix at all curing ages, with a slight decrease of 2.5% on day 56 of curing. SEM images showed the voids in the mortar samples with high percentages of cotton fibres. Overall, 0.8% by the weight of the cement is the optimum quantity of cotton fibres for improving the flexural performance of cementitious composites. This could enhance the recycling of textile waste and would promote the use of natural fibres rather than synthetic fibres. In future studies, the effect of the addition of cotton fibres on the durability of cementitious composites could be investigated.

Author Contributions: Conceptualization, W.T.; methodology, W.T. and R.M.; formal analysis, W.T., R.M. and U.S.; investigation, R.M.; resources, W.T.; data curation; W.T. and U.S.; writing—original draft preparation, R.M.; writing—review and editing, W.T. and U.S.; supervision, W.T.; funding acquisition, W.T. All authors have read and agreed to the published version of the manuscript.

Funding: This research was funded by Cotton Research Development Corporation, grant number: UON2201.

Institutional Review Board Statement: Not applicable.

Informed Consent Statement: Not applicable.

Data Availability Statement: Not applicable.

Acknowledgments: The authors gratefully acknowledge Cotton Research Development Corporation (CRDC) for financial support in this research. We also thank the guidance provided by Electron Microscope and X-Ray Unit (EMX) and the concrete laboratory staff at the University of Newcastle.

Conflicts of Interest: The authors declare no conflict of interest.

References

1. Department of Climate Change, Energy, the Environment and Water. Clothing Textiles. Available online: https://www.dcceew.gov.au/environment/protection/waste/product-stewardship/textile-waste-roundtable (accessed on 6 June 2022).
2. Roth, G. *Australian Grown Cotton Sustainability Report*; Cotton Australia: Mascot, NSW, Australia, 2014.
3. Taylor, L. Tackling Australia's Textile Waste. Available online: https://www.planetark.org/newsroom/news/tackling-australias-textile-waste (accessed on 10 June 2022).
4. Dadkhah, M.; Tulliani, J.-M. Damage Management of Concrete Structures with Engineered Cementitious Materials and Natural Fibers: A Review of Potential Uses. *Sustainability* **2022**, *14*, 3917. [CrossRef]
5. Yan, L.; Chouw, N.; Huang, L.; Kasal, B. Effect of alkali treatment on microstructure and mechanical properties of coir fibres, coir fibre reinforced-polymer composites and reinforced-cementitious composites. *Constr. Build. Mater.* **2016**, *112*, 168–182. [CrossRef]
6. Brahami, Y.; Saeidi, A.; Fiset, M.; Ba, K. The Effects of the Type and Quantity of Recycled Materials on Physical and Mechanical Properties of Concrete and Mortar: A Review. *Sustainability* **2022**, *14*, 14752. [CrossRef]
7. Francioso, V.; Lopez-Arias, M.; Moro, C.; Jung, N.; Velay-Lizancos, M. Impact of Curing Temperature on the Life Cycle Assessment of Sugarcane Bagasse Ash as a Partial Replacement of Cement in Mortars. *Sustainability* **2023**, *15*, 142. [CrossRef]
8. Saleh, H.; Eskander, S.; Fahmy, H. Mortar composite based on wet oxidative degraded cellulosic spinney waste fibers. *Int. J. Environ. Sci. Technol.* **2014**, *11*, 1297–1304. [CrossRef]
9. Jawaid, M.; Khalil, H.A. Cellulosic/synthetic fibre reinforced polymer hybrid composites: A review. *Carbohydr. Polym.* **2011**, *86*, 1–18. [CrossRef]
10. Yan, L.; Kasal, B.; Huang, L. A review of recent research on the use of cellulosic fibres, their fibre fabric reinforced cementitious, geo-polymer and polymer composites in civil engineering. *Compos. Part B Eng.* **2016**, *92*, 94–132. [CrossRef]
11. Nilsson, J.; Sargenius, P. Effect of Microfibrillar Cellulose on Concrete Equivalent Mortar Fresh and Hardened Properties. 2011. Available online: https://www.semanticscholar.org/paper/Effect-of-microfibrillar-cellulose-on-concrete-and-Nilsson-Sargenius/1ae7b6346824fd2bf9beff82a22d46e013e5a373 (accessed on 4 June 2022).
12. Peña-Pichardo, P.; Martínez-Barrera, G.; Martínez-López, M.; Ureña-Núñez, F.; dos Reis, J.M.L. Recovery of cotton fibers from waste Blue-Jeans and its use in polyester concrete. *Constr. Build. Mater.* **2018**, *177*, 409–416. [CrossRef]
13. Jiang, P.; Chen, Y.; Wang, W.; Yang, J.; Wang, H.; Li, N.; Wang, W. Flexural behavior evaluation and energy dissipation mechanisms of modified iron tailings powder incorporating cement and fibers subjected to freeze-thaw cycles. *J. Clean. Prod.* **2022**, *351*, 131527. [CrossRef]
14. Kesikidou, F.; Stefanidou, M. Natural fiber-reinforced mortars. *J. Build. Eng.* **2019**, *25*, 100786. [CrossRef]
15. Onuaguluchi, O.; Banthia, N. Plant-based natural fibre reinforced cement composites: A review. *Cem. Concr. Compos.* **2016**, *68*, 96–108. [CrossRef]
16. Eskander, S.; Saleh, H. Cement mortar-degraded spinney waste composite as a matrix for immobilizing some low and intermediate level radioactive wastes: Consistency under frost attack. *J. Nucl. Mater.* **2012**, *420*, 491–496. [CrossRef]
17. Edmunds, R.; Wadee, M.A. On kink banding in individual PPTA fibres. *Compos. Sci. Technol.* **2005**, *65*, 1284–1298. [CrossRef]
18. Claramunt, J.; Ventura, H.; Fernández-Carrasco, L.J.; Ardanuy, M. Tensile and flexural properties of cement composites reinforced with flax nonwoven fabrics. *Materials* **2017**, *10*, 215. [CrossRef]
19. Selvaraj, R.; Priyanka, R. Study on recycled waste cloth in concrete. *Int. J. Eng. Res* **2015**, *4*, 891–895. [CrossRef]
20. Alomayri, T.; Shaikh, F.; Low, I.M. Thermal and mechanical properties of cotton fabric-reinforced geopolymer composites. *J. Mater. Sci.* **2013**, *48*, 6746–6752. [CrossRef]
21. Affan, M.; Ali, M. Experimental investigation on mechanical properties of jute fiber reinforced concrete under freeze-thaw conditions for pavement applications. *Constr. Build. Mater.* **2022**, *323*, 126599. [CrossRef]

22. Candamano, S.; Crea, F.; Coppola, L.; De Luca, P.; Coffetti, D. Influence of acrylic latex and pre-treated hemp fibers on cement based mortar properties. *Constr. Build. Mater.* **2021**, *273*, 121720. [CrossRef]
23. Zhou, C.; Cai, L.; Chen, Z.; Li, J. Effect of kenaf fiber on mechanical properties of high-strength cement composites. *Constr. Build. Mater.* **2020**, *263*, 121007. [CrossRef]
24. Rahimi, M.; Hisseine, O.A.; Tagnit-Hamou, A. Effectiveness of treated flax fibers in improving the early age behavior of high-performance concrete. *J. Build. Eng.* **2022**, *45*, 103448. [CrossRef]
25. *AS3972-2010*; General Purpose and Blended Cements. Standards Australia: Sydney, Australia, 2010.
26. Solutions, M.B. MasterGlenium SKY 8100. Available online: https://www.master-builders-solutions.com/en-au/products/mastergleniumsky/masterglenium-sky-8100 (accessed on 20 June 2022).
27. Islam, M.S.; Ahmed, S.J. Influence of jute fiber on concrete properties. *Constr. Build. Mater.* **2018**, *189*, 768–776. [CrossRef]
28. *AS2350.12-2006*; Methods of Testing Portland, Blended and Masonry Cements Preparation of a Standard Mortar and Moulding of Specimens. Standards Australia: Sydney, Australia, 2006.
29. *ASTMC1437*; Standard Test Method for Flow of Hydraulic Cement Mortar. American Society of Testing Materials: West Conshohocken, PA, USA, 2020.
30. *AS1012.12.1-1998*; Methods of Testing Concrete Determination of Mass per Unit Volume of Hardened Concrete—Rapid Measuring Method. Standards Australia: Sydney, Australia, 1998.
31. *AS1012.9:2014*; Methods of Testing Concrete Compressive Strength Tests—Concrete, Mortar and Grout Specimens. Standards Australia: Sydney, Australia, 2014.
32. *ASTMC348*; Standard Test Method for Flexural Strength of Hydraulic-Cement Mortars. ASTM: West Conshohocken, PA, USA, 2021.
33. *AS1012.21-1999*; Methods of Testing Concrete Determination of Water Absorption and Apparent Volume of Permeable Voids in Hardened Concrete. Standards Australia: Sydney, Australia, 1999.
34. Zhang, K.; Pan, L.; Li, J.; Lin, C.; Cao, Y.; Xu, N.; Pang, S. How does adsorption behavior of polycarboxylate superplasticizer effect rheology and flowability of cement paste with polypropylene fiber? *Cem. Concr. Compos.* **2019**, *95*, 228–236. [CrossRef]
35. Banfill, P.; Starrs, G.; Derruau, G.; McCarter, W.; Chrisp, T. Rheology of low carbon fibre content reinforced cement mortar. *Cem. Concr. Compos.* **2006**, *28*, 773–780. [CrossRef]
36. Zhang, L.; Miao, X.; Kong, X.; Zhou, S. Retardation effect of PCE superplasticizers with different architectures and their impacts on early strength of cement mortar. *Cem. Concr. Compos.* **2019**, *104*, 103369. [CrossRef]
37. Wydra, M.; Dolny, P.; Sadowski, G.; Fangrat, J. Flexural behaviour of cementitious mortars with the addition of basalt fibres. *Materials* **2021**, *14*, 1334. [CrossRef]
38. Vailati, M.; Mercuri, M.; Angiolilli, M.; Gregori, A. Natural-fibrous lime-based mortar for the rapid retrofitting of heritage masonry buildings. *Fibers* **2021**, *9*, 68. [CrossRef]
39. Kubica, J.; Galman, I. Investigations on flexural and compressive strengths of mortar dedicated to clinker units—Influence of mixing water content and curing time. *Materials* **2022**, *15*, 347. [CrossRef]
40. Ozerkan, N.G.; Ahsan, B.; Mansour, S.; Iyengar, S.R. Mechanical performance and durability of treated palm fiber reinforced mortars. *Int. J. Sustain. Built Environ.* **2013**, *2*, 131–142. [CrossRef]
41. Sebaibi, N.; Benzerzour, M.; Abriak, N.E. Influence of the distribution and orientation of fibres in a reinforced concrete with waste fibres and powders. *Constr. Build. Mater.* **2014**, *65*, 254–263. [CrossRef]
42. Hamada, H.M.; Shi, J.; Al Jawahery, M.S.; Majdi, A.; Yousif, S.T.; Kaplan, G. Application of Natural Fibres in Cement Concrete: A Critical Review. *Mater. Today Commun.* **2023**, *35*, 105833. [CrossRef]
43. Ziada, M.; Erdem, S.; Tammam, Y.; Kara, S.; Lezcano, R.A.G. The Effect of Basalt Fiber on Mechanical, Microstructural, and High-Temperature Properties of Fly Ash-Based and Basalt Powder Waste-Filled Sustainable Geopolymer Mortar. *Sustainability* **2021**, *13*, 12610. [CrossRef]
44. Zhao, K.; Xue, S.; Zhang, P.; Tian, Y.; Li, P. Application of natural plant fibers in cement-based composites and the influence on mechanical properties and mass transport. *Materials* **2019**, *12*, 3498. [CrossRef] [PubMed]
45. Riofrio, A.; Cornejo, M.; Baykara, H. Environmental performance of bamboo fibers and sugarcane bagasse reinforced metakaolin-based geopolymers. *Case Stud. Constr. Mater.* **2022**, *17*, e01150. [CrossRef]
46. Li, T.; Takkellapati, S. The current and emerging sources of technical lignins and their applications. *Biofuels Bioprod. Biorefining* **2018**, *12*, 756–787. [CrossRef]
47. Ma, B.; Qi, H.; Tan, H.; Su, Y.; Li, X.; Liu, X.; Li, C.; Zhang, T. Effect of aliphatic-based superplasticizer on rheological performance of cement paste plasticized by polycarboxylate superplasticizer. *Constr. Build. Mater.* **2020**, *233*, 117181. [CrossRef]
48. The Major Cements: Composition and Properties. Available online: https://www.britannica.com/technology/portland-cement (accessed on 2 July 2022).

sustainability

MDPI

Article

Explainable Ensemble Learning Models for the Rheological Properties of Self-Compacting Concrete

Celal Cakiroglu [1], Gebrail Bekdaş [2,*], Sanghun Kim [3] and Zong Woo Geem [4,*]

1 Department of Civil Engineering, Turkish-German University, 34820 Istanbul, Turkey
2 Department of Civil Engineering, Istanbul University-Cerrahpasa, 34320 Istanbul, Turkey
3 Department of Civil and Environmental Engineering, Temple University, Philadelphia, PA 19122, USA
4 Department of Smart City & Energy, Gachon University, Seongnam 13120, Korea
* Correspondence: bekdas@iuc.edu.tr (G.B.); geem@gachon.ac.kr (Z.W.G.)

Abstract: Self-compacting concrete (SCC) has been developed as a type of concrete capable of filling narrow gaps in highly reinforced areas of a mold without internal or external vibration. Bleeding and segregation in SCC can be prevented by the addition of superplasticizers. Due to these favorable properties, SCC has been adopted worldwide. The workability of SCC is closely related to its yield stress and plastic viscosity levels. Therefore, the accurate prediction of yield stress and plastic viscosity of SCC has certain advantages. Predictions of the shear stress and plastic viscosity of SCC is presented in the current study using four different ensemble machine learning techniques: Light Gradient Boosting Machine (LightGBM), Extreme Gradient Boosting (XGBoost), random forest, and Categorical Gradient Boosting (CatBoost). A new database containing the results of slump flow, V-funnel, and L-Box tests with the corresponding shear stress and plastic viscosity values was curated from the literature to develop these ensemble learning models. The performances of these algorithms were compared using state-of-the-art statistical measures of accuracy. Afterward, the output of these ensemble learning algorithms was interpreted with the help of SHapley Additive exPlanations (SHAP) analysis and individual conditional expectation (ICE) plots. Each input variable's effect on the predictions of the model and their interdependencies have been illustrated. Highly accurate predictions could be achieved with a coefficient of determination greater than 0.96 for both shear stress and plastic viscosity.

Keywords: plastic viscosity; self-compacting concrete; yield stress; V-funnel flow; slump flow; L-Box; XGBoost; LightGBM; CatBoost; SHAP

Citation: Cakiroglu, C.; Bekdaş, G.; Kim, S.; Geem, Z.W. Explainable Ensemble Learning Models for the Rheological Properties of Self-Compacting Concrete. *Sustainability* **2022**, *14*, 14640. https://doi.org/10.3390/su142114640

Academic Editors: Woubishet Zewdu Taffese and Sandra Barbosa Nunes

Received: 6 October 2022
Accepted: 3 November 2022
Published: 7 November 2022

1. Introduction

Self-compacting concrete (SCC) is broadly used in the construction industry due to its good mechanical properties, high fluidity, and ability to pass through and fill the gaps between reinforcing bars without vibrations [1,2]. Self-compactibility and resistance to segregation can be achieved by using superplasticizers, lowering the water-cement ratio, and reducing coarse aggregate content [3]. The flowability of SCC is correlated with the plastic viscosity, yield stress, and the outcome of empirical test procedures. The workability of concrete can be defined as its ability to properly fill its molding while having sufficient strength in its final hardened form [4]. To have good workability, a balance must be maintained between the mechanical properties of concrete and its fluidity. Therefore, it is important to have accurate procedures for the prediction of yield stress and plastic viscosity as these properties determine the workability of concrete. The rheological properties of fresh concrete can be investigated using various test procedures, including slump-flow, L-box, and V-funnel tests. Past studies have aimed at finding correlations between the properties, such as the correlation of slump flow diameter and V-funnel flow time with the yield stress and plastic viscosity of fresh concrete. The yield stress of fresh concrete is defined as the minimum amount of stress that causes permanent deformation and flows [5]. In addition to yield stress, another property of fresh concrete that determines its workability is the plastic viscosity, which is defined as the resistance of concrete to flow when the shear stress is higher than yield stress. Due to the practical significance of these rheological properties, there have been studies aimed at developing models that predict these properties. Neophytou et al. [6] proposed an empirical model that related the yield stress of SCC to the non-dimensional final spread of concrete obtained from a slump flow test. A linear relationship between the non-dimensional final spread and yield stress was demonstrated. Schowalter and Christensen [7] developed an analytical equation that related the slump of fresh concrete to its yield stress. Pashias et al. [4] investigated the relationship between slump height and yield stress in flocculated materials. An approximating equation that related yield stress to slump height was proposed. Le et al. [8] showed that the yield stress of self-consolidating concrete could be predicted by considering concrete as an aggregate suspension in cement paste. Based on excess paste theory and percolation theory, it was shown that the yield stress could be predicted as a function of the excess paste layer thickness, volume fraction of the aggregates, and percolation.

In recent years, machine learning techniques have been applied in various areas of engineering such as bridge construction [9–13], reinforced concrete and steel frames [14–16], modeling of concrete and masonry structures [17–20], modeling of pavement foundations [21], and design of columns [22–24]. Machine learning models have also been applied to plastic viscosity and yield stress prediction for concrete. Different predictive models have been proposed that relates the yield stress and plastic viscosity of concrete to its rheological properties. Benaicha et al. [25] developed an artificial neural network and multi-variable regression models that predicted the yield stress and plastic viscosity of concrete. The predictive models were developed using a dataset of 59 samples. The slump flow diameter, V-funnel flow time, and L-Box ratio were the input variables in these models. The increase in the slump flow diameter and V-funnel flow time was found to positively impact yield stress and plastic viscosity. Alyamac and Ince [26] carried out an experimental study including slump-flow, L-box, and V-funnel tests on SCC with the addition of three different types of marble powder. Compression strength and split tension tests were conducted on hardened concrete specimens. Based on these experiments, a concrete mix design monogram was created that described the relationships between compressive strength, water-cement ratio, aggregate-cement ratio, and cement content.

Data-driven prediction methodologies have also been applied to estimating the chloride permeability and mechanical properties of SCC. Carbon dioxide and chloride penetration is a major factor leading to corrosion in reinforced concrete structures [27]. Yuan et al. [28] investigated the chloride penetration in SCC using single and hybrid regression methods. Cement content, fly ash, silica fume, fine and coarse aggregate percentages, and temperature were the input variables in this study. The predictive model accuracies were measured using the root mean square error, mean absolute error, mean absolute percentage error, and performance index. Kumar et al. [29] demonstrated the applicability of the multivariate adaptive regression spline and minimax probability machine regression models to predict the results of rapid chloride penetration tests. The effects of fly ash and silica fume contents and temperature on the test results were investigated. Ge et al. [30] presented optimized random forest models developed using particle swarm optimization, whale optimization algorithm, and Harris hawks optimization. Input features included cement, fly ash, silica fume, fine and soft aggregate contents, water-to-cement ratio, and temperature. Amin et al. [31] used a gene expression programming algorithm to investigate the effects of fine and coarse aggregate contents, water-to-binder ratio, compressive strength, and metakaolin content on rapid chloride penetration. Aggarwal et al. [32] developed predictive models using random forest, random tree, multilayer perceptron, M5P, and support vector regression algorithms, based on the contents of cement, fine and coarse aggregates, metakaolin, rice husk ash, water, and superplasticizers as input features to predict the 28-day compressive strength of SCC. The models were trained using a dataset of 159 samples. Farooq et al. [33] applied artificial neural network, support vector machine, and gene expression programming techniques to predict the mechanical properties of SCC. The coefficient of determination was used as the metric for accuracy. Zhu et al. [34] developed hybrid predictive models by combining random forest, support vector regression, and multi-layer perceptron techniques with the grey wolf optimization algorithm. The splitting tensile strength of concrete was predicted based on a dataset of 168 samples. De-Prado-Gil et al. [35] compared the performances of extra trees regression, gradient boosting, CatBoost, and XGBoost algorithms in predicting the splitting tensile strength of SCC. The performances of these algorithms were compared on a database consisting of 381 samples using coefficient of determination, root mean square error, and mean absolute error as the metrics for accuracy. The XGBoost algorithm was reported to perform best according to all error metrics.

The current study presents the application of four different ensemble machine learning models to predict the yield stress and plastic viscosity of SCC. The ensemble ML models have been trained on a newly curated database that combines different experimental results from past literature. Furthermore, the SHapley Additive exPlanations (SHAP) methodology has been utilized to make the ensemble learning models explainable. The impact of each rheological property on the predicted yield stress and viscosity was visualized according to the SHAP algorithm. Finally, the effect of changing individual input variables on the output of the ML models were shown using individual conditional expectation (ICE) plots.

2. Materials and Methods

The procedures for developing explainable predictive ensemble learning models for the rheological properties of SCC are presented in this section. After a summary of the test techniques used for generating the machine learning dataset, the applied ensemble learning methods are described.

2.1. Test Procedures

This section briefly describes the experiments for the measurement of the rheological properties of SCC.

2.1.1. Slump Flow Test

The slump flow test quantifies the filling ability of SCC. The test procedure followed the EFNARC guidelines for SCC [36]. A mold with the shape of a truncated cone was placed on a flow table, as shown in Figure 1, and filled with SCC. The mold had a top and bottom diameter of 100 mm and 200 mm, respectively. The height of the cone was 300 mm. A circle of 500 mm diameter with the cone was drawn on the flow table. After the mold was lifted, SCC started to spread. The time needed for concrete to reach 500 mm diameter was recorded as T_{500}. Once SCC reached its final shape, the diameter of the spread was measured in two perpendicular directions (D1 and D2, shown in Figure 1), and the average value of these two measurements was recorded as the slump flow diameter.

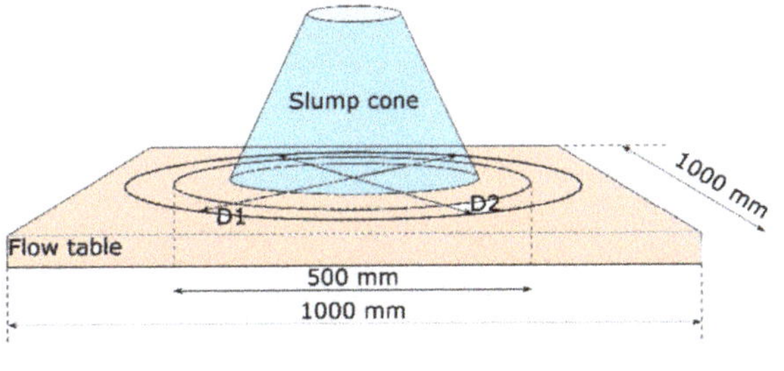

Figure 1. Slump flow test setup.

2.1.2. V-Funnel Test

The V-funnel test quantifies the ability of SCC to pass through narrow openings. The test was carried out using a V-shaped funnel (Figure 2), through which SCC passed under its weight. The standard dimensions of the equipment, shown in Figure 2, were adopted from the JSCE guidelines [37].

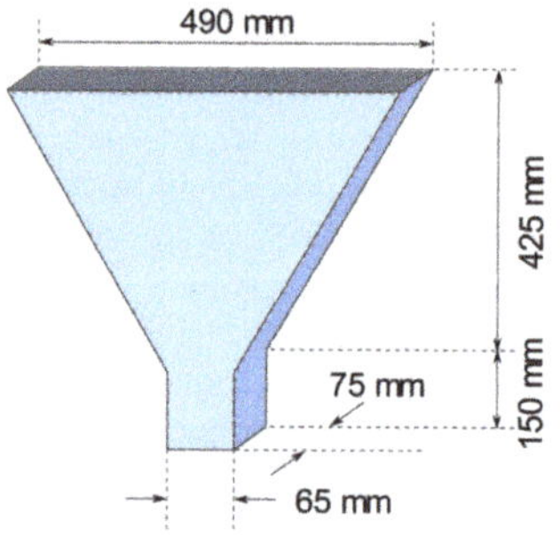

Figure 2. V-funnel test equipment [1].

The V-funnel test was carried out with 11.2 L of concrete that filled the V-shaped funnel. The time it took for this volume of SCC to completely flow out of the V-funnel was recorded. V-funnel flow time is an indication of how quickly SCC can fill narrow voids. Furthermore, greater V-funnel flow times indicate greater plastic viscosity [38].

2.1.3. L-Box Test

The equipment for the L-Box test consists of a vertical and horizontal section. Initially, the vertical section is filled with SCC, and then the SCC is allowed to flow into the horizontal section, passing through steel bars. A description of this test setup is shown in Figure 3.

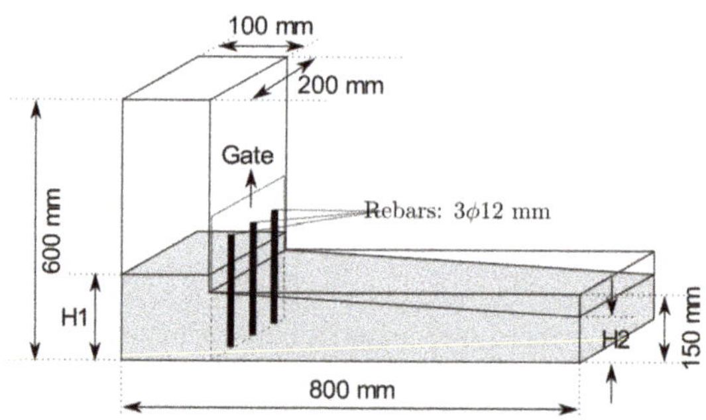

Figure 3. L-Box test setup [11].

After stabilization of the concrete, the heights H1 and H2 are measured. The ratio of these two values (H1/H2) is used as an indicator of the ability of SCC to pass through narrow spaces. The ratio PA = H1/H2 is used as the passing ability parameter [39].

2.2. Ensemble Machine Learning Process

A dataset consisting of 170 samples was collected from the literature [3,25,39–41]. In this dataset, each sample contained the measurements for SCC from the V-funnel, slump flow, and L-Box tests. Furthermore, the shear stress and plastic viscosity values corresponding to each data sample was recorded using a rheometer. In this dataset, the slump flow diameter, V-funnel flow time, and passing ability parameter were the input features for the machine learning models, whereas shear stress and plastic viscosity were the predicted output variables. In addition to these three input features, some models have been developed where the input data has been augmented with the plastic viscosity values when predicting shear stress, and vice versa. As the algorithms of predictive modeling, Extreme Gradient Boosting (XGBoost), Categorical Gradient Boosting (CatBoost), random forest, and Light Gradient Boosting Machine (LightGBM) were selected. The models were implemented using the Scikit-learn package available in the Python programming language. This selection was based on various studies in the recent years that reported these algorithms as some of the best-performing algorithms among state-of-the art machine learning techniques. Rahman et al. [42] presented the results of eleven different machine learning algorithms, including linear regression, ridge regression, lasso regression, decision tree, random forest, support vector regression, k-nearest neighbors (KNN), artificial neural network (ANN), XGBoost, AdaBoost, and CatBoost, in the prediction of the shear strength of steel fiber-reinforced concrete beams. The XGBoost algorithm was shown to deliver the best accuracy among these methods, followed by random forest, AdaBoost, and CatBoost. Cakiroglu et al. [24] developed machine learning models for the prediction of the axial load carrying capacity of concrete filled steel tubular stub columns. The performances of Lasso regression, random forest, AdaBoost, LightGBM, Gradient Boosting Machine, XGBoost, and CatBoost models were compared. The CatBoost, LightGBM, and XGBoost models were demonstrated to perform better than the remaining models. Somala et al. [43] demonstrated the application of linear regression, KNN, support vector machine, random forest, and XGBoost models in predicting the peak ground acceleration and peak ground velocity during an earthquake. The best predictions were obtained using the random forest and XGBoost models for the prediction of peak ground acceleration and peak ground velocity, respectively. Degtyarev and Naser [44] developed predictive models for the estimation of the shear strength and elastic shear buckling load of cold-formed steel channels, using the gradient boosting regressor, XGBoost, LightGBM, AdaBoost, and CatBoost methods. The CatBoost method was observed to deliver the highest accuracy among these models. Sun et al. [9] used the ANN, KNN, decision tree, random forest, AdaBoost, gradient boosting regression tree, and XGBoost algorithms for the prediction of tuned mass damper accelerations. Among these algorithms, the random forest model was shown to achieve the best accuracy.

The maximum and minimum values, as well as the distribution of the input and output features used in this study, have been visualized in Figure 4. In Figure 4, each feature has been split into four compartments, with different colors based on the magnitude of the features. The length of each compartment depends on the number of samples in that compartment. The minimum and maximum values of each compartment are written above the boundaries of the compartments.

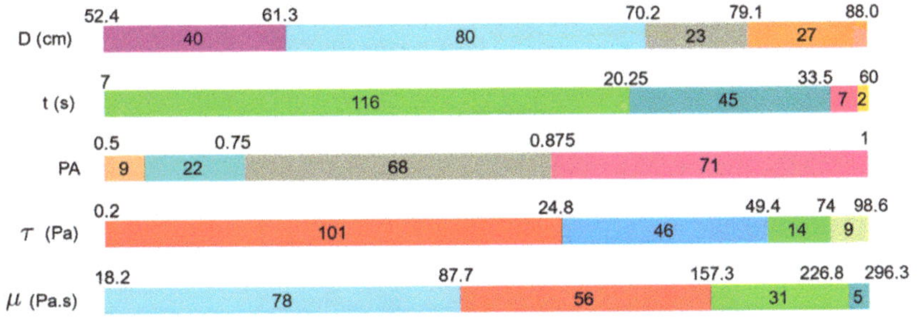

Figure 4. Variable ranges and distributions.

In Figure 4, the slump flow diameter, V-funnel flow time, L-Box H2/H1 ratio, shear stress, and plastic viscosity are shown with D, t, PA, τ, and μ, respectively. The majority of cases (68%) have a V-funnel flow time between 7 and 20.25 s, whereas only 5.3% of the samples have a V-funnel flow time between 33.5 and 60 s. In addition, 70.6% of the samples have a slump flow diameter of less than 70.2 cm and 54.1% of the samples have a plastic viscosity higher than 87.7 Pa.s. A visualization of the correlation between the different variables in this study is shown in Figure 5.

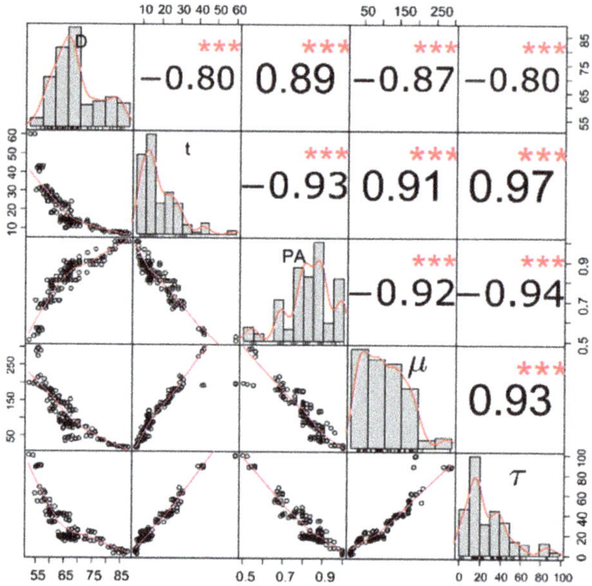

Figure 5. The correlation matrix with variable distributions (three stars indicate the significance of correlation).

The correlation of all the variables in the dataset and their statistical distributions are shown in Figure 5. Each tile on the diagonal shows the frequency distribution of a variable, whereas the lower triangular area contains bivariate scatter plots with regression lines. The upper triangular part of the correlation matrix contains the Pearson correlation values between the variables, where the font size and number of stars indicate the strength of the correlation. The Pearson correlation value r_{xy} was computed as shown in Equation (1), and an r_{xy} value close to 1 indicates a strong correlation between the variables x and y.

$$r_{xy} = \frac{n\sum_{i=1}^{n} x_i y_i - \sum_{i=1}^{n} x_i \sum_{i=1}^{n} y_i}{\sqrt{n\sum_{i=1}^{n} x_i^2 - \left(\sum_{i=1}^{n} x_i\right)^2}\sqrt{n\sum_{i=1}^{n} y_i^2 - \left(\sum_{i=1}^{n} y_i\right)^2}} \quad (1)$$

In Equation (1), x and y are two sets of values containing n samples each. For each variable in Figure 5, a scale showing the magnitude of this variable is available on both the horizontal and vertical axes. According to Figure 5, the passing ability (PA) and V-funnel flow time (t) were correlated with a Pearson correlation coefficient of −0.93. The strongest correlation between any two variables was observed between the V-funnel flow time (t) and shear stress (τ). On the other hand, relatively weak correlations were found between the slump flow diameter (D) and t, as well as between D and τ.

The dataset was randomly split into a training and test set for the training and testing of the predictive models. Based on a consensus in the machine learning-related literature, a 70% to 30% ratio between the training and test sets was adopted. Some of the notable examples were studies carried out by Feng et al. [45] and Nguyen et al. [46]. Nguyen et al. [46] carried out a comprehensive parametric study to demonstrate the effect of changing training/test set split ratio on the predictive model output. A total of nine split ratios ranging between 10/90 and 90/10 were tested, and the 70/30 split ratio was demonstrated to be the most suitable split ratio. Table 1 shows the statistical properties of the training and test sets where SD, As, and K denote standard deviation, skewness, and kurtosis, respectively. The grid search approach was adopted to tune the hyperparameters of the predictive models using the training sets. Table 2 provides an overview of the hyperparameters used in the predictive models. The hyperparameters of Table 2 were obtained based on 10-fold cross-validation using the training set. In this process, the training set was split into 10 equal-sized segments, and each of these segments was used as a test set once, while the model was trained on the remaining part of the training set.

Table 1. Statistical properties of the training and test sets.

Dataset	Property	D	t	PA	μ	τ
Training (119 samples)	Unit	cm	s	-	[Pa · s]	[Pa · s]
	Min	52.4	7.0	0.5	18.2	0.2
	Max	88.0	60.0	1.0	296.3	98.6
	Mean	68.19	18.10	0.83	107.40	29.23
	SD	9.08	9.94	0.11	67.34	21.54
	As	0.65	1.30	−0.69	0.57	1.18
	K	−0.54	2.09	0.35	−0.10	1.28
Test (51 samples)	Min	54.0	7.0	0.52	18.3	0.8
	Max	88.0	60.0	1.0	274.65	97.8
	Mean	69.30	16.04	0.85	93.26	25.34
	SD	9.12	9.57	0.11	58.22	20.71
	As	0.59	2.38	−0.83	0.81	1.63
	K	−0.55	7.45	0.99	0.52	2.78

Table 2. Hyperparameters for the predictive models.

Model	Parameter	Value
Random Forest	Number of estimators (n_estimators)	5
	Minimum samples for split (min_samples_split)	3
	Minimum samples of leaf node (min_samples_leaf)	1
	Maximum tree depth (max_depth)	None
	Number of features (max_features)	"sqrt"
XGBoost	Number of estimators (n_estimators)	50
	Step size shrinkage (eta)	0
	Learning rate	0.1
	Subsample ratio of the training instances (subsample)	0.5
	Maximum depth of a tree	6
LightGBM	Number of estimators (n_estimators)	500
	Maximum number of decision leaves (num_leaves)	5
	Maximum depth of a tree (max_depth)	4
	Learning rate	0.2
	use extremely randomized trees (extra_trees)	True
CatBoost	Bagging temperature (bagging_temperature)	10
	Learning rate	0.3
	Depth	8
	Tree growing policy (grow_policy)	"Depthwise"

Figure 6 shows the learning curves of the predictive models. In each subplot of Figure 6, the development of the R^2 scores obtained from the training and test sets are shown in red and green colors, respectively. For each model, the training samples were fed into the algorithm in 30 batches and the performance of the model was plotted after the model parameter updates. Figure 6 shows that the prediction performances of the test set improves as the number of training samples increases. The learning curves converge to their best performance, which indicates that the models have a good fit.

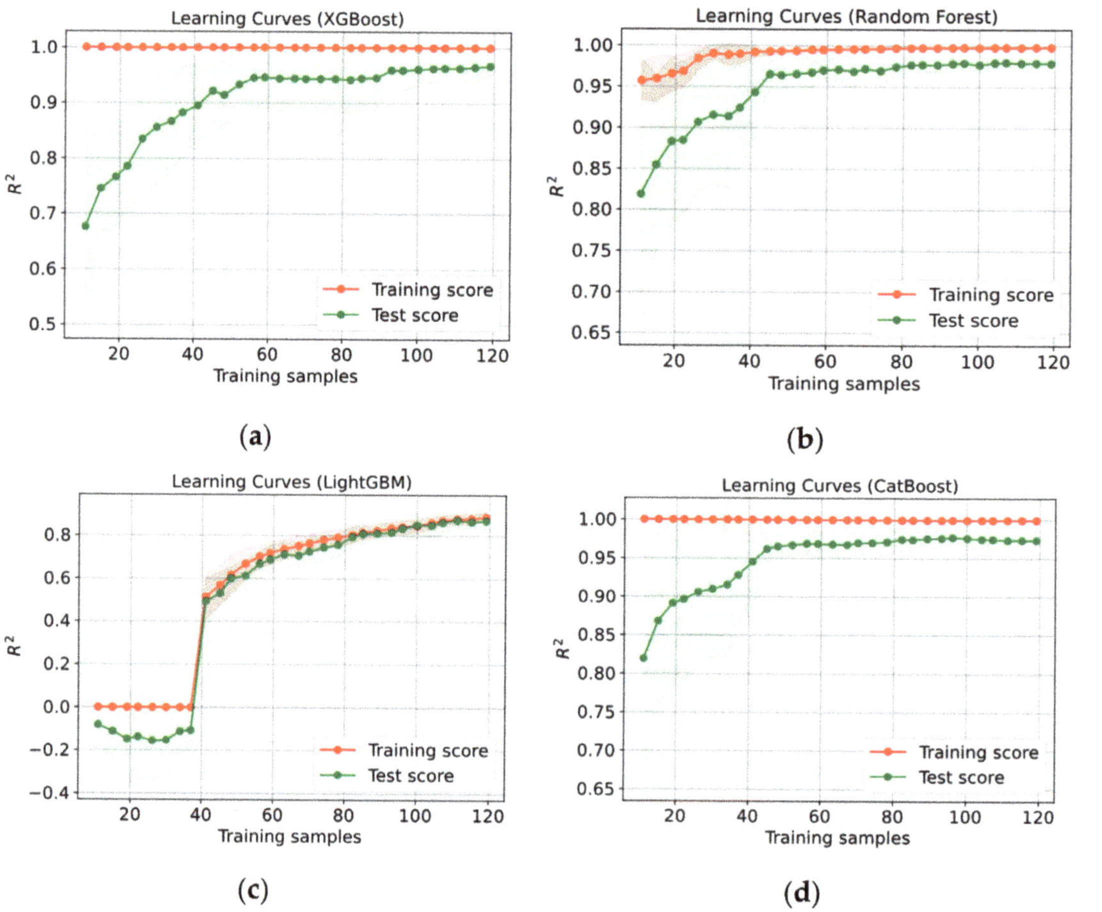

Figure 6. Learning curves of the (**a**) XGBoost, (**b**) random forest, (**c**) LightGBM, and (**d**) CatBoost models.

Figures 7 and 8 show the variations of different variables used in the predictions for shear stress (τ) and plastic viscosity (μ). In these plots, the colors of the dots represent the changes in the magnitudes of the output variables. In Figure 7, as the magnitude of τ increases, the colors of the dots change from blue to red. Similarly, in Figure 8, the colors of the dots lighten as the magnitude of μ increases. Both Figures 7 and 8 show that as the V-funnel flow time (t) decreases, the slump flow diameter rapidly increases for $t < 15$ s. A similar relationship can be observed between D and passing ability (PA). Particularly, for $PA > 0.9$, a rapid increase in D takes place. Both τ and μ are observed to decrease with increasing D values, and increase with increasing t values. A nearly linear relationship is observed between these variables (τ and μ) and the V-funnel flow time t, particularly for $t < 30$ s.

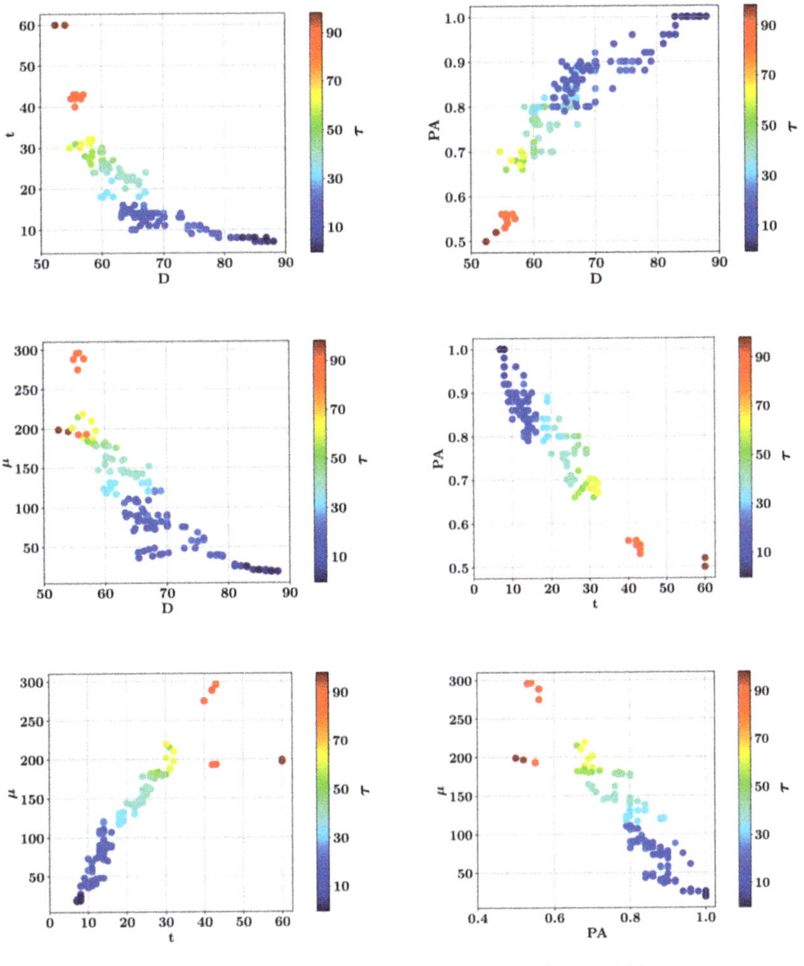

Figure 7. Relationships between the input variables and dependent variable τ.

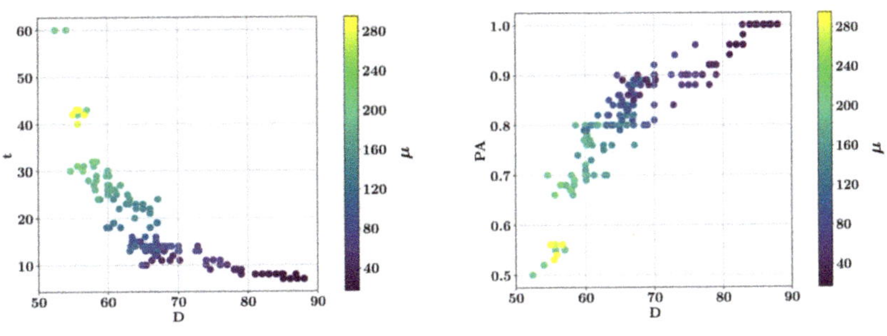

Figure 8. *Cont.*

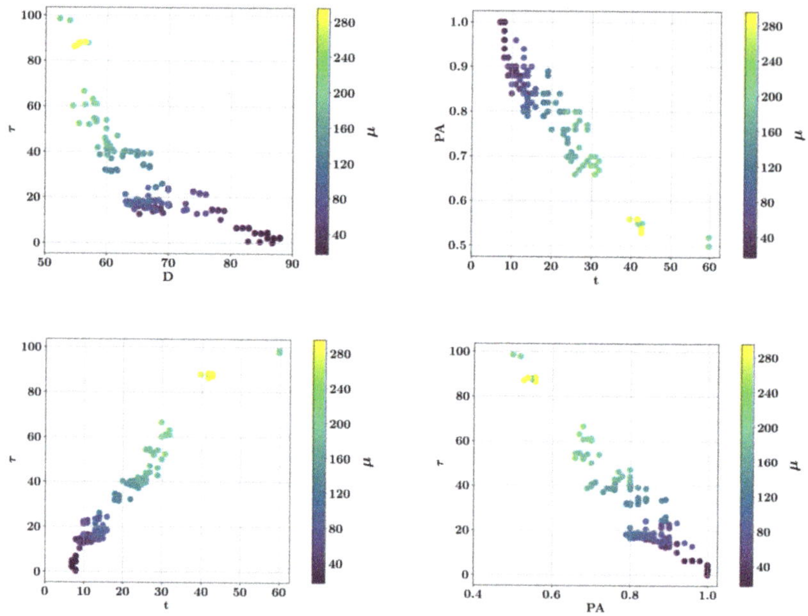

Figure 8. Relationships between the input variables and dependent variable μ.

2.3. Gradient Boosting Algorithms

The gradient boosting algorithms in this study were built on the technique of combining weak decision trees to generate strong learners. This procedure is also called ensemble learning. Among the algorithms in this category, the eXtreme Gradient Boosting (XGBoost) algorithm stands out as one of the most successful and frequently used algorithms. Equation (2) describes the process of generating the predicted values, $\hat{y}_i$, for the i-th data point in an XGBoost model. In Equation (2), the predictive model output is described as a linear combination of the outputs of individual regression trees, $f_k(\mathbf{x}_i)$, where K is the total number of regression trees. Here, $\mathbf{x}_i$ denotes a vector of input features for a data point with index i. Equation (2) also shows the regularized objective function, $\mathcal{L}(\Phi)$, the minimization of which yields the individual regression trees, f_k. The leaf weights of these regression trees are combined in the vectors $\mathbf{w}_k$, the total number of the leaves is denoted with T, and the loss function, which depends on the difference between the target and predicted values, is shown with $L(y_i, \hat{y}_i)$. In Equation (2), w_j^* stands for the optimal values of the leaf weights that minimize the loss function, and I_j is the set that contains the sample indices of the j-th leaf [47,48].

$$\begin{gathered}\hat{y}_i = \sum_{k=1}^{K} f_k(\mathbf{x}_i)\\ \mathcal{L}(\Phi) = \sum_i L(y_i, \hat{y}_i) + \sum_k \Omega(f_k) = \sum_i L(y_i, \hat{y}_i) + \sum_k \gamma T + \tfrac{1}{2}\lambda \|\mathbf{w}_k\|^2 \\ w_j^* = -\frac{\sum_{i\in I_j} g_i}{\sum_{i\in I_j} h_i + \lambda}, \quad g_i = \frac{\partial L\left(y_i, \hat{y}_i^{(t-1)}\right)}{\partial \hat{y}_i^{(t-1)}}, \quad h_i = \frac{\partial^2 L\left(y_i, \hat{y}_i^{(t-1)}\right)}{\partial \left(\hat{y}_i^{(t-1)}\right)^2}\end{gathered} \tag{2}$$

Another algorithm that combines weak decision trees to generate strong learners is the random forest algorithm. Bagging and random feature selection methodologies are utilized to train each decision tree on a randomly selected subset of the training set. The random forest model forecast is determined by the average value of the individual decision tree predictions, as indicated in Equation (3). In Equation (3), $\hat{m}_j(\mathbf{x})$ and $\hat{m}(\mathbf{x})$ denote the predictions of a single decision tree and the entire random forest model, respectively, for the input vector **x**, where K is the total number of the decision trees in the model [49].

$$\hat{m}(\mathbf{x}) = \frac{1}{K}\sum_{j=1}^{K}\hat{m}_j(\mathbf{x}) \quad (3)$$

A more developed version of the gradient boosting machine algorithm with better accuracy, more efficient memory usage, and increased speed is called the LightGBM algorithm. Due to gradient-based one-side sampling (GOSS), parallel learning, and exclusive feature bundling (EFB) techniques, the LightGBM algorithm is capable of processing large datasets. The GOSS algorithm is based on the ranking of data instances to the magnitudes of their gradients. Using this methodology, the predictive model can be trained on a smaller training subset with greater gradient magnitudes, which increases model efficiency [50].

The CatBoost algorithm was mainly designed for the efficient processing of datasets with categorical features. Better prediction accuracy is achieved through the ordered boosting algorithm. Furthermore, using ordered target statistics, CatBoost eliminates the prediction shift that is observed in the other ensemble learning methods [51].

3. Results

In this section, the predictions of the ensemble learning algorithms are compared with the actual experimental values of shear stress and plastic viscosity. The accuracy and computational speed of the predictive models were quantified and tabulated. For each predictive model, the coefficient of determination (R^2), root mean square error (RMSE), and mean absolute error were used as the metrics for accuracy. Separate predictive models were developed for the shear stress and plastic viscosity of SCC. The impact of each input variable on model prediction was investigated using SHAP methodology. Individual conditional expectation (ICE) plots were generated for each input feature.

Figure 9 shows the comparison between the predicted and actual experimental shear stress values for the four different ensemble machine learning models. The predictions of the training and test sets are shown separately, with different colors and symbols. The predictive models in Figure 9 were developed using the slump flow diameter, V-funnel flow time, L-Box passing ability, and plastic viscosity as the input features affecting the shear stress τ. In Table 3, the best accuracy values are shown in bold font. Table 3 shows that the XGBoost model delivered the best accuracy on the test set in terms of the R^2 score, MAE, VAF, and RMSE metrics, followed by the random forest and CatBoost algorithms. On the other hand, the LightGBM algorithm was observed to have the worst performance in terms of all four metrics. A comparison of the model performances on the test set revealed that the XGBoost model performed best according to the R^2, MAE, and RMSE metrics, whereas the random forest algorithm performed best according to the VAF metric. The XGBoost, random forest, and LightGBM algorithms fared almost equally in terms of processing speed, whereas the CatBoost technique was significantly slower. Figure 10a shows the comparison of the predicted and target shear stress values for the training and test sets, where a close overlap between these two quantities can be observed. Figure 10b shows the prediction errors in the training and test sets in terms of percentages. The error percentages were reduced to near zero values for the entire training set, whereas up to 40% of over- or underestimation of the shear stress could be observed for certain samples in the test set. Figure 10c,d shows the distributions of the error percentages for the training and test sets, respectively. Figure 10c shows that the error percentages of the training set are clustered around the zero value, whereas a wider distribution can be observed with error percentages that are an order of magnitude greater, as shown in Figure 10d.

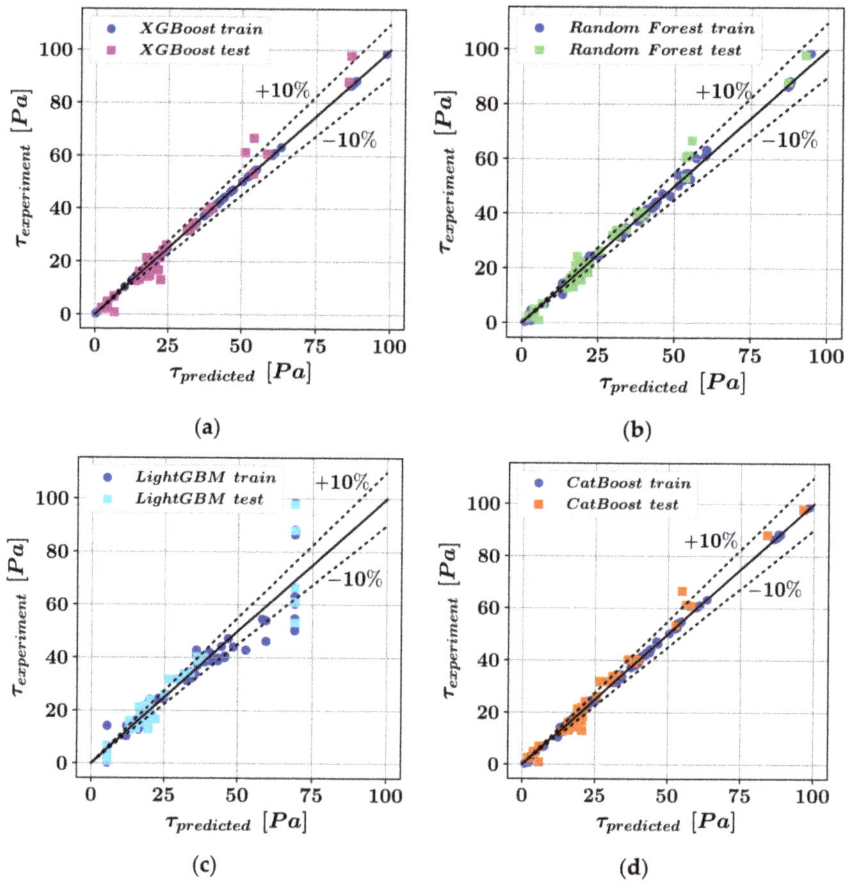

Figure 9. Comparison of the experimental and predicted shear stress values for (**a**) XGBoost, (**b**) random forest, (**c**) LightGBM, and (**d**) CatBoost models.

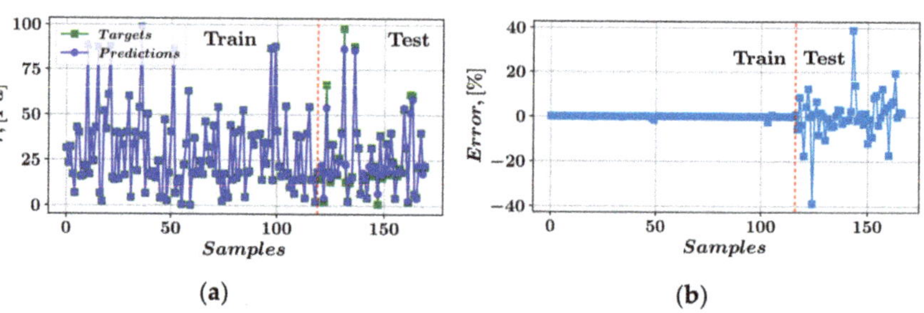

Figure 10. *Cont.*

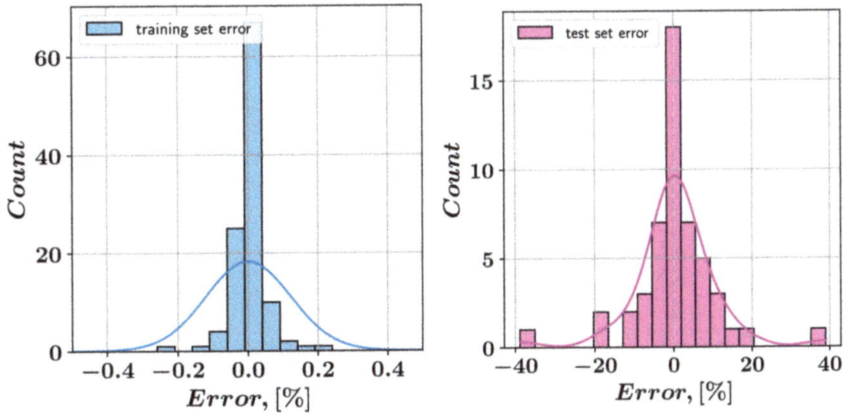

(c) (d)

Figure 10. (**a**) Predictions of the shear stress, (**b**) error percentages, (**c**) error distribution of the training set, and (**d**) error distribution of the test set for the XGBoost model.

Table 3. Prediction accuracy of the machine learning models for shear stress.

Algorithm	R^2		MAE		VAF		RMSE		Duration [s]
	Train	Test	Train	Test	Train	Test	Train	Test	
XGBoost	**0.9997**	**0.9802**	**0.094**	**1.712**	**99.99**	97.04	**0.397**	**2.885**	4.54
Random Forest	0.9977	0.9797	0.658	1.795	99.76	**98.05**	1.037	2.924	3.24
LightGBM	0.8968	0.9111	4.104	3.624	90.08	90.80	6.888	6.114	4.04
CatBoost	0.9988	0.9779	0.572	2.120	99.92	97.98	0.747	3.047	22.69

Figure 11 shows the comparison between the predicted and target plastic viscosity values. Similar to Figure 9, the accuracies of the models in predicting the plastic viscosity values are visualized by the positions of the predicted/target value pairs. The slump flow diameter, V-funnel flow time, L-Box passing ability, and shear stress were used as input variables to predict the plastic viscosities (μ) in Figure 11. The model performances in predicting plastic viscosity were quantified using the R^2 score, MAE, VAF, and RMSE metrics, as shown in Table 4. In Table 4, the best accuracy values have been shown in bold font. The performance values in Table 4 show that the CatBoost model was the most accurate for all four metrics, followed by random forest. On the other hand, the random forest algorithm was the fastest in terms of computational speed, whereas the CatBoost algorithm was significantly slower. Figure 12 shows the variation in the prediction error percentages throughout the training and test sets. An overlap of the predicted and target plastic viscosity values can be observed in Figure 12a. Error percentages increased in the transition from training set to test set, shown in Figure 12b. The distribution of these error percentages are shown as histogram plots in Figure 12c,d for the training and test sets, respectively. According to Figure 12c, the error percentages are mainly clustered in the $\pm 1\%$ range, whereas the error percentages are distributed on a wider range in Figure 12d.

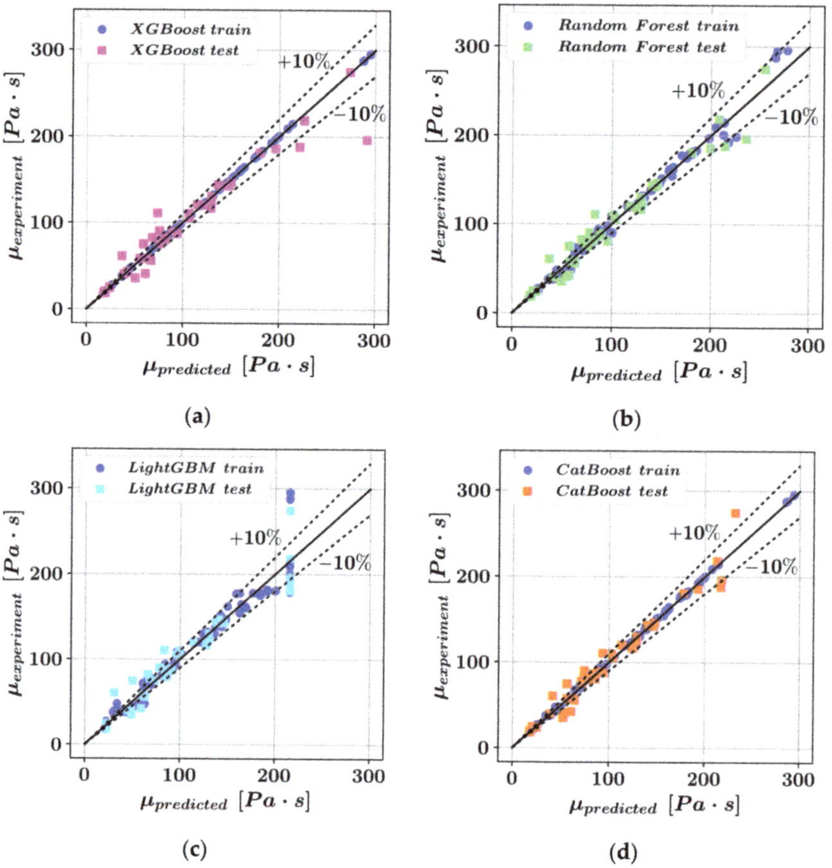

Figure 11. Comparison of the experimental and predicted plastic viscosity values for (**a**) XGBoost, (**b**) random forest, (**c**) LightGBM, and (**d**) CatBoost models.

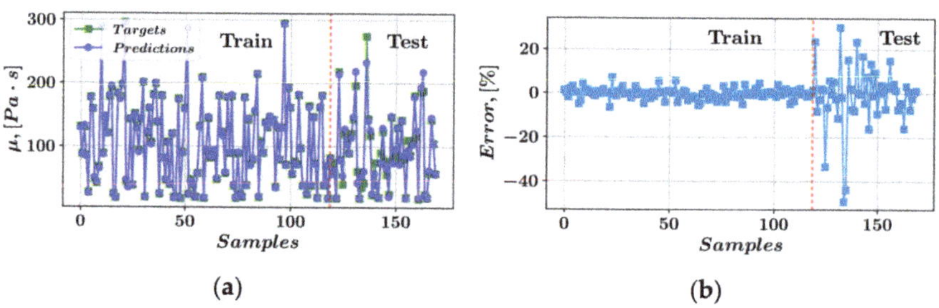

Figure 12. *Cont.*

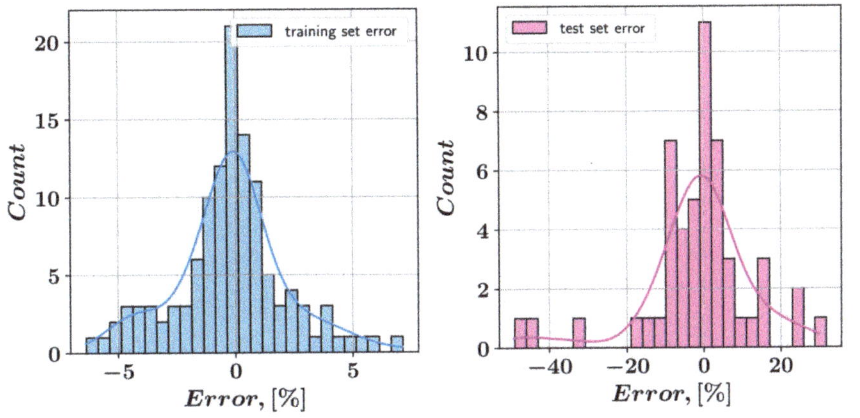

(c) (d)

Figure 12. (**a**) Predictions of the plastic viscosity, (**b**) error percentages, (**c**) error distribution of the training set, and (**d**) error distribution of the test set for the CatBoost model.

Table 4. Prediction accuracy of the machine learning models for plastic viscosity.

Algorithm	R^2		MAE		VAF		RMSE		Duration [s]
	Train	Test	Train	Test	Train	Test	Train	Test	
XGBoost	**0.9999**	0.9132	**0.041**	8.274	**99.99**	91.56	**0.084**	16.986	4.66
Random Forest	0.9896	0.9570	3.665	7.703	98.74	95.44	6.846	11.961	3.15
LightGBM	0.9324	0.9387	9.527	9.286	93.61	93.18	17.437	14.270	3.65
CatBoost	0.9986	**0.9654**	1.764	**7.602**	99.95	**96.32**	2.487	**10.727**	19.82

SHAP Analysis

The SHAP methodology is an effective way of clarifying the impacts different input features have on the predictions of a model. The SHAP technique explains complex ML models using simpler explanation models that approximate the original model. This method can be summarized through Equation (4), where s is the explanation model, $x\prime \in \{0,1\}^M$ is a binary variable connected to the actual input feature values x through a mapping function, such that $x = h_x(x\prime)$, and M is the total number of input features [52].

$$\begin{gathered} s(x\prime) = \Phi_0 + \sum_{j=1}^{M} \Phi_j x'_j \\ \Phi_j = \sum_{S \subseteq F \setminus \{i\}} \frac{|S|!(|F|-|S|-1)!}{|F|!} \left[f_{S\cup\{i\}}\left(\mathbf{x}_{S\cup\{i\}}\right) - f_S(\mathbf{x}_S) \right] \end{gathered} \tag{4}$$

In Equation (4), Φ_j represents the effect of each input feature. The set of all features is denoted with F, and S is a subset of F that does not contain the i-th feature. The values of the input features in the subset S are contained in the vector $\mathbf{x}_S$. The output of the explanation model for when all input features are missing is denoted with Φ_0. In Equation (4), the function f represents the actual predictive model. The Shapley regression values are computed based on the differences between the model predictions with and without the i-th input feature. The SHAP process can also be visually explained, as in Figure 13, where Φ_1 and Φ_2 have an increasing effect and Φ_3 has a decreasing effect on the model prediction. Further details of this procedure can be found in [52].

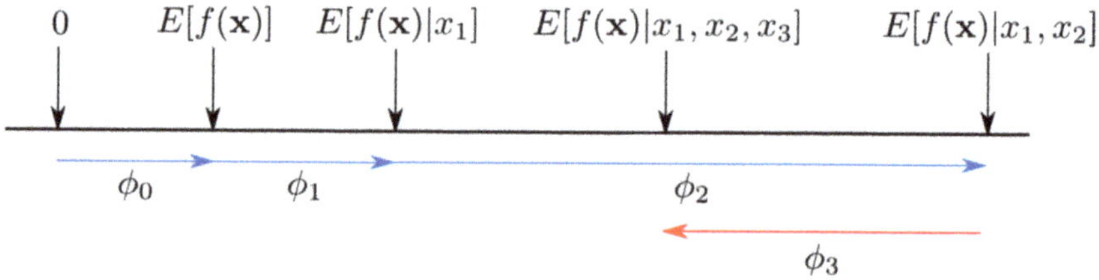

Figure 13. SHAP (SHapley Additive exPlanation) values [53].

The SHAP summary plots in Figures 14 and 15 visualize the impact of each input variable on predicted shear stress and plastic viscosity, respectively. In Figures 14 and 15, each data sample is shown with a dot for each input feature. The horizontal positions of these dots are determined by their SHAP values. The positive SHAP values indicate an increasing impact of the input feature on the model output, whereas negative SHAP values indicate a decreasing impact. Furthermore, the magnitude of each feature in a certain sample is represented by the color of its dot in the diagram, where shades of red correspond to high feature values and shades of blue correspond to low feature values. Figure 14 is based on the XGBoost model, whereas Figure 15 is generated using the CatBoost model, as these were the models with the highest accuracy in the prediction of τ and μ, respectively. According to Figure 14, the plastic viscosity had the highest impact on the XGBoost model prediction, followed by V-funnel flow time, passing ability, and slump flow diameter. Figure 15 shows that shear stress had the highest impact on the predicted plastic viscosity, and the remaining features had the same ranking (D < PA < t) as in Figure 14. Figure 14 shows that high values of t and μ have an increasing effect on the predicted value of τ, whereas high values of PA and D have a decreasing effect on τ. Similarly, Figure 15 shows that high values of τ and t have an increasing effect on μ, and high values of PA and D have a decreasing effect on μ. These observations are also supported by Figures 7 and 8.

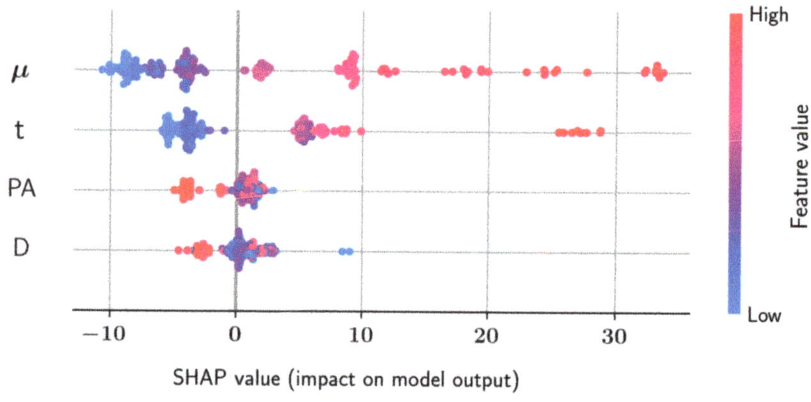

Figure 14. SHAP summary plot for the prediction of shear stress (τ) (XGBoost).

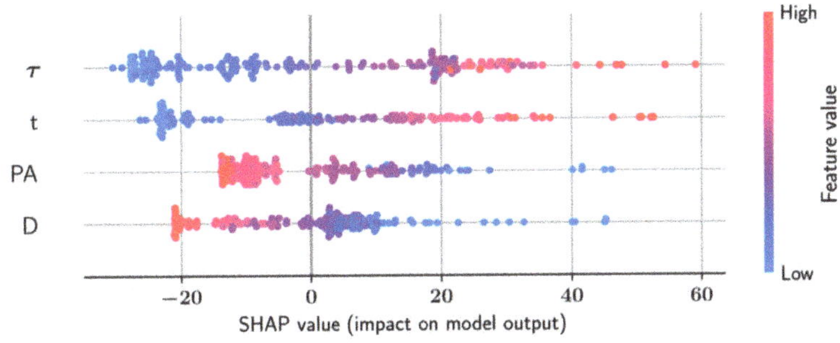

Figure 15. SHAP summary plot for the prediction of plastic viscosity (μ) (CatBoost).

The feature dependence plots in Figures 16 and 17 were generated from the XGBoost and CatBoost models, respectively. The feature dependence plots were generated for each input feature, and each feature dependence plot contains information about the variation of the SHAP values of a feature, with respect to the magnitude of that feature. For each input feature, another feature most dependent on that feature was included in the plots. The magnitudes of the most dependent features are presented with colors. Figure 16a shows that as the value of D increases, this variable begins to have negative SHAP values and the V-funnel flow time t, which is the most dependent variable on D, begins to have smaller values. From Figure 16b, as the value of t increases, passing ability (PA) decreases and the SHAP value for t increases, which indicates that increases in t increases the predicted shear stress (τ). Figure 16c shows that increasing the PA variable also increases D. On the other hand, increases in both D and PA have a decreasing effect on the predicted τ. According to Figure 16d, increasing the value of the plastic viscosity (μ) decreases D, but has an increasing effect on τ. Figure 17 shows the feature dependence plots obtained during the prediction of μ. Figure 17a shows that for the larger values of D, the SHAP value becomes negative and shear stress decreases. The variation of t and corresponding SHAP values in Figure 17b resembles Figure 16b, such that PA appears to be the input feature most dependent on t in both plots, and is adversely affected by t. Figure 17c shows that an increase in the passing ability of concrete reduces plastic viscosity (negative SHAP values) and corresponds to lower τ values. According to Figure 17d, the SHAP values of τ almost linearly increase, up to when τ reaches around 50 MPa. From this point, more irregular variation in the SHAP value is observed, whereas the D values are on the lower side of the spectrum for these larger values of τ.

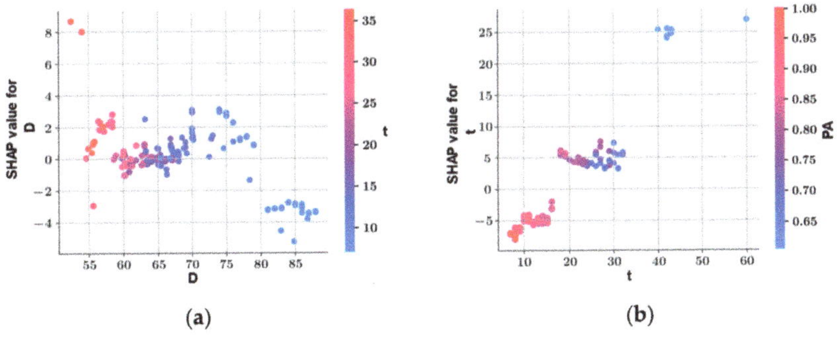

Figure 16. *Cont.*

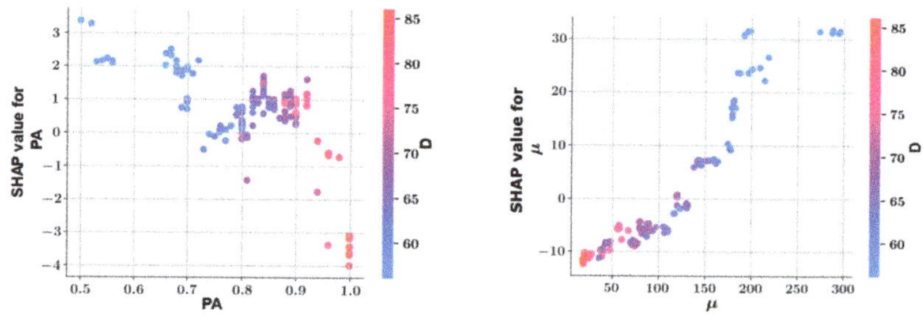

(c) (d)

Figure 16. Feature dependence plots for the variables in the prediction of τ: (**a**) D, (**b**) t, (**c**) PA, and (**d**) μ.

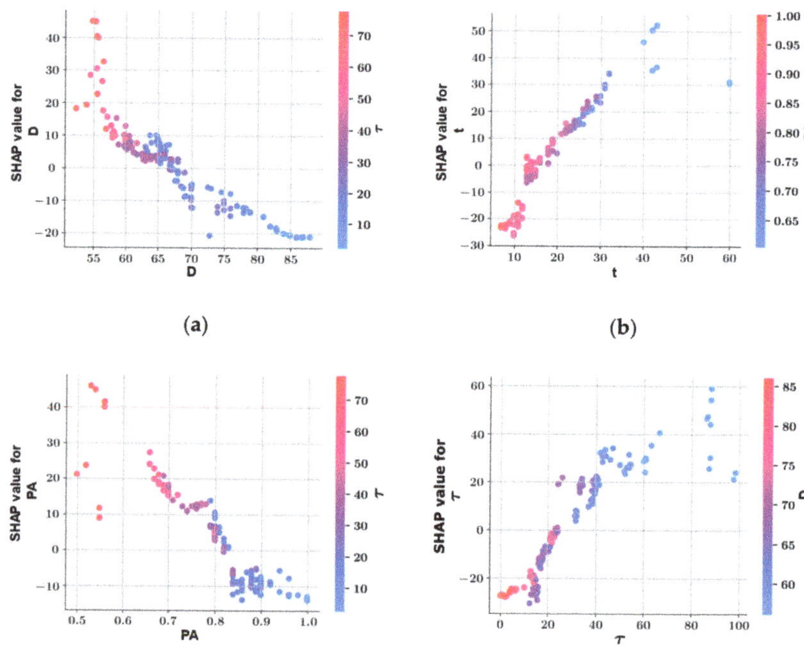

(a) (b)

(c) (d)

Figure 17. Feature dependence plots for the variables in the prediction of μ: (**a**) D, (**b**) t, (**c**) PA, and (**d**) τ.

The plots of prediction consistency in Figures 18 and 19 were generated using the predictions of the models on the test sets, which comprised 30% of the entire data set. The vertical axes in these plots show the ratio of the actual experimental values to predicted values. For each predictive model, a linear curve was fitted onto the τ_{exp}/τ_{pre} and μ_{exp}/μ_{pre} values in Figures 18 and 19, respectively. In addition to the linear curves, the τ_{exp}/τ_{pre} and μ_{exp}/μ_{pre} values from the XGBoost and CatBoost plots have been presented as scatter plots. The perfect match between the experimental and predicted values, as well as $\pm$10% deviation from the perfect match, has been shown with horizontal solid and dashed lines in black in each subplot of Figures 18 and 19. The consistency plots present information about the tendencies of each predictive model to overestimate or underestimate the target value. In this sense, a perfectly horizontal curve would indicate that the model predictions are equally good for the entire range of input features. Figure 18 shows that the LightGBM predictions shown with the solid blue line had the least consistency among the predictive models. For D values less than 65 cm, the LightGBM predictions underestimated the τ values, and for D values less than 58 cm, the LightGBM model underestimated τ over 10%. Similarly, for D values greater than 75 cm, the LightGBM model overestimated τ more than 10%. A similar pattern of consistency could also be observed for other input features during the estimation of τ. On the other hand, Figure 18 shows that the most consistent model in the prediction of τ was the CatBoost model. A different pattern of consistency was observed in the prediction such that LightGBM was observed to be the most consistent model. The curve fit for the μ_{exp}/μ_{pre} values obtained through the LightGBM model exhibit a near horizontal course for all input features. On the other hand, the least consistent model was the XGBoost model in the prediction of μ. Overall, the predictive models exhibited better consistency in the prediction of μ. XGBoost was the only model that over-or underestimated the target value by more than 10% for τ > 75 MPa and t > 40 s.

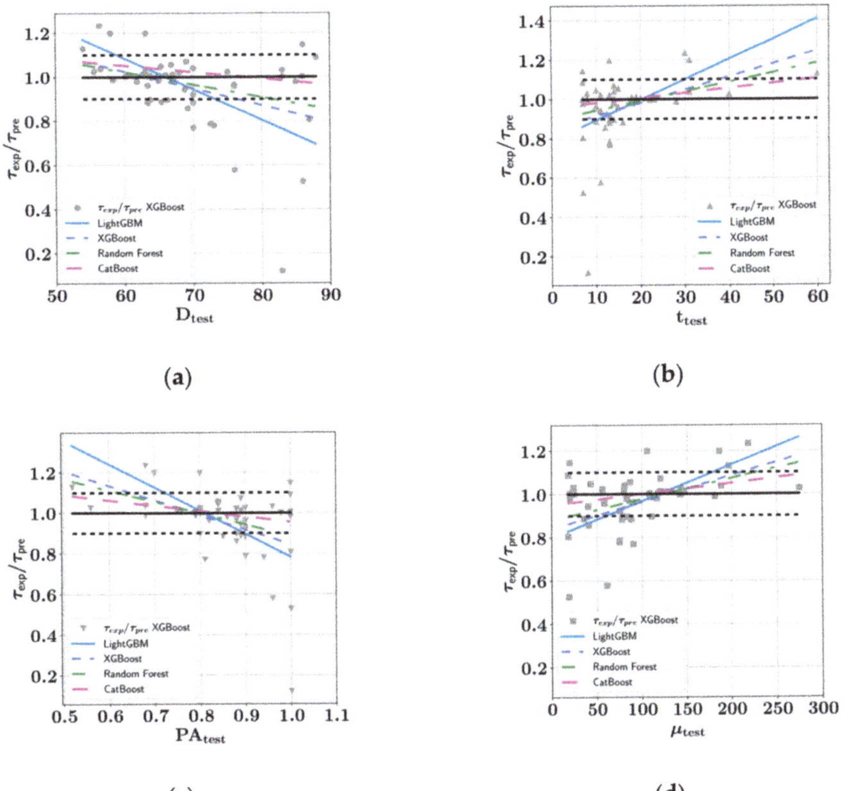

Figure 18. Prediction consistency plots for the variables in the prediction of τ: (**a**) D, (**b**) t, (**c**) PA, and (**d**) τ.

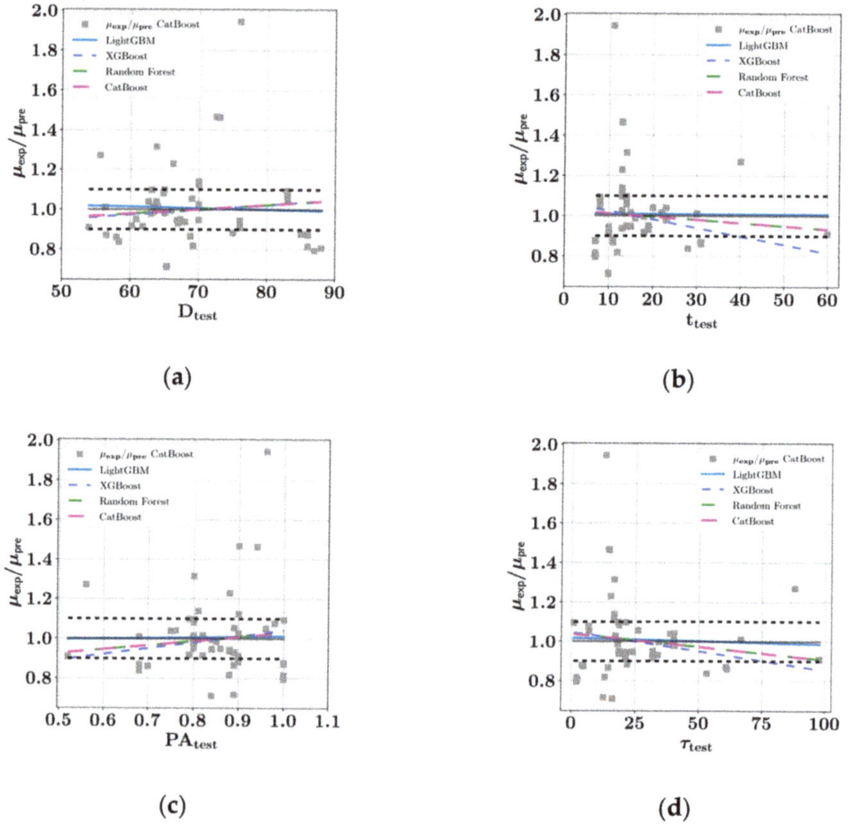

Figure 19. Prediction consistency plots for the variables in the prediction of μ: (**a**) D, (**b**) t, (**c**) PA, and (**d**) τ.

The ICE plots in Figures 20 and 21 show the variations in the predictive model output to each input feature for each data sample in the dataset. For any given input feature and data sample, all the values of the remaining input features were kept constant while the model predictions (denoted by f(x)) were calculated for different values of that particular input feature. As a result, for each data sample, a different curve was generated in the ICE plots. These plots show whether there are differences in the interactions between the feature values and model predictions for different samples. In addition to the individual curves, Figures 20 and 21 also contain darker blue curves, which represent the average of all the curves in a plot. Both Figures 20 and 21 show that most curves have a similar course in all subplots, whereas only minor deviations from the average curve pattern can be observed in some of the curves. For example, in Figure 21 for D > 70 cm, most curves exhibit a near horizontal course, whereas in some of the samples, a slight increase in predicted μ values can be observed for 70 < D < 75 cm. Similarly, in Figure 20, most of the curves exhibit a near horizontal course for PA > 0.8, whereas some of the samples in the 0.88 < PA < 0.92 range show a slight increase in the predicted τ value. Also, for PA > 0.95, a slight drop in the predicted τ value can be observed in some of the samples.

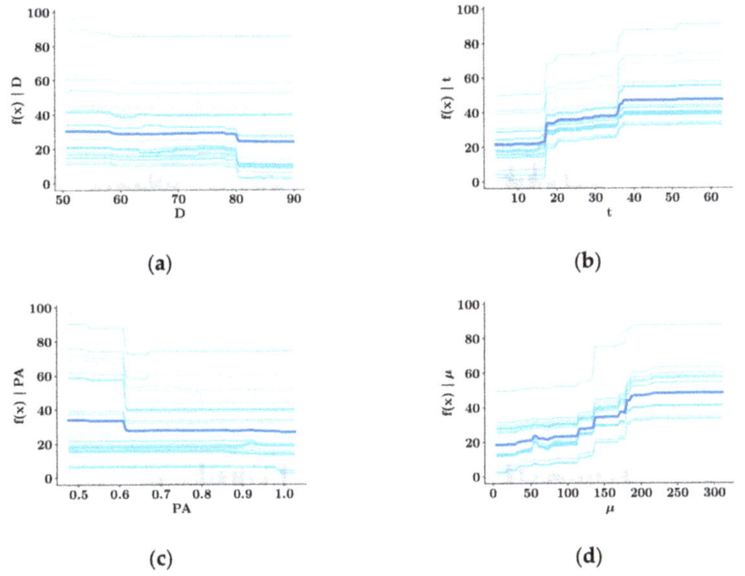

Figure 20. Individual conditional expectation (ICE) plots for the variables in the prediction of τ: (**a**) D, (**b**) t, (**c**) PA, and (**d**) μ.

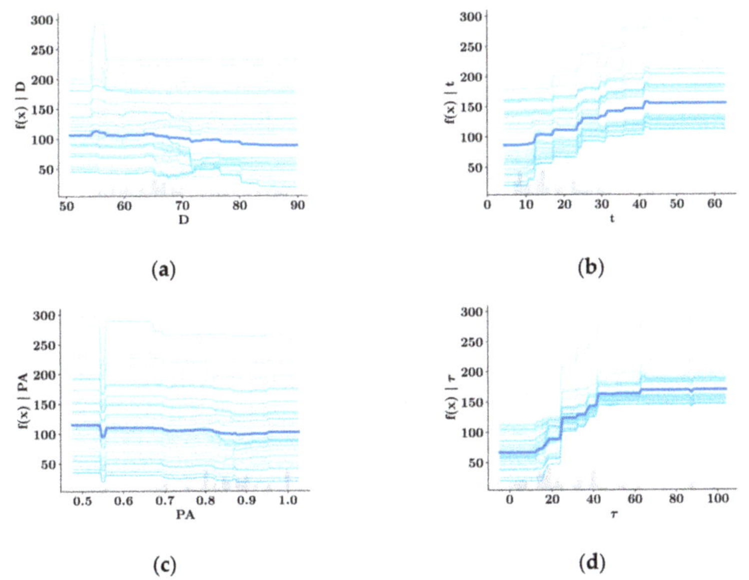

Figure 21. Individual conditional expectation (ICE) plots for the variables in the prediction of μ: (**a**) D, (**b**) t, (**c**) PA, and (**d**) τ.

4. Discussion and Conclusions

Yield stress and plastic viscosity are significant indicators of the workability of SCC. Therefore, having the appropriate tools to accurately predict its material properties is a great advantage. In recent years, machine learning techniques have been increasingly used to investigate different engineering systems to predict their structural behavior. The current study demonstrates the performances of four ensemble learning techniques on a newly curated data set that contains information about the relationships between the slump flow diameter, V-funnel flow time, passing ability from L-Box tests, plastic viscosity, and shear strength of SCC. By splitting the dataset into training and testing sets, the ensemble learning models were developed on the training set, and then their performances were measured on the test set. As the metrics for predictive model accuracy, the commonly used coefficient of determination, root mean square error, and mean absolute error were adopted. The current work includes extensive model interpretations based on the SHAP algorithm, consistency plots, and individual conditional expectation (ICE) plots. The interactions between the different input features and predictive model output values have been presented in detail using feature dependence plots. The SHAP analysis revealed that plastic viscosity had the greatest influence on model output when predicting shear stress, and shear stress had the highest impact when predicting plastic viscosity. Furthermore, slump flow diameter was found to have the lowest impact on the model output for the predictions of yield stress and plastic viscosity of SCC. Among all ensemble learning models, the CatBoost model was the most consistent model in the prediction of shear stress, and LightGBM was the most consistent model in the prediction of plastic viscosity. The relationships between the predictive model outputs and different input features were further clarified using ICE plots, in which the model predictions were visualized for the entire range of input features for each data sample. The average values of these predictions are presented in the ICE plots. The main conclusions of the study can be summarized as follows:

- The XGBoost model performed best on the test set during the prediction of shear stress as a function of the variables D, t, PA, and μ, with an R^2 score of 0.9802, followed by random forest (R^2 = 0.9797), CatBoost (R^2 = 0.9779), and LightGBM (R^2 = 0.9111).
- The CatBoost model performed best on the test set during the prediction of plastic viscosity as a function of D, t, PA, and τ, with an R^2 score of 0.9654, followed by random forest (R^2 = 0.9570), LightGBM (R^2 = 0.9387), and XGBoost (R^2 = 0.9132).
- Shear strength and plastic viscosity features were found to have the highest impact on the predictive model output during prediction of each other, based on the SHAP analysis. In the prediction of both shear stress and plastic viscosity, the slump flow diameter was found to have the lowest impact on the model output.
- In the prediction of shear stress, the most consistent predictions were made by the CatBoost model, whereas the LightGBM model was most consistent in predicting plastic viscosity.

Future research should include the study of the compressive strength and split tensile strength of SCC as a function of different rheological properties. The results of sieve segregation resistance tests could be included in predictive model development. Furthermore, closed-form equations could be developed with the help of optimization techniques that relate different rheological parameters to compressive strength, split tension strength, shear stress, and plastic viscosity of SCC. Overall, the availability of open-source machine learning techniques and predictive models is a great benefit for practical engineers and researchers working in the field of concrete research.

Author Contributions: Writing—original draft preparation, C.C.; conceptualization, C.C. and G.B.; data curation, C.C.; visualization, C.C.; writing—review and editing, G.B., S.K. and Z.W.G.; supervision G.B., S.K. and Z.W.G. All authors have read and agreed to the published version of the manuscript.

Funding: This research received no external funding.

Data Availability Statement: Data will be made available upon reasonable request.

Conflicts of Interest: The authors declare no conflict of interest.

References

1. Bartos, P.J.M.; Sonebi, M.; Tamimi, A.K. *Report of Rilem Technical Committee TC145 WSM, Compendium of Tests, Workability and Rheology of Fresh Concrete*; RILEM (The International Union of Testing and Research Laboratories for Materials and Structures): Paris, France, 2002.
2. Ben Aicha, M.; Burtschell, Y.; Alaoui, A.H.; El Harrouni, K.; Jalbaud, O. Correlation between Bleeding and Rheological Characteristics of Self-Compacting Concrete. *J. Mater. Civ. Eng.* **2017**, *29*, 05017001. [CrossRef]
3. Okamura, H.; Ouchi, M. Self-compacting high performance concrete. *Prog. Struct. Eng. Mater.* **1998**, *1*, 378–383. [CrossRef]
4. Pashias, N.; Boger, D.V.; Summers, J.; Glenister, D.J. A fifty cent rheometer for yield stress measurement. *J. Rheol.* **1996**, *40*, 1179–1189. [CrossRef]
5. Roussel, N. Correlation between Yield Stress and Slump: Comparison between Numerical Simulations and Concrete Rheometers Results. *Mater. Struct.* **2006**, *39*, 501–509. [CrossRef]
6. Neophytou, M.K.A.; Pourgouri, S.; Kanellopoulos, A.D.; Petrou, M.F.; Ioannou, I.; Georgiou, G.; Alexandrou, A. Determination of the rheological parameters of self-compacting concrete matrix using slump flow test. *Appl. Rheol.* **2010**, *20*. [CrossRef]
7. Schowalter, W.R.; Christensen, G. Toward a rationalization of the slump test for fresh concrete: Comparisons of calculations and experiments. *J. Rheol.* **1998**, *42*, 865–870. [CrossRef]
8. Lee, J.H.; Kim, J.H.; Yoon, J.Y. Prediction of the yield stress of concrete considering the thickness of excess paste layer. *Constr. Build. Mater.* **2018**, *173*, 411–418. [CrossRef]
9. Sun, Z.; Feng, D.-C.; Mangalathu, S.; Wang, W.-J.; Su, D. Effectiveness Assessment of TMDs in Bridges under Strong Winds Incorporating Machine-Learning Techniques. *J. Perform. Constr. Facil.* **2022**, *36*, 04022036. [CrossRef]
10. Todorov, B.; Billah, A.M. Post-earthquake seismic capacity estimation of reinforced concrete bridge piers using Machine learning techniques. *Structures* **2022**, *41*, 1190–1206. [CrossRef]
11. Todorov, B.; Billah, A.M. Machine learning driven seismic performance limit state identification for performance-based seismic design of bridge piers. *Eng. Struct.* **2022**, *255*, 113919. [CrossRef]
12. Xu, J.; Feng, D.; Mangalathu, S.; Jeon, J. Data-driven rapid damage evaluation for life-cycle seismic assessment of regional reinforced concrete bridges. *Earthq. Eng. Struct. Dyn.* **2022**, *51*, 2730–2751. [CrossRef]
13. Chen, M.; Mangalathu, S.; Jeon, J.-S. Machine Learning–Based Seismic Reliability Assessment of Bridge Networks. *J. Struct. Eng.* **2022**, *148*, 06022002. [CrossRef]
14. Somala, S.N.; Karthikeyan, K.; Mangalathu, S. Time period estimation of masonry infilled RC frames using machine learning techniques. *Structures* **2021**, *34*, 1560–1566. [CrossRef]
15. Hwang, S.-H.; Mangalathu, S.; Shin, J.; Jeon, J.-S. Estimation of economic seismic loss of steel moment-frame buildings using a machine learning algorithm. *Eng. Struct.* **2022**, *254*, 113877. [CrossRef]
16. Hwang, S.-H.; Mangalathu, S.; Shin, J.; Jeon, J.-S. Machine learning-based approaches for seismic demand and collapse of ductile reinforced concrete building frames. *J. Build. Eng.* **2020**, *34*, 101905. [CrossRef]
17. Moghaddas, S.A.; Nekoei, M.; Golafshani, E.M.; Nehdi, M.; Arashpour, M. Modeling carbonation depth of recycled aggregate concrete using novel automatic regression technique. *J. Clean. Prod.* **2022**, *371*, 133522. [CrossRef]
18. Safayenikoo, H.; Nejati, F.; Nehdi, M.L. Indirect Analysis of Concrete Slump Using Different Metaheuristic-Empowered Neural Processors. *Sustainability* **2022**, *14*, 10373. [CrossRef]
19. Shah, H.A.; Nehdi, M.L.; Khan, M.I.; Akmal, U.; Alabduljabbar, H.; Mohamed, A.; Sheraz, M. Predicting Compressive and Splitting Tensile Strengths of Silica Fume Concrete Using M5P Model Tree Algorithm. *Materials* **2022**, *15*, 5436. [CrossRef]
20. Zhao, G.; Wang, H.; Li, Z. Capillary water absorption values estimation of building stones by ensembled and hybrid SVR models. *J. Intell. Fuzzy Syst.* **2022**, 1–13. [CrossRef]
21. Benemaran, R.S.; Esmaeili-Falak, M.; Javadi, A. Predicting resilient modulus of flexible pavement foundation using extreme gradient boosting based optimised models. *Int. J. Pavement Eng.* **2022**. [CrossRef]
22. Bekdaş, G.; Cakiroglu, C.; Kim, S.; Geem, Z.W. Optimization and Predictive Modeling of Reinforced Concrete Circular Columns. *Materials* **2022**, *15*, 6624. [CrossRef] [PubMed]
23. Cakiroglu, C.; Islam, K.; Bekdaş, G.; Kim, S.; Geem, Z.W. Interpretable Machine Learning Algorithms to Predict the Axial Capacity of FRP-Reinforced Concrete Columns. *Materials* **2022**, *15*, 2742. [CrossRef] [PubMed]
24. Cakiroglu, C.; Islam, K.; Bekdaş, G.; Isikdag, U.; Mangalathu, S. Explainable machine learning models for predicting the axial compression capacity of concrete filled steel tubular columns. *Constr. Build. Mater.* **2022**, *356*, 129227. [CrossRef]
25. BEN Aicha, M.; Al Asri, Y.; Zaher, M.; Alaoui, A.H.; Burtschell, Y. Prediction of rheological behavior of self-compacting concrete by multi-variable regression and artificial neural networks. *Powder Technol.* **2022**, *401*, 117345. [CrossRef]
26. Alyamaç, K.E.; Ince, R. A preliminary concrete mix design for SCC with marble powders. *Constr. Build. Mater.* **2009**, *23*, 1201–1210. [CrossRef]
27. Taffese, W. Data-Driven Method for Enhanced Corrosion Assessment of Reinforced Concrete Structures. *arXiv* **2020**, arXiv:2007.01164. [CrossRef]
28. Yuan, J.; Zhao, M.; Esmaeili-Falak, M. A comparative study on predicting the rapid chloride permeability of self-compacting concrete using meta-heuristic algorithm and artificial intelligence techniques. *Struct. Concr.* **2022**, *23*, 753–774. [CrossRef]
29. Kumar, S.; Rai, B.; Biswas, R.; Samui, P.; Kim, D. Prediction of rapid chloride permeability of self-compacting concrete using Multivariate Adaptive Regression Spline and Minimax Probability Machine Regression. *J. Build. Eng.* **2020**, *32*, 101490. [CrossRef]

30. Ge, D.-M.; Zhao, L.-C.; Esmaeili-Falak, M. Estimation of rapid chloride permeability of SCC using hyperparameters optimized random forest models. *J. Sustain. Cem. Mater.* **2022**, 1–19. [CrossRef]
31. Amin, M.N.; Raheel, M.; Iqbal, M.; Khan, K.; Qadir, M.G.; Jalal, F.E.; Alabdullah, A.A.; Ajwad, A.; Al-Faiad, M.A.; Abu-Arab, A.M. Prediction of Rapid Chloride Penetration Resistance to Assess the Influence of Affecting Variables on Metakaolin-Based Concrete Using Gene Expression Programming. *Materials* **2022**, *15*, 6959. [CrossRef]
32. Aggarwal, S.; Bhargava, G.; Sihag, P. Prediction of compressive strength of scc-containing metakaolin and rice husk ash using machine learning algorithms. In *Computational Technologies in Materials Science*; CRC Press: Boca Raton, FL, USA, 2021; pp. 193–205. [CrossRef]
33. Farooq, F.; Czarnecki, S.; Niewiadomski, P.; Aslam, F.; Alabduljabbar, H.; Ostrowski, K.A.; Śliwa-Wieczorek, K.; Nowobilski, T.; Malazdrewicz, S. A Comparative Study for the Prediction of the Compressive Strength of Self-Compacting Concrete Modified with Fly Ash. *Materials* **2021**, *14*, 4934. [CrossRef]
34. Zhu, Y.; Huang, L.; Zhang, Z.; Bayrami, B. Estimation of splitting tensile strength of modified recycled aggregate concrete using hybrid algorithms. *Steel Compos. Struct.* **2022**, *44*, 375–392. [CrossRef]
35. De-Prado-Gil, J.; Palencia, C.; Jagadesh, P.; Martínez-García, R. A Comparison of Machine Learning Tools That Model the Splitting Tensile Strength of Self-Compacting Recycled Aggregate Concrete. *Materials* **2022**, *15*, 4164. [CrossRef]
36. EFNARC. Specification and Guidelines for Self-Compacting Concrete. 2002. Available online: https://wwwp.feb.unesp.br/pbastos/c.especiais/Efnarc.pdf (accessed on 16 September 2022).
37. JSCE, Japan Society of Civil Engineers. Recommendations for Self-Compacting Concrete, Concrete Library of JSCE. 1999, 31. 77p. Available online: http://www.jsce.or.jp/committee/concrete/e/newsletter/newsletter01/recommendation/selfcompact/4.pdf (accessed on 16 September 2022).
38. Yang, S.; Zhang, J.; An, X.; Qi, B.; Li, W.; Shen, D.; Li, P.; Lv, M. The Effect of Sand Type on the Rheological Properties of Self-Compacting Mortar. *Buildings* **2021**, *11*, 441. [CrossRef]
39. Sahraoui, M.; Bouziani, T. Effects of fine aggregates types and contents on rheological and fresh properties of SCC. *J. Build. Eng.* **2019**, *26*, 100890. [CrossRef]
40. EL Asri, Y.; Benaicha, M.; Zaher, M.; Alaoui, A.H. Prediction of plastic viscosity and yield stress of self-compacting concrete using machine learning technics. *Mater. Today: Proc.* **2022**, *59*, A7–A13. [CrossRef]
41. Benaicha, M.; Roguiez, X.; Jalbaud, O.; Burtschell, Y.; Alaoui, A.H. Influence of silica fume and viscosity modifying agent on the mechanical and rheological behavior of self compacting concrete. *Constr. Build. Mater.* **2015**, *84*, 103–110. [CrossRef]
42. Rahman, J.; Ahmed, K.S.; Khan, N.I.; Islam, K.; Mangalathu, S. Data-driven shear strength prediction of steel fiber reinforced concrete beams using machine learning approach. *Eng. Struct.* **2021**, *233*, 111743. [CrossRef]
43. Somala, S.N.; Chanda, S.; Karthikeyan, K.; Mangalathu, S. Explainable Machine learning on New Zealand strong motion for PGV and PGA. *Structures* **2021**, *34*, 4977–4985. [CrossRef]
44. Degtyarev, V.; Naser, M. Boosting machines for predicting shear strength of CFS channels with staggered web perforations. *Structures* **2021**, *34*, 3391–3403. [CrossRef]
45. Feng, D.-C.; Cetiner, B.; Kakavand, M.R.A.; Taciroglu, E. Data-Driven Approach to Predict the Plastic Hinge Length of Reinforced Concrete Columns and Its Application. *J. Struct. Eng.* **2021**, *147*, 04020332. [CrossRef]
46. Nguyen, Q.H.; Ly, H.-B.; Ho, L.S.; Al-Ansari, N.; Van Le, H.; Tran, V.Q.; Prakash, I.; Pham, B.T. Influence of Data Splitting on Performance of Machine Learning Models in Prediction of Shear Strength of Soil. *Math. Probl. Eng.* **2021**, *2021*, 1–15. [CrossRef]
47. Bakouregui, A.S.; Mohamed, H.M.; Yahia, A.; Benmokrane, B. Explainable extreme gradient boosting tree-based prediction of load-carrying capacity of FRP-RC columns. *Eng. Struct.* **2021**, *245*, 112836. [CrossRef]
48. Chen, T.; Guestrin, C. Xgboost: A scalable tree boosting system. In Proceedings of the 22nd Acm Sigkdd International Conference on Knowledge Discovery and Data Mining, San Francisco, CA, USA, 13–17 August 2016; pp. 785–794.
49. Feng, D.-C.; Wang, W.-J.; Mangalathu, S.; Hu, G.; Wu, T. Implementing ensemble learning methods to predict the shear strength of RC deep beams with/without web reinforcements. *Eng. Struct.* **2021**, *235*, 111979. [CrossRef]
50. Ke, G.; Meng, Q.; Finley, T.; Wang, T.; Chen, W.; Ma, W.; Ye, Q.; Liu, T.Y. Lightgbm: A highly efficient gradient boosting decision tree. In Proceedings of the 31st International Conference on Neural Information Processing Systems, Long Beach, CA, USA, 4–9 December 2017; Curran Associates Inc.: New York, NY, USA, 2017.
51. Prokhorenkova, L.; Gusev, G.; Vorobev, A.; Dorogush, A.V.; Gulin, A. CatBoost: Unbiased boosting with categorical features. In Proceedings of the 32nd International Conference on Neural Information Processing Systems, Montréal, QC, Canada, 3–8 December 2018; Curran Associates Inc.: New York, NY, USA, 2018.
52. Lundberg, S.M.; Lee, S.-I. A unified approach to interpreting model predictions. In Proceedings of the 31st Conference on Neural Information Processing Systems (NIPS 2017), Long Beach, CA, USA, 4–9 December 2017. [CrossRef]
53. Mangalathu, S.; Hwang, S.-H.; Jeon, J.-S. Failure mode and effects analysis of RC members based on machine-learning-based SHapley Additive exPlanations (SHAP) approach. *Eng. Struct.* **2020**, *219*, 110927. [CrossRef]

sustainability MDPI

Article

Impact of Curing Temperature on the Life Cycle Assessment of Sugarcane Bagasse Ash as a Partial Replacement of Cement in Mortars

Vito Francioso [1], Marina Lopez-Arias [1], Carlos Moro [1,2], Nusrat Jung [1] and Mirian Velay-Lizancos [1,*]

1 Lyles School of Civil Engineering, Purdue University, West Lafayette, IN 47907, USA
2 Department of Engineering Technology, Texas State University, San Marcos, TX 78666, USA
* Correspondence: mvelayli@purdue.edu

Abstract: Sugarcane bagasse ash (SCBA), a biomass waste resulting from sugarcane bagasse burning for electricity production, has shown to be a viable alternative option as a partial cement replacement due to its chemical composition and physical properties. Besides, previous research indicates that higher curing temperature may improve the mechanical properties of mixes containing SCBA as cement replacement. However, the environmental assessment of those mixes is lacking in the literature. This study aims to understand how curing temperature impacts the Life Cycle Assessment (LCA) of SCBA as a partial replacement of cement in mortars. An LCA was performed from the extraction of the raw materials to the material production part of the life cycle, including transport. This study shows that the reduction of environmental impact when using SCBA highly depends on the curing temperature. When mortars were cured at 45 °C, the use of SCBA reduced the environmental impact of mortars two times with respect to the reduction at 21 °C (31% reduction when cured at 45 °C vs. 14% at 21 °C, with a 20% replacement). This difference is mainly related to the fact that the higher the curing temperature, the better SCBA mortars perform in terms of strength, thus, net savings of cement required to achieve a given performance are higher. Results indicate that the sustainability of SCBA utilization as a partial replacement of cement will be better when mortar is poured in hot regions or during days with higher ambient temperatures. Likewise, the advantages of using SCBA in terms of sustainability will decrease if the external temperature is low.

Keywords: sugar cane bagasse ash; cement mortar; curing temperature; life cycle assessment; environmental footprint

Citation: Francioso, V.; Lopez-Arias, M.; Moro, C.; Jung, N.; Velay-Lizancos, M. Impact of Curing Temperature on the Life Cycle Assessment of Sugarcane Bagasse Ash as a Partial Replacement of Cement in Mortars. *Sustainability* **2023**, *15*, 142. https://doi.org/10.3390/su15010142

Academic Editors: Woubishet Zewdu Taffese and Sandra Barbosa Nunes

Received: 29 November 2022
Revised: 16 December 2022
Accepted: 17 December 2022
Published: 22 December 2022

1. Introduction

1.1. Cement Industry and the Need of Waste Valorization

The demand for construction materials has surged due to the enormous development in the construction industry and the increase in the world's population. This has been translated into a worldwide increase of 175% in cement production over the last 20 years [1]. Regrettably, cement manufacturing consumes a great deal of energy and negatively impacts the environment by consuming raw materials and releasing a massive amount of CO_2 [2]. Furthermore, besides the consumption of natural resources for building materials production, continuous waste generation has become one of our society's most prominent environmental issues. Hence, much endeavor is being made to design viable solutions to solve these problems, including developing more sustainable cementitious composites with the inclusion of industrial waste, by-products, or alternative aggregates [3,4]. The partial or total replacement of ordinary Portland cement with supplementary cementitious materials (SCMs) and natural aggregate by recycled aggregates are potential pivotal approaches to reduce the environmental concerns and carbon footprint deriving from the construction industry [4].

Thus, based on previous research, higher curing temperature may improve the mechanical properties of mixes containing SCBA as cement replacement.

If curing temperature changes the effect of SCBA on the composites' compressive strength, it can be an important factor that affects the actual change in sustainability produced by the use of biomass ash in cementitious composite production and must be included in the environmental assessments. Therefore, this research aims to understand and quantify the impact of curing temperature on the environmental performance of mortars with different replacements of cement by SCBA.

2. Materials and Methods

2.1. Materials

Mortars were produced with Portland cement Type I (Buzzi Unicem, Greencastle, IN US with a specific gravity of 3.02. The fine aggregate used was natural siliceous sand compliant with ASTM C33/C33M [48] and had a fineness modulus of 2.99, an absorption of 1.97%, and a relative density (specific gravity in saturated surface dry-SSD-condition) of 2.45. The sugar cane bagasse ash (SCBA) was supplied by a sugar plant located in Valle del Cauca (Colombia), and is a waste generated after the combustion of sugarcane bagasse for energy production. The received ash was ground with a RETSCH ZM-1 Lab Benchtop Ultra Centrifuge Mill Grinder (passing an 80 μm mesh) for 2 min to reduce the particle size and improve homogenization and reactivity. The original and final particle size was obtained with a PSA 1090 (Anton Paar, Graz, Austria) and is reported in Table 1.

Table 1. SCBA particle size.

Type	D10 (μm)	D50 (μm)	D90 (μm)	Mean Size (μm)
As received	8.0	31.9	71.6	37.6
Ground	3.7	18.7	47.5	23.6

The chemical configuration of the ash was obtained by a Lab X500 XRF analyzer (Hitachi, Tokyo, Japan) and is reported in Table 2. The SCBA used in this study satisfied the minimum requirement of 70% in the sum of SiO_2 + Al_2O_3 + Fe_2O_3 pozzolanic oxides conform to the ASTM C618 [49]. However, the loss of ignition (LOI) was about 7% higher than the minimum 10% requirement [49].

Table 2. Chemical composition of SCBA.

Compound	Percentage (%)	Standard Deviation—σ
SiO_2	54.97	0.026
Al_2O_3	13.85	0.019
CaO	9.98	0.013
Fe_2O_3	8.568	0.0054
MgO	2.074	0.0123
TiO_2	1.595	0.0064
K_2O	1.385	0.0017
Mn_2O_3	1.245	0.0067
SO_3	0.926	0.0015
Na_2O	0.509	0.0052
P_2O_5	0.213	0.006
SrO	0.158	0.0005
Cr_rO_3	0.089	0.0006
ZnO	0.087	0.0003
LOI	16.5	-

2.2. Mix Design, Sample Manufacturing, and Curing Conditions

The mortar formulation used as reference was prepared with a 1:3:0.5 (cement:sand:water) weight ratio. Two percentages of cement replacement with SCBA were selected for this study based on previous literature [10]. The amount of cement replaced was 10% and 20% (by mass of cement). Water to binder (cement + SCBA) and binder to aggregate ratios were constant for all mixes at 0.50 and 3, respectively. Mix proportions of the composites are presented in Table 3.

Table 3. Mix proportions (per m^3 of mortar).

Component	Reference	10% SCBA	20% SCBA
Cement (kg)	486.39	437.75	389.12
SCBA (kg)	0.00	48.64	97.28
Sand (SSD) (kg)	1459.18	1459.18	1459.18
w/b	0.50	0.50	0.50

Six cubes with dimensions of 50.8 × 50.8 × 50.8 mm^3 were cast per each mortar mixture. All the samples were cured in their molds at room temperature (21.5 °C and 60% RH) covered with a plastic sheet during the first day. After demolding, three samples per mixture were cured at 21 °C, while another three were cured at 45 °C, until the testing day.

2.3. Methods

2.3.1. Compressive Strength Test

Compressive strength tests were carried out according to ASTM standards C349 [50]. An MTS (Eden Prairie, MN, USA) machine with a load capacity of 300 kN, under a displacement control of 0.05 mm/s, was used. The mortar compressive strength was examined at 28 days.

2.3.2. Life Cycle Assessment (LCA) of Mortars

The framework of the LCA performed in this study is presented in Figure 1. The LCA methodology of this study follows the ISO standards 14040 and 14044 [29,51]. According to these standards, four steps are identified:

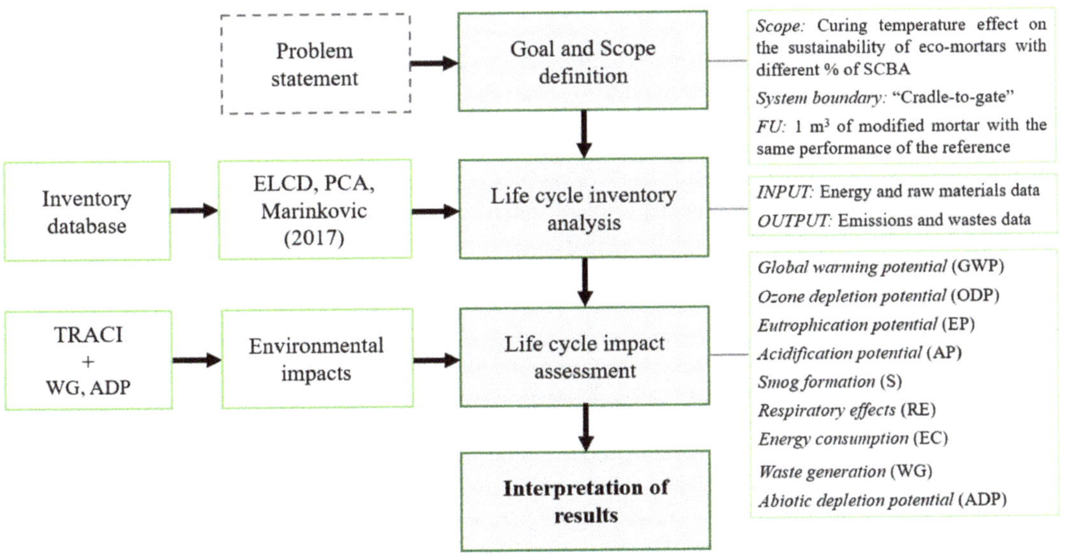

Figure 1. LCA framework for this study.

- Goal and scope

This research aims to understand and quantify how curing temperature affects the environmental performance of eco-mortars with different percentages of sugarcane bagasse ash (SCBA). Three mixes with 0%, 10%, and 20% replacement of SCBA by mass of cement and two different curing temperatures (21 °C and 45 °C) were studied. A "cradle-to-gate" analysis was considered as the system boundary, which allows for the quantification of the embodied environmental impacts of the material from the extraction of raw materials (cradle) to the mortar production (gate) stages. The boundary of the LCA is illustrated in Figure 2. The functional unit (FU) selected for this study was 1 m^3 of mortar with the same compressive strength as the reference mixture (plain Portland cement mortar without SCBA cured at the same curing temperature). To compare the environmental impact of the investigated mortars, an experimental campaign was performed beforehand to obtain a curve exhibiting the development of compressive strength as a function of cement content for the reference mortar. This curve estimates the variations in the cement content needed for achieving the same compressive strength as the reference mortar when SCBA is used to replace part of the cement.

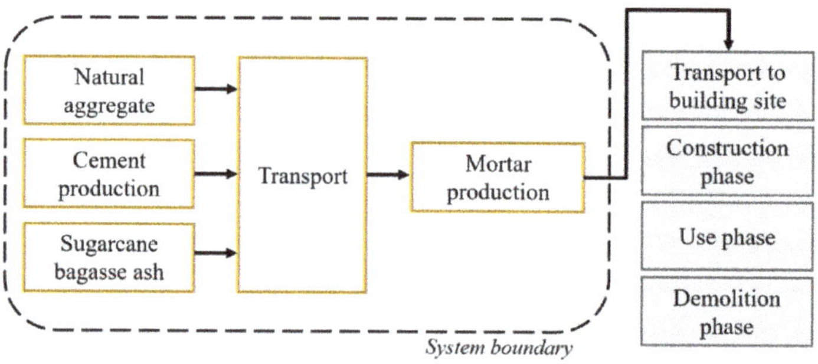

Figure 2. LCA system boundary.

- Life cycle inventory (LCI)

In this stage, energy and raw materials data (input) and emissions and wastes data (output) must be identified and allocated for each material production. In order to facilitate the LCI phase, numerous databases have been developed. These databases provide essential inventory data such as raw materials, electricity generation, transport, and waste generation data that are essential in every LCA and are based on average data representing average production and supply conditions for products and services. Since no global database exists, different sources were used in this study depending on their availability and reliability. Cement and natural fine aggregate production and total transportation impact data were obtained from the European Platform on Life Cycle Assessment (ELCD database) [52]. Mortar production input data was assumed to be equivalent to concrete production based on their essentially identical composition (aggregate, cement, and water) and because of the more abundant and reliable data on concrete production. The main difference between the two products is the aggregate size (with mortar using only the fraction passing through a 4.75 mm sieve). These data were obtained from a previous study [53]. Waste generation was estimated considering the density of the studied mortars, where the average density of all hardened samples was 2220 kg/m^3 (with a standard deviation lower than 2.1%). Finally, transport distances were collected from the Life Cycle Inventory of Portland Cement Concrete by the Portland Cement Association (PCA) [54]. Table 4 presents the LCI data considered in this study.

Table 4. Life inventory data used for this study.

	Cement (kg)	Sand (kg)	SCBA (kg)	Mortar (m^3)	Transport (t·km)
INPUTS Fossil fuels (kg)					
Diesel	3.56×10^{-2}	3.29×10^{-4}	0.00	2.00×10^{-2}	2.06×10^{-2}
Gas	8.53×10^{-3}	1.24×10^{-4}	0.00	5.30×10^{-2}	1.13×10^{-3}
Soft coal	2.67×10^{-2}	3.13×10^{-4}	0.00	5.25×10^{0}	7.31×10^{-5}
Hard coal	4.83×10^{-2}	2.18×10^{-4}	0.00	4.01×10^{-2}	9.01×10^{-5}
OUTPUTS Emissions in the air (kg)					
CO_2	8.85×10^{-1}	2.34×10^{-3}	0.00	4.59×10^{0}	6.40×10^{-2}
CO	2.14×10^{-3}	4.19×10^{-6}	0.00	8.81×10^{-4}	1.10×10^{-4}
CH_4	5.80×10^{-4}	3.72×10^{-6}	0.00	2.19×10^{-3}	6.25×10^{-5}
C_2H_4	3.95×10^{-10}	9.24×10^{-12}	0.00	7.36×10^{-8}	3.70×10^{-10}
CFC-11	5.22×10^{-9}	1.75×10^{-10}	0.00	2.09×10^{-15}	6.08×10^{-11}
CFC-114	5.35×10^{-9}	1.79×10^{-10}	0.00	3.67×10^{-9}	6.23×10^{-11}
SO_x	1.05×10^{-3}	9.49×10^{-6}	0.00	5.34×10^{-2}	3.41×10^{-5}
NO_x	1.79×10^{-3}	1.52×10^{-5}	0.00	8.01×10^{-2}	5.39×10^{-4}
N_2O	2.22×10^{-6}	3.81×10^{-8}	0.00	2.20×10^{-5}	7.32×10^{-7}
NH_3	3.91×10^{-2}	7.24×10^{-9}	0.00	3.27×10^{-7}	4.00×10^{-7}
NMVOC	2.26×10^{-1}	1.37×10^{-6}	0.00	9.20×10^{-5}	3.20×10^{-5}
HCl	1.99×10^{-2}	1.80×10^{-7}	0.00	4.00×10^{-4}	8.20×10^{-8}
N (water)	1.16×10^{-4}	4.23×10^{-9}	0.00	4.81×10^{-6}	2.35×10^{-7}
PO_4^{-3} (groundwater)	5.14×10^{-7}	1.23×10^{-8}	0.00	3.85×10^{-3}	5.88×10^{-7}

As shown in Table 4, no allocation (waste status) was applied to the SCBA. This scenario is based on the following assumptions: (i) SCBA results from a by-product (sugarcane bagasse) of a multifunctional process (sugarcane processing) and thus is a waste; (ii) even if considered as a by-product, the resulting impact is likely to be very small, therefore, the environmental impact of its production will not be relevant; (iii) it is believed that the CO_2 emissions from biomass burning do not contribute to the greenhouse effect, since the carbon released from crops already existed in the atmosphere and was absorbed during the plant growth, thus, the net CO_2 associated with the whole process can be considered neutral [31,55]. Besides, the authors deliberately avoided estimating of the cost for the industrial implementation of SCBA because it was out of the scope of the present paper, which was to study the effect of curing temperature on the embodied environmental impacts of the material. An economic analysis should consider market variables such as availability, local cost (which in the very first moment would be zero–waste value), and demand (like other residual pozzolans, the cost of the material will increase with the increasing demand), as well as the cost of further processing, but also potential revenues from carbon credits.

- Life cycle impact assessment (LCIA)

The LCIA phase is crucial to evaluate the importance and relevance of the environmental impact of a product based on the LCI results. In this study, the TRACI methodology (mid-point approach) was employed to assess the total environmental impact of the investigated mortars. The life cycle impacts were assessed by analyzing nine environmental impact categories. The categories (from TRACI) used to analyze and compare the environmental impact of the studied mixes, and their units are displayed in Table 5. Among these, ecotoxicity was excluded from the analysis since no evidence of this impact has been found in the main components of mortars [56,57]. Moreover, the characterization factors, which estimate the relative contribution of each substance (inputs and outputs in the LCI) to each impact category in its corresponding unit, were considered. The characterization factors were obtained from the study of Ryberg et al. [58].

Table 5. Designated impact categories from TRACI (last two added).

Category	Units
(i) Global warming potential (GWP)	kg CO_2 eq
(ii) Ozone depletion potential (ODP)	kg CFC-11 eq
(iii) Eutrophication potential (EP)	kg N eq
(iv) Acidification potential (AP)	kg SO_2 eq
(v) Smog formation (S)	kg O_3 eq
(vi) Respiratory effects (RE)	kg $PM_{2.5}$ eq
(vii) Energy consumption (EC)	MJ surplus
(viii) Waste generation (WG)	kg
(ix) Abiotic depletion potential (ADP)	kg Sb eq

Besides, two other categories (not measured in TRACI) were considered to account for the potential reduction of raw materials due to the replacement of cement by SCBA: (i) waste generation (WG) and (ii) abiotic depletion potential (ADP). Waste generation quantifies the net kilograms of material produced that can become waste at the product's end of life. Abiotic depletion potential is defined as the consumption of non-renewable mineral (raw materials) resources. Therefore, if ADP and WG are not considered in the environmental assessment of mixes with SCBA, the real environmental benefits of using this waste in mortars will not be fully reflected.

- Interpretation of the results

The last phase of the LCA is the interpretation of the results obtained from the impact assessment. This phase is not strictly defined as the previous ones, allowing practitioners to apply findings from the LCA to various situations, knowing the uncertainty and the assumptions used to generate the results. In this study, the analysis was conducted following two different approaches:

- Single categories analysis: Single impact categories were analyzed, and the results were discussed to appreciate the potential implications of replacing cement with SCBA as a function of curing temperature.
- Normalized unified index: Environmental impacts of single categories were normalized using the normalization factors (NFs) recommended by Ryberg et al. [58] to relate the environmental impact results of each category to a common reference. The goal of this normalization is to put each environmental impact in relation to the impact of society's production and consumption activities. As a result, normalized values will better reflect the product system's contribution to each category's environmental impact compared to those of the reference system. The reference system used was the environmental impacts of each category per year in the US [58]. The normalized values were calculated by dividing the environmental impact of each category by its corresponding normalization factor. Note that this section did not consider WG and ADP since there are no NFs available for these two categories.

3. Results and Discussion

3.1. Compressive Strength and Binder Content Variation Based on Designated FU

Figure 3a displays the compressive strength results of the studied mortars after 28 days of curing at 21 °C and 45 °C. Results showed that the replacement of cement with SCBA at standard curing temperature (21 °C) had a slightly negative effect on the compressive strength. The higher the replacement level, the higher the reduction of compressive strength. Nevertheless, the recorded loss was lower than 8% as shown in Figure 3b (6.3% and 7.3% for 10 and 20% replacement, respectively) compared to the reference mortar cured at the same temperature. However, when the curing temperature was higher (45 °C), the replacement of SCBA positively influenced the compressive strength when the results were compared to the reference mortar cured at 45 °C. Mortars with 10 and 20% SCBA replacement exhibited an increase in compressive strength of 24.5% and 22.2% (Figure 3b) compared to the plain mortar cured at the same temperature, respectively. Besides, the use of SCBA mitigated the high-temperature crossover effect observed in the reference mortar. It means that replacing Portland cement with SCBA may be beneficial when the external curing temperature is high or when a lower heat of hydration is required.

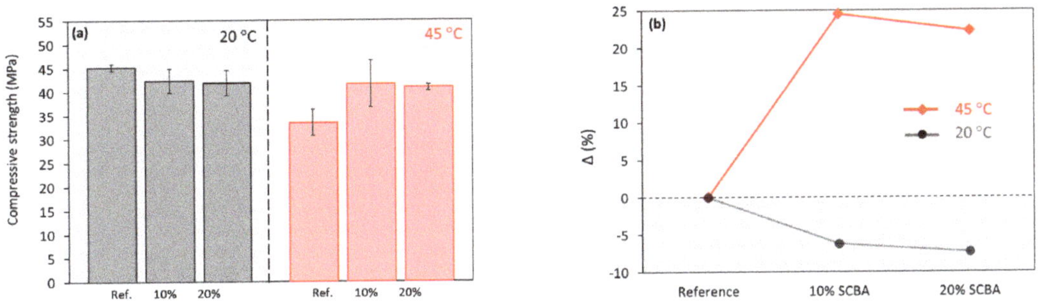

Figure 3. Compressive strength of mortars at 28 days (**a**) and its variation compared to the reference (**b**).

Thus, results show that curing temperature modifies the effect of SCBA on the mortar's compressive strength. Therefore, a secondary experimental campaign was performed to assess the influence of partial replacement of cement by SCBA on the compressive strength of the mortar with different water to cement ratios. The objective of this secondary experimental campaign was to estimate the variation of cement content required to obtain a compressive strength equal to the compressive strength of the reference mortar (no SCBA) for each SCBA replacement level. Thus, the savings or additions of cement content needed for each case can be estimated. The results of this additional experimental campaign contributed as input data to the LCA.

Two extra reference mixtures (besides the one listed in Table 3) were made, varying the cement content (w/c ratio). Figure 4 displays the compressive strength of the reference plain mortar (no SCBA) as a function of the cement content for two different curing temperatures. This graph was used to estimate the cement content required in the modified mixes to obtain the same compressive strength as the reference mortar (at the same curing temperature). First, the compressive strength obtained from the modified mixtures was introduced in the graph to estimate the corresponding cement content of a mixture with the same strength if SCBA was not used. Next, the difference between this estimate and the cement content needed to achieve the reference strength was calculated. Then, this variation was added to the initial cement content used for the specific mortar. The result is the estimated cement content required to achieve the same compressive strength as the reference mortar (without SCBA). The binder content of each modified mixture is the estimated cement content plus the SCBA content (applicable). The results are presented in Table 6, which displays the initial and the required binder content to achieve the same compressive strength as the reference mortar at 28 days when SCBA is used, as a function of the curing temperature. The aggregate content was kept constant, and therefore, the changes in the binder content to achieve the desired strength only affected the water-to-binder ratio.

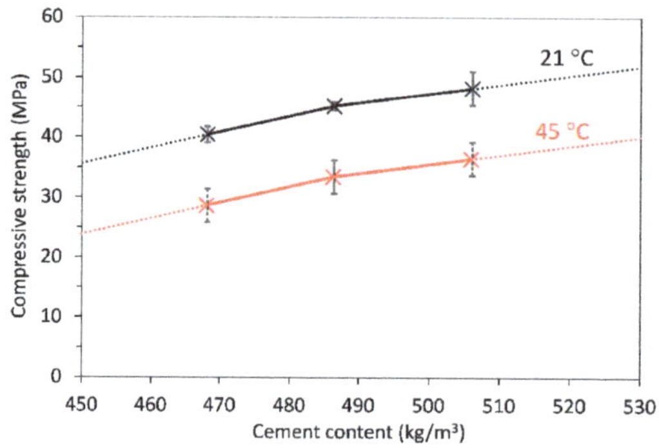

Figure 4. 28-day compressive strength as a function of cement content (no SCBA).

Table 6. Estimation of cement content to obtain the FU.

Mixture	Curing Temperature (°C)	Compressive Strength, f_c (MPa)	Initial Cement Content (kg/m^3)	Estimated Binder [Cement, SCBA] Content to Achieve Reference f_c (kg/m^3)
Reference	21 °C	45.20	486.4	486.4 [486.4, 0]
BA10		42.37	437.8	497.1 [448.5, 48.64]
BA20		41.88	389.1	499.0 [401.7, 97.28]
Reference	45 °C	33.44	486.4	486.4 [486.4, 0]
BA10		41.62	437.8	414.1 [365.5, 48.64]
BA20		40.88	389.1	420.6 [323.4. 97.28]

3.2. *Life Cycle Assessment (LCA)*

3.2.1. Environmental Impact of Single Categories

Figure 5 displays the environmental impacts of each LCA category considered for the studied mortars as a function of curing temperature differentiating the contribution of each component. In the graphs, for each material (including mortar production), the transport impact data was included in the results. The amount of component materials (cement, natural aggregate, and SCBA) used in the mix design is according to tests performed for determining the mix proportions of the modified mortars that would have the same compressive strength as their reference (without SCBA).

Table 7 reports the contribution of each material and process to the total environmental impact of the mortars cured at 21 °C. In the table, the transportation contribution to the environmental impact was separated from the impact of the materials.

Results show that cement is the most significant contributor to all impact categories. The main reason is the great deal of CO_2 emissions due to the calcination of limestone and fossil fuel and energy use required for clinker production. Thus, as expected, the reduction of the cement content produced by the utilization of SCBA reduced the environmental impact of mortars, not only on the global warming potential (which is expressed in kg CO_2 eq) but also on all other categories, as can be seen in Table 7. The impacts from cement production are, for each category, at least one order of magnitude higher than the other two largest contributors (natural aggregate and mortar production). The contribution of aggregate and mortar production does not change, regardless of the amount of cement replaced by SCBA, because the amount does not change in the mortar formulation considered for the LCA in this study.

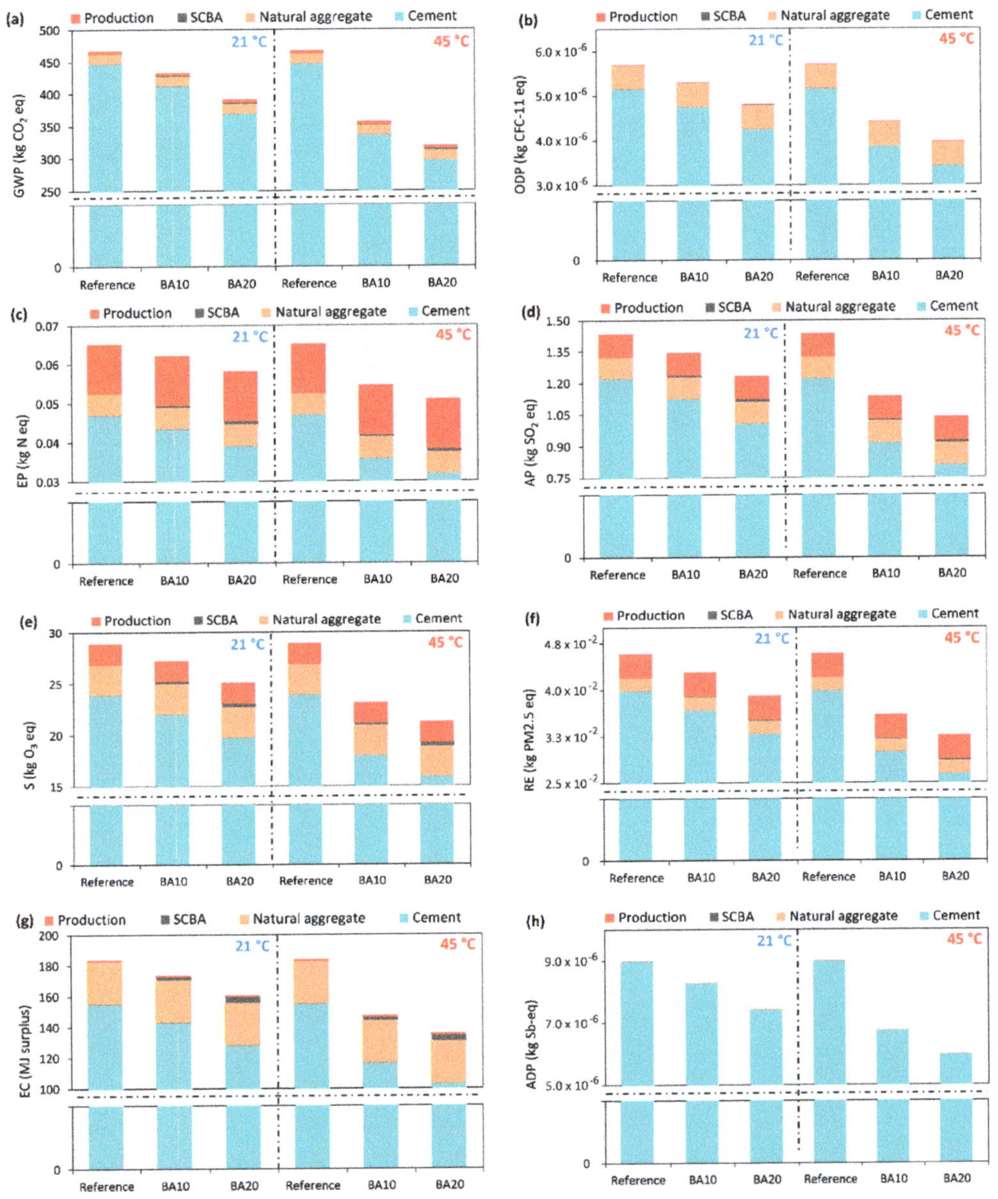

Figure 5. Environmental impact of the investigated mortars cured at 21 °C (blue background) and cured at 45 °C (red background). (**a**) Global warming potential, (**b**) Ozone depletion potential, (**c**) Eutrophication potential, (**d**) Acidification potential, (**e**) Smog formation, (**f**) Respiratory effects, (**g**) Energy consumption, (**h**) Abiotic depletion potential.

Table 7. Environmental impacts for 1 m^3 of mortars cured at 21 °C (* modified mortar that would have the same compressive strength as their reference).

Mortar	Cement	Natural Aggregate	SCBA	Production	Transportation of Components	Total
			GWP—Global warming potential (kg CO_2 eq)			
Reference	437.84	3.57	0.00	4.65	21.04	467.10
BA10 *	403.73	3.57	0.00	4.65	21.24	433.19
BA20 *	361.62	3.57	0.00	4.65	21.28	391.12
			ODP—Ozone depletion potential (kg CFC-11 eq)			
Reference	5.14×10^{-6}	5.17×10^{-7}	0.00	3.67×10^{-9}	3.94×10^{-8}	5.70×10^{-6}
BA10	4.74×10^{-6}	5.17×10^{-7}	0.00	3.67×10^{-9}	3.98×10^{-8}	5.30×10^{-6}
BA20	4.24×10^{-6}	5.17×10^{-7}	0.00	3.67×10^{-9}	3.98×10^{-8}	4.81×10^{-6}
			EP—Eutrophication potential (kg N eq)			
Reference	4.13×10^{-2}	1.03×10^{-3}	0.00	1.27×10^{-2}	8.16×10^{-3}	6.32×10^{-2}
BA10	3.81×10^{-2}	1.03×10^{-3}	0.00	1.27×10^{-2}	8.24×10^{-3}	6.01×10^{-2}
BA20	3.41×10^{-2}	1.03×10^{-3}	0.00	1.27×10^{-2}	8.25×10^{-3}	5.61×10^{-2}
			AP—Acidification potential (kg SO_2 eq)			
Reference	1.16	2.96×10^{-2}	0.00	1.10×10^{-1}	1.32×10^{-1}	1.43
BA10	1.07	2.96×10^{-2}	0.00	1.10×10^{-1}	1.33×10^{-1}	1.35
BA20	0.96	2.96×10^{-2}	0.00	1.10×10^{-1}	1.34×10^{-1}	1.23
			S—Smog formation (kg O_3 eq)			
Reference	22.02	0.56	0.00	1.99	4.31	28.88
BA10	20.31	0.56	0.00	1.99	4.36	27.21
BA20	18.19	0.56	0.00	1.99	4.36	25.09
			RE—Respiratory effects (kg PM2.5 eq)			
Reference	3.91×10^{-2}	1.01×10^{-3}	0.00	3.84×10^{-3}	1.93×10^{-3}	4.58×10^{-2}
BA10	3.60×10^{-2}	1.01×10^{-3}	0.00	3.84×10^{-3}	1.95×10^{-3}	4.28×10^{-2}
BA20	3.23×10^{-2}	1.01×10^{-3}	0.00	3.84×10^{-3}	1.95×10^{-3}	3.91×10^{-2}
			EC—Energy consumption (MJ surplus)			
Reference	136.48	4.09	0.00	0.96	42.44	183.98
BA10	125.84	4.09	0.00	0.96	42.86	173.76
BA20	112.72	4.09	0.00	0.96	42.93	160.71
			ADP—Abiotic depletion potential (kg Sb eq)			
Reference	9.00×10^{-6}	1.69×10^{-9}	0.00	0.00	0.00	9.00×10^{-6}
BA10	8.30×10^{-6}	1.69×10^{-9}	0.00	0.00	0.00	8.30×10^{-6}
BA20	7.43×10^{-6}	1.69×10^{-9}	0.00	0.00	0.00	7.43×10^{-6}

Besides cement production, transportation of component materials is the major contributor to the environmental burden of cementitious materials. Transport impacts directly depend on the transportation type and distance. While transportation impact data was obtained from the European Platform on Life Cycle Assessment (ELCD database) [52], transport distances were obtained from the Life Cycle Inventory of Portland Cement Concrete by the PCA [54] and multiplied by two (round-trip). In Figure 5, transportation was added up to the impact of each material. Therefore, the impacts coming from SCBA are none other than the impact of their transportation to the mixing location. That leaves the transport phase as a possible source of varying the environmental impact of these eco-friendly mortars. In this analysis, the transportation distance for SCBA was assumed to be the same as the transportation distance for Portland cement, fly ash, and other SCMs indicated in the Life Cycle Inventory of Portland Cement Concrete [54].

The use of SCBA produced important changes in the environmental impact of several impact categories, especially on the ADP and the GWP. The GWP quantifies the greenhouse gases released into the atmosphere (carbon dioxide (CO_2), methane (CH_4), and nitrous oxides (N_2O)) as a function of the amount of CO_2 that would have the same impact over a 100-year period [59]. When the curing temperature was 21 °C, BA20* presented a decrease of over 16% in GWP compared to the reference mortar while keeping the same estimated strength. Moreover, the ODP impact was reduced by 7% and 16% for BA10* and BA20*, respectively, for the same curing temperature. The ODP refers to a decrease in the ozone density through the thinning of the stratospheric ozone layer due to anthropogenic pollutants (e.g., halocarbons). That implies an increased exposure of human skin to UV light, which may lead to a higher risk of melanoma [60]. Besides, for the same curing temperature, the total energy demand was reduced by 6% and 13% in BA10* and BA20*, respectively, compared to the energy demand for the reference mortar.

The S, AP, RE, and EP impact categories were also reduced by 13%, 14%, 15%, and 11%, respectively, when SCBA was used (BA20*) and when the curing temperature was 21 °C. It is important because of the consequences that those impact categories have on the environment. The S impact category ponders the air pollution caused by the reaction between sunlight and emissions from fossil fuel combustion during the production of raw materials. As a result, the formation of other chemicals is promoted (e.g., ozone and peroxide), which leads to an increase in ground-level ozone concentration [61]. The AP impact category is also associated with the combustion of fossil fuels (e.g., nitrous and sulfide oxides), which may leach in the presence of oxygen and water and endanger the surrounding ecosystems [62]. The EP impact category relates to the natural or artificial discharge of nutritional elements to a body of water. These compounds decrease the oxygen available in aquatic systems, consequently reducing the water quality. The water contaminated by these organisms can be a threat to public health and biodiversity, thus, it must be monitored [63].

Finally, regarding abiotic depletion and waste generation, the replacement of cement by SCBA also reduced the impact of these two categories. The ADP impact category reflects the exhaustion of non-renewable resources and consequent environmental impacts, thus, a reduction of cement used for mortar production will have a positive effect. ADP and WG were reduced by 18% and 4% for BA20*, respectively, compared with the corresponding values for the reference mortar, both with 21 °C curing temperature.

Table 8 presents the environmental impacts of the production of 1 m^3 of the studied modified mortars cured at 45 °C, with the same target compressive strength as the reference. The environmental results of the mortars cured at 45 °C followed the same trend of mortars cured at 21 °C, with the environmental impact decreasing as the replacement of cement increased. However, at 45 °C, the use of SCBA reduced the environmental impact of mortars to a greater extent than at 21 °C, as is clearly shown in Figure 6.

Considering the results for each analyzed impact category, Figure 6 displays the reduction of the environmental impacts of modified mortars in comparison to the reference (no SCBA) for both 21 °C and 45 °C. Besides, the LCA accounted for the deviation of compressive strength obtained by the tested samples (Figure 3a). The environmental impact for each impact category was calculated assuming the average compressive strength (solid lines) as well as the minimum and maximum strength results to determine the range of uncertainty (shadowed regions). The results of each impact category for mortars containing SCBA were normalized to the corresponding results for the reference mortar (no SCBA) assuming its average compressive strength in the calculation. The result is a dimensionless value equal to 1 for the reference–the lower the value, the lower the environmental impact.

Table 8. Environmental impacts for 1 m^3 of mortars cured at 45 °C (* modified mortar that would have the same compressive strength as their reference).

Mortar	Cement	Natural Aggregate	SCBA	Production	Transportation	Total
		GWP—Global warming potential (kg CO_2 eq)				
Reference	437.84	3.57	0.00	4.65	21.04	467.10
BA10 *	328.98	3.57	0.00	4.65	19.65	356.85
BA20 *	291.08	3.57	0.00	4.65	19.77	319.08
		ODP—Ozone depletion potential (kg CFC-11 eq)				
Reference	5.14×10^{-6}	5.17×10^{-7}	0.00	3.67×10^{-9}	3.94×10^{-8}	5.70×10^{-6}
BA10	3.86×10^{-6}	5.17×10^{-7}	0.00	3.67×10^{-9}	3.68×10^{-8}	4.42×10^{-6}
BA20	3.42×10^{-6}	5.17×10^{-7}	0.00	3.67×10^{-9}	3.70×10^{-8}	3.97×10^{-6}
		EP—Eutrophication potential (kg N eq)				
Reference	4.13×10^{-2}	1.03×10^{-3}	0.00	1.27×10^{-2}	8.16×10^{-3}	6.32×10^{-2}
BA10	3.11×10^{-2}	1.03×10^{-3}	0.00	1.27×10^{-2}	7.62×10^{-3}	5.24×10^{-2}
BA20	2.75×10^{-2}	1.03×10^{-3}	0.00	1.27×10^{-2}	7.67×10^{-3}	4.89×10^{-2}
		AP—Acidification potential (kg SO_2 eq)				
Reference	1.16	2.96×10^{-2}	0.00	1.10×10^{-1}	1.32×10^{-1}	1.43
BA10	0.87	2.96×10^{-2}	0.00	1.10×10^{-1}	1.23×10^{-1}	1.14
BA20	0.77	2.96×10^{-2}	0.00	1.10×10^{-1}	1.24×10^{-1}	1.04
		S—Smog formation (kg O_3 eq)				
Reference	22.02	0.56	0.00	1.99	4.31	28.88
BA10	16.55	0.56	0.00	1.99	4.03	23.12
BA20	14.64	0.56	0.00	1.99	4.05	21.24
		RE—Respiratory effects (kg PM2.5 eq)				
Reference	3.91×10^{-2}	1.01×10^{-3}	0.00	3.84×10^{-3}	1.93×10^{-3}	4.58×10^{-2}
BA10	2.93×10^{-2}	1.01×10^{-3}	0.00	3.84×10^{-3}	1.81×10^{-3}	3.60×10^{-2}
BA20	2.60×10^{-2}	1.01×10^{-3}	0.00	3.84×10^{-3}	1.82×10^{-3}	3.26×10^{-2}
		EC—Energy consumption (MJ surplus)				
Reference	136.48	4.09	0.00	0.96	42.44	183.98
BA10	102.55	4.09	0.00	0.96	39.64	147.25
BA20	90.73	4.09	0.00	0.96	39.90	135.69
		ADP—Abiotic depletion potential (kg Sb eq)				
Reference	9.00×10^{-6}	1.69×10^{-9}	0.00	0.00	0.00	9.00×10^{-6}
BA10	6.76×10^{-6}	1.69×10^{-9}	0.00	0.00	0.00	6.76×10^{-6}
BA20	5.98×10^{-6}	1.69×10^{-9}	0.00	0.00	0.00	5.98×10^{-6}

Figure 6 shows that replacing cement with SCBA can help reduce the environmental impact of cementitious composites and that the higher the replacement, the lower the environmental footprint of mortars in the studied range of SCBA replacement. While WG impact did not change, all other categories showed important differences produced by curing temperature. Whereas at 21 °C, BA20* presented a reduction of single impact categories ranging between 11.23% and 17.40% (in comparison to reference mortar), at 45 °C, the reduction of the environmental impacts due to the use of SCBA doubled, ranging from 22.61% to 33.51%. The LCA results can be explained by looking at the compressive strength results and the functional unit chosen for the LCA. The reference mortar cured at 45 °C presented a compressive strength 26% lower than the same mixture cured at 21 °C. This tendency of high curing temperatures to induce lower strength after 28 days is well documented in the literature as a consequence of the poorer quality of the hydration products [64], and it is known as the crossover effect. However, the data of the present study showed that SCBA was effective in mitigating the crossover effect.

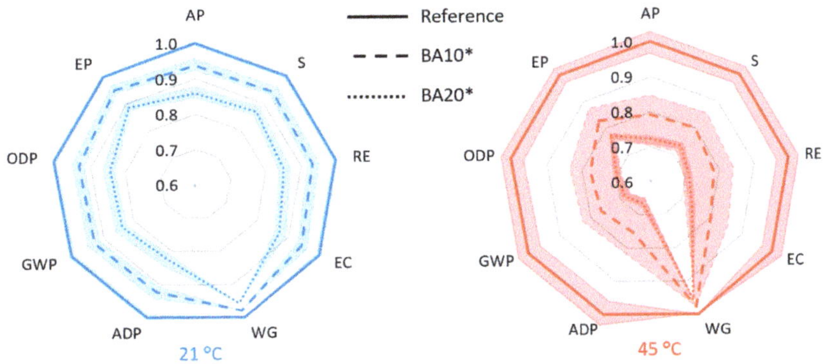

Figure 6. LCA results of SCBA mortars normalized by the reference mortar (AP—Acidification potential, S—Smog formation, RE—respiratory effects, EC—energy consumption, WG—waste generation, ADP—abiotic depletion potential, GWP—global warming potential, ODP—ozone depletion potential, EP—eutrophication potential).

3.2.2. Environmental Impact Assessed with Normalized Unified Index

Normalization factors (NFs) may be used as an optional aid for the interpretation of the results of the LCA. LCIA results for each impact category were first normalized by the NFs. A unified index was calculated by adding the normalized impacts of each category. Next, the normalized unified index was calculated as the unified index of each mixture divided by the unified index of their reference mortar. While previous results considered ADP and WG, as mentioned in the methods section, these two impact categories were not considered in the analysis to obtain the normalized unified index due to the lack of NFs.

The normalized unified index represents with a single score the relative environmental impact of each mixture compared to the reference one. Figure 7 displays the results of this approach. For a curing temperature of 21 °C, BA10* and BA20* reduced their environmental impact by 6% and 14%, respectively, compared to the reference mortar cured at the same temperature. Nevertheless, the best results in terms of environmental performance were obtained when the curing temperature was higher (45 °C). At 45 °C curing temperature, the reduction in the environmental impact was more pronounced on BA10* and BA20*, with values of 23% and 31%, respectively, compared to their reference mortar cured at 45 °C.

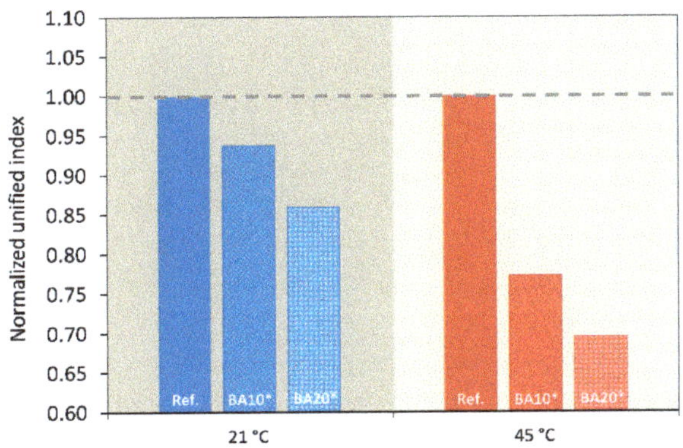

Figure 7. Normalized unified index as a function of the curing temperature. Dotted line represents the threshold (where 1.00 means same environmental impact as the reference mortar).

4. Conclusions

The impact of curing temperature on the sustainability of sugarcane bagasse ash as a partial replacement of cement in mortars was assessed in this study. Nine impact categories were analyzed for each mixture and curing temperature. Based on the results of this study it was concluded that:

- Among the impact categories, waste generation is the only impact category that was not affected by curing temperature. For all of the others analyzed categories, when mortars were cured at 45 °C, the use of SCBA reduced the environmental impact of mortars two times with respect to the reduction at 21 °C.
- At 45 °C, a replacement of 97 kg of cement with SCBA (per m^3 of mortar) produced a reduction of the environmental impact (presented with the normalized unified index) of 31%, while the reduction produced by the same amount of SCBA with a curing temperature of 21 °C was 14%.
- The reduction of environmental impact when using SCBA as a partial replacement for cement highly depends on the curing temperature. The results clearly indicate that the sustainability of SCBA utilization as a partial replacement of cement will be better when mortar is poured in hot regions or during days with higher ambient temperatures.
- The advantages of using SCBA in terms of sustainability will decrease if the external temperature is low. Therefore, external curing temperature is an important factor that should be considered when the sustainability of cementitious composites containing SCBA is assessed.

5. Future Directions and Limitations

This study focused on the effect of curing temperature on the environmental impact of using SCBA as a partial replacement of cement in the production of mortars (cradle-to-gate). Future directions are (i) extending the analysis considering a cradle-to-grave analysis, thus, including durability as a parameter, and (ii) assessing the effect of temperature on the sustainability of other alternative SCMs and wastes. Besides, this study was focused on mortar. Therefore, the analysis of other cementitious composites such as concrete or plaster, as well as a comparison between them to assess the best usage of this waste, will be of great interest. Finally, the principle of this research should be applied to other types of waste.

Author Contributions: Conceptualization, V.F. and M.V.-L.; Methodology, V.F., C.M., N.J. and M.V.-L.; Validation, V.F., M.L.-A., C.M., N.J. and M.V.-L.; Formal analysis, V.F., C.M. and M.V.-L.; Investigation, V.F., M.L.-A. and M.V.-L.; Resources, M.V.-L.; Data curation, V.F. and M.V.-L.; Writing—original draft, V.F. and M.V.-L.; Writing—review & editing, V.F., M.L.-A., C.M., N.J. and M.V.-L.; Visualization, V.F., M.L.-A., C.M. and N.J.; Supervision, M.V.-L.; Funding acquisition, M.V.-L.. All authors have read and agreed to the published version of the manuscript.

Funding: The authors gratefully acknowledge start-up funding from Purdue University (V.F.), (C.M.), (M.L.-A.), and (M.V.-L.). The experiments reported in this study were performed in the Pankow Materials Laboratories at Lyles School of Civil Engineering (Purdue University).

Institutional Review Board Statement: Not applicable.

Data Availability Statement: The data that support the findings of this study are available from the corresponding author upon request.

Conflicts of Interest: The authors declare no conflict of interest.

References

1. Statista. Cement Production Global 2021. 2021. Available online: https://www.statista.com/statistics/1087115/global-cement-production-volume/ (accessed on 31 May 2022).
2. Andrew, R.M.; Peters, G.P. *The Global Carbon Project's Fossil CO_2 Emissions Dataset*; Zenodo: Geneva, Switzerland, 2021. [CrossRef]
3. Imbabi, M.S.; Carrigan, C.; McKenna, S. Trends and developments in green cement and concrete technology. *Int. J. Sustain. Built Environ.* **2012**, *1*, 194–216. [CrossRef]

4. Adesina, A. Recent advances in the concrete industry to reduce its carbon dioxide emissions. *Environ. Chall.* **2020**, *1*, 100004. [CrossRef]
5. Paris, J.M.; Roessler, J.G.; Ferraro, C.C.; Deford, H.D.; Townsend, T.G. A review of waste products utilized as supplements to Portland cement in concrete. *J. Clean. Prod.* **2016**, *121*, 1–18. [CrossRef]
6. Juenger, M.C.G.; Snellings, R.; Bernal, S.A. Supplementary cementitious materials: New sources, characterization, and performance insights. *Cem. Concr. Res.* **2019**, *122*, 257–273. [CrossRef]
7. Li, G.; Zhou, C.; Ahmad, W.; Usanova, K.I.; Karelina, M.; Mohamed, A.M.; Khallaf, R. Fly Ash Application as Supplementary Cementitious Material: A Review. *Materials* **2022**, *15*, 2664. [CrossRef]
8. U.S. Energy Information Administration—Independent Statistics and Analysis. Nearly a Quarter of the Operating U.S. Coal-Fired Fleet Scheduled to Retire by 2029. Available online: https://www.eia.gov/todayinenergy/detail.php?id=54559# (accessed on 10 November 2022).
9. Thomas, B.S.; Yang, J.; Mo, K.H.; Abdalla, J.A.; Hawileh, R.A.; Ariyachandra, E. Biomass ashes from agricultural wastes as supplementary cementitious materials or aggregate replacement in cement/geopolymer concrete: A comprehensive review. *J. Build. Eng.* **2021**, *40*, 102332. [CrossRef]
10. Kolawole, J.T.; Babafemi, A.J.; Fanijo, E.; Chandra Paul, S.; Combrinck, R. State-of-the-art review on the use of sugarcane bagasse ash in cementitious materials. *Cem. Concr. Compos.* **2021**, *118*, 103975. [CrossRef]
11. Asocaña. *Más Que Azúcar, una Fuente de Energía Renovable Para el País*; Sect. Agroindustral la Caña: Cali, Colombia, 2017; Volume 10.
12. Pratap, S.; Jawaid, M.; Chandrasekar, M.; Senthilkumar, K.; Yadav, B.; Saba, N.; Siengchin, S. Sugarcane wastes into commercial products: Processing methods, production optimization and challenges. *J. Clean. Prod.* **2021**, *328*, 129453. [CrossRef]
13. Cherubin, M.R.; Franchi, M.R.A.; de Lima, R.P.; de Moraes, M.T.; da Luz, F.B. Sugarcane straw effects on soil compaction susceptibility. *Soil Tillage Res.* **2021**, *212*, 105066. [CrossRef]
14. FAO. FAOSTAT—Production Quantities of Sugar Cane by Country. 2020. Available online: https://www.fao.org/faostat/en/#data/QCL/visualize (accessed on 31 May 2022).
15. Statista. Global Sugar Cane Production 2020. 2020. Available online: https://www.statista.com/statistics/249604/sugar-cane-production-worldwide/ (accessed on 31 May 2022).
16. de Almeida, G.M.; Pereira, G.T.; Bahia, A.S.R.d.S.; Fernandes, K.; Marques Júnior, J. Machine learning in the prediction of sugarcane production environments. *Comput. Electron. Agric.* **2021**, *190*, 106452. [CrossRef]
17. Staples, M.; Malina, R.; Barrett, S. The limits of bioenergy for mitigating global life-cycle greenhouse gas emissions from fossil fuels. *Nat. Energy* **2017**, *2*, 16202. [CrossRef]
18. de Souza, N.R.D.; Duft, D.G.; Bruno, K.M.B.; Henzler, D.d.S.; Junqueira, T.L.; Cavalett, O.; Hernandes, T.A.D. Unraveling the potential of sugarcane electricity for climate change mitigation in Brazil. *Resour. Conserv. Recycl.* **2021**, *175*, 105878. [CrossRef]
19. Jittin, V.; Minnu, S.N.; Bahurudeen, A. Potential of sugarcane bagasse ash as supplementary cementitious material and comparison with currently used rice husk ash. *Constr. Build. Mater.* **2021**, *273*, 121679. [CrossRef]
20. Montakarntiwong, K.; Chusilp, N.; Tangchirapat, W.; Jaturapitakkul, C. Strength and heat evolution of concretes containing bagasse ash from thermal power plants in sugar industry. *Mater. Des.* **2013**, *49*, 414–420. [CrossRef]
21. Medina, J.M.; Sáez del Bosque, I.F.; Frías, M.; Sánchez de Rojas, M.I.; Medina, C. Design and properties of eco-friendly binary mortars containing ash from biomass-fuelled power plants. *Cem. Concr. Compos.* **2019**, *104*, 103372. [CrossRef]
22. Yogitha, B.; Karthikeyan, M.; Muni Reddy, M.G. Progress of sugarcane bagasse ash applications in production of Eco-Friendly concrete—Review. *Mater. Today Proc.* **2020**, *33*, 695–699. [CrossRef]
23. Katare, V.D.; Madurwar, M.V. Process standardization of sugarcane bagasse ash to develop durable high-volume ash concrete. *J. Build. Eng.* **2021**, *39*, 102151. [CrossRef]
24. González-Kunz, R.N.; Pineda, P.; Bras, A.; Morillas, L. Plant biomass ashes in cement-based building materials. Feasibility as eco-efficient structural mortars and grouts. *Sustain. Cities Soc.* **2017**, *31*, 151–172. [CrossRef]
25. Athira, G.; Bahurudeen, A.; Sahu, P.K.; Santhanam, M.; Nanthagopalan, P.; Lalu, S. Effective utilization of sugar industry waste in Indian construction sector: A geospatial approach. *J. Mater. Cycles Waste Manag.* **2020**, *22*, 724–736. [CrossRef]
26. Quedou, P.G.; Wirquin, E.; Bokhoree, C. Sustainable concrete: Potency of sugarcane bagasse ash as a cementitious material in the construction industry. *Case Stud. Constr. Mater.* **2021**, *14*, e00545. [CrossRef]
27. Jha, P.; Sachan, A.K.; Singh, R.P. Agro-waste sugarcane bagasse ash (ScBA) as partial replacement of binder material in concrete. *Mater. Today Proc.* **2021**, *44*, 419–427. [CrossRef]
28. Barbosa, F.L.; Cordeiro, G.C. Partial cement replacement by different sugar cane bagasse ashes: Hydration-related properties, compressive strength and autogenous shrinkage. *Constr. Build. Mater.* **2021**, *272*, 8–11. [CrossRef]
29. *ISO 14040:2006*; Environmental Management—Life Cycle Assessment—Principles and Framework. International Organization for Standardization (ISO): Geneva, Switzerland, 2006. Available online: https://www.iso.org/standard/37456.html (accessed on 31 May 2022).
30. Science Inventory. US EPA. LIFE-CYCLE ASSESSMENT. 2008. Available online: https://cfpub.epa.gov/si/si_public_record_report.cfm?Lab=NRMRL&count=10000&dirEntryId=156704&searchall=&showcriteria=2&simplesearch=0&timstype= (accessed on 31 May 2022).

31. Fairbairn, E.M.R.; De Paula, T.P.; Cordeiro, G.C.; Americano, B.B.; Toledo Filho, R.D. Evaluation of partial clinker replacement by sugar cane bagasse ash: CO2 emission reductions and potential for carbon credits. *Rev. IBRACON Estrut. Mater.* **2012**, *5*, 229–251. [CrossRef]
32. Jamora, J.B.; Gudia, S.E.L.; Go, A.W.; Giduquio, M.B.; Orilla, J.W.A.; Loretero, M.E. Potential reduction of greenhouse gas emission through the use of sugarcane ash in cement-based industries: A case in the Philippines. *J. Clean. Prod.* **2019**, *239*, 118072. [CrossRef]
33. Sinoh, S.S.; Ibrahim, Z.; Othman, F.; Kuang, L.M.; Zaki, A. Life Cycle Assessment of Sugarcane Bagasse Ash as Partial Cement Replacement in Concrete. In Proceedings of the 4th International Conference on Sustainable Innovation 2020–Technology, Engineering and Agriculture (ICoSITEA 2020), Yogyakarta, Indonesia, 13–14 October 2020; Volume 199, pp. 144–150. [CrossRef]
34. Martínez-Lage, I.; Vázquez-Burgo, P.; Velay-Lizancos, M. Sustainability evaluation of concretes with mixed recycled aggregate based on holistic approach: Technical, economic and environmental analysis. *Waste Manag.* **2020**, *104*, 9–19. [CrossRef] [PubMed]
35. Moro, C.; Francioso, V.; Schager, M.; Velay-Lizancos, M.; Schrager, M. TiO_2 nanoparticles influence on the environmental performance of natural and recycled mortars: A life cycle assessment. *Environ. Impact Assess. Rev.* **2020**, *84*, 106430. [CrossRef]
36. Velay-Lizancos, M.; Martinez-Lage, I.; Azenha, M.; Vázquez-Burgo, P. Influence of temperature in the evolution of compressive strength and in its correlations with UPV in eco-concretes with recycled materials. *Constr. Build. Mater.* **2016**, *124*, 276–286. [CrossRef]
37. Chajec, A.; Chowaniec, A.; Królicka, A.; Sadowski, Ł.; Żak, A.; Piechowka-Mielnik, M.; Šavija, B. Engineering of green cementitious composites modified with siliceous fly ash: Understanding the importance of curing conditions. *Constr. Build. Mater.* **2021**, *313*, 125209. [CrossRef]
38. Francioso, V.; Moro, C.; Martinez-Lage, I.; Velay-Lizancos, M. Curing temperature: A key factor that changes the effect of TiO_2 nanoparticles on mechanical properties, calcium hydroxide formation and pore structure of cement mortars. *Cem. Concr. Compos.* **2019**, *104*, 103374. [CrossRef]
39. Yang, J.; Fan, J.; Kong, B.; Cai, C.S.; Chen, K. Theory and application of new automated concrete curing system. *J. Build. Eng.* **2018**, *17*, 125–134. [CrossRef]
40. Zhang, Z.; Wang, Q.; Chen, H. Properties of high-volume limestone powder concrete under standard curing and steam-curing conditions. *Powder Technol.* **2016**, *301*, 16–25. [CrossRef]
41. Hamzah, S.; Aprianti, E. The Effect of Supplementary Cementitious Material Using Thermal Method. In *Sustainable Future for Human Security*; Springer: Berlin/Heidelberg, Germany, 2017. [CrossRef]
42. Kim, T.; Rens, K.L. Concrete Maturity Method Using Variable Temperature Curing for Normal and High-Strength Concrete. I: Experimental Study. *J. Mater. Civ. Eng.* **2008**, *20*, 727–734. [CrossRef]
43. Yousuf, S.; Shafigh, P.; Ibrahim, Z.; Hashim, H.; Panjehpour, M. Crossover effect in cement-based materials: A review. *Appl. Sci.* **2019**, *9*, 2776. [CrossRef]
44. Jiang, P.; Jiang, L.; Zha, J.; Song, Z. Influence of temperature history on chloride diffusion in high volume fly ash concrete. *Constr. Build. Mater.* **2017**, *144*, 677–685. [CrossRef]
45. Wang, M.; Xie, Y.; Long, G.; Ma, C.; Zeng, X. Microhardness characteristics of high-strength cement paste and interfacial transition zone at different curing regimes. *Constr. Build. Mater.* **2019**, *221*, 151–162. [CrossRef]
46. Murugesan, T.; Vidjeapriya, R.; Bahurudeen, A. Reuse of Silica Rich Sugarcane Bagasse Ash in Concrete and Influence of Different Curing on the Performance of Concrete. *Silicon* **2021**, *14*, 3069–3080. [CrossRef]
47. Rajasekar, A.; Arunachalam, K.; Kottaisamy, M.; Saraswathy, V. Durability characteristics of Ultra High Strength Concrete with treated sugarcane bagasse ash. *Constr. Build. Mater.* **2018**, *171*, 350–356. [CrossRef]
48. *ASTM International C33/C33M-18*; Standard Specification for Concrete Aggregates. ASTM International: West Conshohocken, PA, USA, 2018. [CrossRef]
49. *ASTM International C618-22*; Standard Specification for Coal Fly Ash and Raw or Calcined Natural Pozzolan for Use in Concrete. ASTM International: West Conshohocken, PA, USA, 2022. [CrossRef]
50. *ASTM International C349-18*; Standard Test Method for Compressive Strength of Hydraulic-Cement Mortars (Using Portions of Prisms Broken in Flexure). ASTM International: West Conshohocken, PA, USA, 2018. [CrossRef]
51. *ISO 14044:2006*; Environmental management—Life Cycle Assessment—Requirements and Guidelines. International Organization for Standardization (ISO): Geneva, Switzerland, 2006. Available online: https://www.iso.org/standard/38498.html (accessed on 31 May 2022).
52. European Platform on Life Cycle Assessment. 2018. Available online: https://eplca.jrc.ec.europa.eu/ (accessed on 31 May 2022).
53. Marinković, S.; Dragaš, J.; Ignjatović, I.; Tošić, N. Environmental assessment of green concretes for structural use. *J. Clean. Prod.* **2017**, *154*, 633–649. [CrossRef]
54. Marceau, M.L.; Nisbet, M.A.; Vangeem, M.G. *Lyfe Cycle Inventory of Portland Cement Concrete*; Portland Cement Association: Skokie, IL, USA, 2007.
55. U.S. Energy Information Administration (EIA). Biomass and the Environment. 2021. Available online: https://www.eia.gov/energyexplained/biomass/biomass-and-the-environment.php (accessed on 31 May 2022).
56. Kurda, R.; Silvestre, J.D.; de Brito, J. Life cycle assessment of concrete made with high volume of recycled concrete aggregates and fly ash. *Resour. Conserv. Recycl.* **2018**, *139*, 407–417. [CrossRef]

57. Rodrigues, P.; Silvestre, J.D.; Flores-Colen, I.; Viegas, C.A.; de Brito, J.; Kurad, R.; Demertzi, M. Methodology for the Assessment of the Ecotoxicological Potential of Construction Materials. *Materials* **2017**, *10*, 649. [CrossRef]
58. Ryberg, M.; Vieira, M.D.M.; Zgola, M.; Bare, J.; Rosenbaum, R.K. Updated US and Canadian normalization factors for TRACI 2.1. *Clean Technol. Environ. Policy* **2014**, *16*, 329–339. [CrossRef]
59. US EPA. Overview of Greenhouse Gases Nitrous Oxide Emissions. pp. 2–3. 2016. Available online: https://www.epa.gov/ghgemissions/overview-greenhouse-gases (accessed on 31 May 2022).
60. Kim, T.; Tae, S.; Chae, C.U.; Tam, V.W.Y.; Le, K.N.; Shen, L. Analysis of Environmental Impact for Concrete Using LCA by Varying the Recycling Components, the Compressive Strength and the Admixture Material Mixing. *Sustainability* **2016**, *8*, 389. [CrossRef]
61. Haagen-Smit, A.J.; Bradley, C.E.; Fox, M.M. Ozone Formation in Photochemical Oxidation of Organic Substances. *Ind. Eng. Chem.* **2002**, *45*, 2086–2089. [CrossRef]
62. Dincer, I.; Abu-Rayash, A. Sustainability modeling. In *Energy Sustainability*; Academic Press: Cambridge, MA, USA, 2020; pp. 119–164. [CrossRef]
63. Wetzel, R. *Limnology—Lake and River Ecosystems*; Academic Press: Cambridge, MA, USA, 2001.
64. Alexander, K.M.; Taplin, J.H. Concrete strength, paste strength, cement hydration, and the maturity rule. *Aust. J. Appl. Sci.* **1962**, *13*, 277–284.

sustainability

MDPI

Article

Remaining Fatigue Life Predictions of Railway Prestressed Concrete Sleepers Considering Time-Dependent Surface Abrasion

Dan Li [1,2,3], Sakdirat Kaewunruen [1,2,*] and Ruilin You [4]

1 Department of Civil Engineering, School of Engineering, University of Birmingham, Birmingham B15 2TT, UK
2 TOFU Lab (Track Engineering and Operations for Future Uncertainties), School of Engineering, University of Birmingham, Birmingham B15 2TT, UK
3 Structural Engineering Institute, East China Architectural Design & Research Institute, Shanghai 200002, China
4 Railway Engineering Institute, China Academy of Railway Sciences, Beijing 100081, China
* Correspondence: s.kaewunruen@bham.ac.uk

Abstract: One of the safety-critical components of ballasted track systems is railway sleepers whose main functions are to (i) transfer vertical load, (ii) maintain rail gauge, and (iii) restrain longitudinal rail movement. Railway sleepers can be manufactured using timber, concrete, steel, composite, and any other engineered materials. Prestressed concrete sleepers are the most commonly used type worldwide because of their superior value-for-money performance. In practice, railway sleepers experience thousands of cycles of aggressive wheel–rail dynamic loads and wear deterioration can be observed over their service life. Not only does the deterioration affect track quality and geometries, but it also undermines the structural integrity of the track structures. The wear and abrasion directly decrease the capacity of railway sleepers, resulting in the reduction in service life. In this paper, the emphasis is placed on the assessment of the fatigue life of prestressed concrete railway sleepers with imperfect geometry. This study is the world's first to establish a new fatigue simulation of railway concrete sleepers considering accumulative non-constant amplitudes, which has been validated using full-scale experimental results and empirical analyses. Parametric studies have been conducted to obtain new insights into the fatigue performance of the worn sleepers. The new findings will improve railway sleeper maintenance and inspection criteria, and will provide a new guideline on track-condition monitoring networks.

Keywords: railway concrete sleepers; fatigue; S–N curve; finite element method (FEM); geometric damage; abrasion; remaining service life

check for updates

Citation: Li, D.; Kaewunruen, S.; You, R. Remaining Fatigue Life Predictions of Railway Prestressed Concrete Sleepers Considering Time-Dependent Surface Abrasion. *Sustainability* **2022**, *14*, 11237. https://doi.org/10.3390/su141811237

Academic Editors: Sandra Barbosa Nunes and Woubishet Zewdu Taffese

Received: 3 August 2022
Accepted: 3 September 2022
Published: 8 September 2022

1. Introduction

At present, railway transport is one of the most sustainable and commonly used forms of transport for passengers and goods [1]. Conventional ballasted track systems can be divided into superstructures and substructures. The superstructure includes the rails, fastening systems, railway sleepers, rail pads, and ballast. The substructure consists of the sub-ballast, formation, subgrade, and structural fills. A railway sleeper is a safety-critical component, which transfers the force from the wheels to the substructure [2–4]. Prestressed concrete sleepers are the most widely used type of sleeper materials because of their good structural performance, low cost, and ease of maintenance [5]. However, repeated loads from wheel–rail interactions could result in accumulated damage on railway sleepers, which reduces the service life. Both the static and dynamic load-carrying capacities of railway sleepers can degrade over time. In addition, early deterioration could lead to premature failure of railway sleepers. Such problems are a form of serviceability limit states, which have become of great concern over the past few decades for railway engineers.

The most common causes of concrete sleeper failures have been ranked using the results obtained from a worldwide survey shown in Table 1 [6,7], where it can be seen that deterioration, abrasion, fatigue, and cracking are the most critical problems for railway sleepers. Such problems can lead to a combined reaction that accelerates the failure of railway sleepers. In limit-states design practice, the fatigue limit state shall be considered.

Table 1. Most common causes of concrete sleeper failures [6].

Main Causes	Problems	Worldwide Response [1]
Lateral load	• Abrasion on rail-seat	3.15
	• Shoulder/fastening system wear or fatigue	5.5
Vertical dynamic load	• Cracking from dynamic loads	5.21
	• Derailment damage	4.57
Manufacturing and maintenance defects	• Cracking from center binding	5.36
	• Tamping damage	6.14
	• Others (e.g., manufactured defects)	4.09
Environmental considerations	• Cracking from environmental or chemical degradation	4.67

[1] Ranked from 1 to 8, with 8 being the most critical.

In order to investigate the serviceability performance of prestressed concrete sleepers, the dynamic condition needs to be investigated. In reality, a dynamic load is usually related to the rail–wheel interaction and train speed [8]. Railway sleepers often experience impact loading due to wheel–rail interactions associated with abnormalities in either a wheel or a rail [9]. The high dynamic impact forces induced by wheel–rail irregularities could greatly exceed the static wheel load, which could cause deterioration. The most common forms of wear are rail-seat or soffit abrasion at mid-span [10]. These deterioration mechanisms can be observed in the field. Extensive experimental and analytical investigations have been conducted by Kaewunruen and Remennikov on the dynamic behavior of prestressed concrete sleepers [11–13]. In their research, the impact damage and failure patterns of prestressed concrete sleepers were studied, which provided a database for life cycle investigation. Chen et al. presented an experimentally validated three-dimensional finite element model of a prestressed concrete sleeper that can be used to study and improve the design and performance [14]. Bastos et al. investigated the deterioration of prestressed concrete sleepers under cyclic loads and exposed to moisture [15]. Parvez et al. investigated the efficiency of using steel fibers to improve the fatigue performance of prestressed concrete sleepers [16]. Riding et al. conducted a study of environmental and track factors that influenced the abrasion damage of prestressed concrete sleepers [17]. Kernes et al. investigated the mechanics of rail-seat deterioration and methods to improve the abrasion resistance of concrete sleepers at rail seats [18]. It is noted that, for high-quality concrete sleepers, most studies did not observe any bond slip in the prestress wires under static and high-cycle fatigue tests. This is due to the fact that, in various modern design standards (such as in Europe, Australia, Japan, and China), there is a restriction in the allowable dimension of the prestress wires which helps it to increase the total bond perimeters and restrict any slip. In recent studies, Ngamkhanong et al. studied the effect of surface abrasion on the impact capacity of prestressed concrete sleepers [19,20]. Meanwhile, You et al. developed a nonlinear finite element model for determining the structural capacity of a prestressed concrete sleeper with rail-seat abrasion [21]. Further, Li et al. investigated creep and shrinkage of prestressed concrete sleepers influenced by surface abrasion [22]. Despite a number of previous studies into concrete sleepers and fatigue life, most studies focus on a general understanding in order to establish an empirical fatigue-life estimation that is relatively specific to the cases. On this ground, many factors related to time-dependent variables (such as wear, creep, and shrinkage, and so on) cannot be properly identified. This study has thus embarked on the development of nonlinear models of concrete sleepers capable of fatigue failure analyses and for virtual tests of the time-dependent factors.

Numerous numerical and experimental investigations have been initially conducted to establish rational railway-sleeper models by the authors [23]. Our studies into the fatigue life of concrete sleepers originated with the damage accumulation method [24,25]. This tailored method is extended from Miner's rule. Traditionally, the cumulative damage under various loads is regarded as equivalent fatigue stress with constant amplitude [26]. Our study is unique in the way that the fatigue rule has been modified to enable iterative and variable stress fields (or non-constant amplitude), which can be more realistic and suitable for track load spectra. The critical literature review reveals that the fatigue life-cycles of prestressed concrete sleepers have not been fully studied, especially when the sleepers are deteriorated by excessive wear [27,28]. This paper is thus the world's first to investigate and present an advanced railway concrete-sleeper model capable of parametric analysis into the effect of surface abrasion on the fatigue life of prestressed concrete sleepers. This model is highly original since it can cope with the accumulative, non-constant amplitudes of fatigue loads. The emphasis of this paper is placed on the fatigue life of prestressed concrete sleepers when subjected to wear or abrasion in comparison with intact railway sleepers. The new findings and insights will help track engineers to have a better understanding of the remaining fatigue life of damaged sleepers and provide a guideline to identify appropriate damage-detection technology for track systems. The new insight into fatigue life will also improve the safety and reliability of railway infrastructure.

2. Theoretical Life-Cycle Assessment Method of Prestressed Concrete Sleepers

2.1. Damage Accumulation Method

You et al. [29,30] demonstrated a new approach based on the damage accumulation method to analyze the life-cycle of prestressed concrete sleepers. The cumulative damage index is calculated by Miner's rule:

$$\sum D_i = \sum_i \frac{n(\Delta\sigma_i)}{N(\Delta\sigma_i)} \tag{1}$$

where $n(\Delta\sigma_i)$ is the applied number of cycles at a stress range $\Delta\sigma_i$; $N(\Delta\sigma_i)$ is the resisting number of cycles at a stress range $\Delta\sigma_i$.

The fatigue life can be determined by:

$$L_f = \frac{1}{D_i} \tag{2}$$

2.2. Material Properties of Fatigue

The S–N curve can be defined as the number of cycles to failure when a material is repeatedly loaded through an alternating stress range. Therefore, the fatigue life of prestressed concrete sleepers can be determined as the maximum applied number of cycles to failure. [24,31]. The typical S–N curve for concrete and prestressing tendons is illustrated in Figures 1 and 2.

$$S_{c,\,max} = \frac{|\sigma_{c,max}|}{f_{ck,fat}} \tag{3}$$

$$S_{c,\ min} = \frac{|\sigma_{c,min}|}{f_{ck,fat}} \quad (4)$$

where $S_{c,\ max}$ is the maximum compressive stress level; $S_{c,\ min}$ is the minimum compressive stress level; $\sigma_{c,max}$ is the maximum compressive stress; $\sigma_{c,min}$ is the minimum compressive stress; $f_{ck,fat}$ is the fatigue-reference compressive strength.

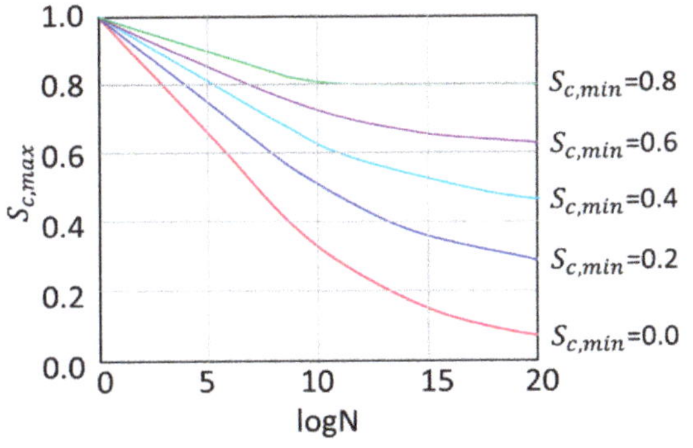

Figure 1. S–N curve for concrete [31].

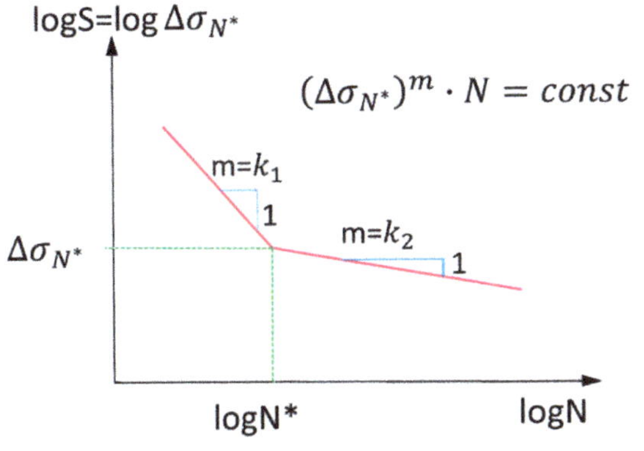

Figure 2. S–N curve for prestressing tendons [31].

If $(\Delta\sigma > \Delta\sigma_{N^*})$

$$\log N_f = \log N^* - k_1[\log(\Delta\sigma) - \log(\Delta\sigma_{N^*})] \quad (5)$$

If $(\Delta\sigma < \Delta\sigma_{N^*})$

$$\log N_f = \log N^* + k_2[\log(\Delta\sigma_{N^*}) - \log(\Delta\sigma)] \quad (6)$$

where $\Delta\sigma$ is the stress range of the prestressing steel; $\Delta\sigma_{N^*}$ is the stress range at N^* cycles; k_1, k_2 are the stress exponents. Table 2 illustrates the parameters of prestressing steel S–N curve.

Table 2. Parameters of S–N curve for prestressing steel [31].

S–N Curve of Prestressing Steel Used for	Stress Exponent			$\Delta\sigma_{N^*}$ at N^* Cycles (MPa)
	N^*	k_1	k_2	
Pre-Tensioning	10^6	5	9	185

2.3. Fatigue Life Assessment

In fatigue life assessment, there are two steps: cracking load calculation and fatigue life calculation. Initially, the stress range at the bottom fiber needs to be analyzed by:

$$\sigma_{cF}^{b} = \frac{nA_{ps}\sigma_{se}}{A_t} + \frac{nA_{ps}\sigma_{se}e}{I_t}y_t \tag{7}$$

where σ_{se} is the effective stress at each prestressing steel; A_{ps} is the cross-section area of a prestressing steel; e is the eccentricity; A_t is the transformed area of the sleeper; I_t is the inertia moment of transformed section before cracking; y_t is the distance of the centroidal axis of the transformed area from the soffit.

The cracking moment is calculated by:

$$M_{cr} = I_t\frac{\sigma_{cF}^{b}+f_{cf}}{y_t} \tag{8}$$

where f_{cf} is the tensile strength.

The neutral axis of the cross-section of the sleeper starts changing when the cracking propagates. Therefore, the distance from the center of gravity of the effective transformed area to the top of the compressed area can be calculated by:

$$y_{CG} = root[[S_{pcII} - n_eA'_{p3}(h-y_{cg}-d_3) - n_eA'_{p2}(h-y_{cg}-d_2) - n_eA'_{p1}(h-y_{cg}-d_1)], y_{cg}] \tag{9}$$

where the S_{pcII} is the first moment at the bottom fiber after cracking; A'_{pi} is the total area of the prestressed steel at layer i; d_i is the distance from the prestressed steel at layer i to the bottom of the steel area; and n_e is the modular ratio.

The effective transformed section can be estimated using the transformed area of the sleeper cross-section A_{cII}:

$$A_{tII} = A_{cII} + n_eA_p \tag{10}$$

The moment of inertia of the cracking section is presented by:

$$I_{cr} = I_{ccr} + n_eA'_{p3}(h-y_{cg}-d_3)^2 + n_eA'_{p2}(h-y_{cg}-d_2)^2 + n_eA'_{p1}(h-y_{cg}-d_1)^2 \tag{11}$$

The effective inertia moment in the lifetime is given by:

$$I_{ef} = I_{cr} + (I_t - I_{cr})\left(\frac{M_{cr}}{M_{max}}\right)^3 \tag{12}$$

where I_t is the inertia moment of transformed section before cracking; M_{cr} is the cracking moment; M_{max} is the maximum bending moment at the section under cyclic loads:

$$\Delta\sigma_{pt1} = n_e\frac{M_{max}-M_{min}}{I_{ef}}(h-y_{cg}-d_1) \tag{13}$$

where M_{min} is the minimum bending moment at the section under cyclic loads.

Using the output value of $\Delta\sigma_{pt1}$, the failure cycles of the prestressing steel under constant cyclic loading can be estimated by Equations (5) or (6).

3. Numerical Life-Cycle Assessment Method of Prestressed Concrete Sleepers

3.1. Fatigue Analysis Decisions

Numerical fatigue analysis can be divided into five common stages listed in Figure 3. This figure illustrates the general flow of decisions required to perform a fatigue analysis.

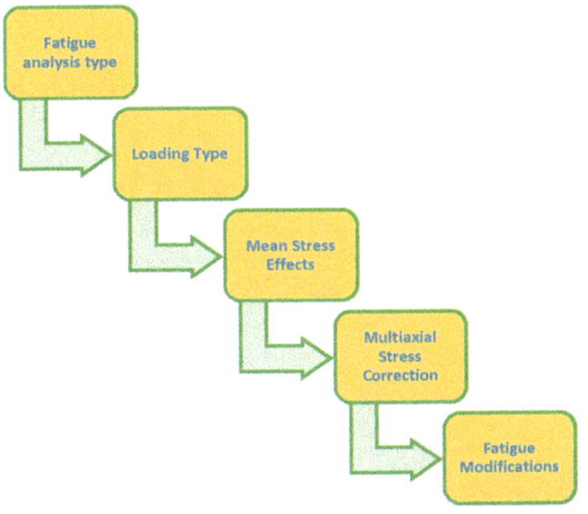

Figure 3. Flowchart for fatigue analysis decision tree.

Fatigue analysis consists of ascertaining the stress life and strain life. Stress life is based on empirical S–N curves and then modified by a variety of factors. Strain life is based on the strain-life relation equation. The strain-life method is concerned with crack initiation and usually deals with relatively low cycles (less than 10^5 cycles). In this study, the stress-life method is used for life-cycle simulation of the railway sleeper. The stress-life method focuses on total fatigue life, which doesn't distinguish between cracking initiation and progression. The stress-life method is also suitable for simulation of relatively high cycles (more than 10^5 cycles) [32]. Normally, railway sleepers experience millions of cyclic loads during service life.

3.2. Types of Cyclic Loading

There are four types of fatigue loading in the simulation:

- Constant amplitude, proportional loading;
- Constant amplitude, non-proportional loading;
- Non-constant amplitude, proportional loading;
- Non-constant amplitude, non-proportional loading.

Each passed wheel is assumed as a single cycle with constant amplitude. The principal stress axes do not change over time [32]. Therefore, constant amplitude with proportional loading is used for fatigue-life simulation of the railway sleeper. Figure 4 shows the types of loading used in the simulation.

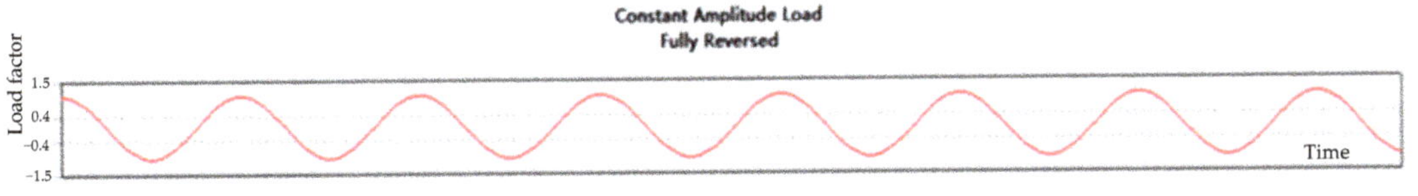

Figure 4. Cyclic loading type for life-cycle simulation of the railway sleeper.

3.3. Fatigue Life Results

In the fatigue analysis, the typical stress analysis can be used to determine the fatigue life of the prestressed concrete sleeper when it is subject to cyclic loads. This method is based on the S–N curve as part of the material definition of railway sleepers. The S–N curve is cyclic stress (S) against the logarithmic scale of cycles to failure (N). In the stress-life method, the alternating stress vs. the number of cycles to failure data are needed to put into the model [32]. It should be noted that the fatigue-life results using the stress-life method represent the available life for the given fatigue analysis without considering crack propagation and factors such as bond-slip, environmental conditions, etc.

4. Prestressed Concrete Sleeper Modelling

4.1. Properties of the Railway Sleeper

This project is a collaborative study between the University of Birmingham and the China Academy of Railway Sciences (CARS). CARS has commissioned full-scale experimental studies to support this collaboration. In this study, the 2600-mm long Chinese Type III prestressed concrete sleeper (Figure 5) was originally designed for high-speed railways, with the properties shown in Table 3. The dimension of the sleeper is approximately 2600 mm × 320 mm × 260 mm including 10 prestressing tendons with 7 mm diameters. The sleeper will be assessed using the numerical and theoretical approaches to calculate the life-cycle.

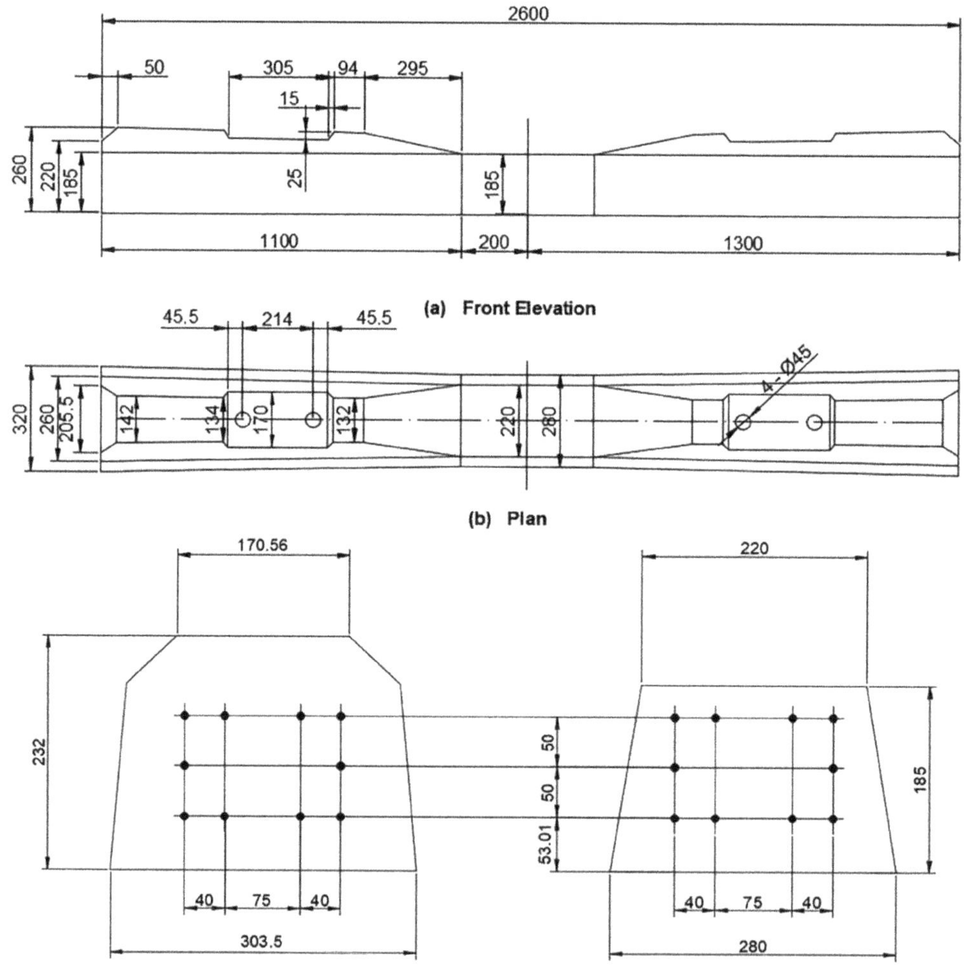

Figure 5. Chinese Type III prestressed concrete sleeper.

Table 3. Material properties of Chinese Type III prestressed concrete sleeper.

Material Properties	Basic Variables	Value
Concrete	• Mean compressive strength	65 MPa
	• Modulus of elasticity	33 GPa
Prestressed wire	• Yield strength	1570 MPa
	• Modulus of elasticity	200 GPa
	• Prestressing force	420 kN

4.2. Finite Element Model

In this study, the finite element method (FEM) is used to analyze the life-cycle of prestressed concrete sleepers with surface abrasion. The prestressed concrete sleeper is modelled in ANSYS Workbench (shown in Figure 6). The finite element model is composed of concrete and prestressed tendons. The Solid65 element is used to model the concrete material. In the model, thermal condition is used to simulate the characteristics of prestressing tendons. The support condition is modelled as fixed support. In the model, the contact condition No Separation is used between concrete and prestressing tendons. Therefore, the concrete and prestressing tendons are considered as being well adhered, while the bond slip and bursting are ignored. The material properties of the FE sleeper model are shown in Table 3. The modelling is performed with the minimum difference from the real sleeper.

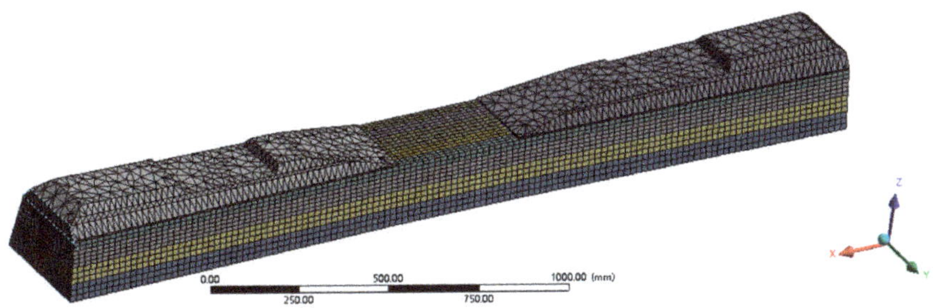

Figure 6. FE sleeper model.

4.3. Experimental Program

The bending moment test at the midspan of the full-scale prestressed concrete sleeper using the digital image correlation method (DIC) conducted by Jing et al. [33], as part of CARS's commissioned tests, is used for the FE model validation. This experiment presented the nonlinear structural performance of Chinese Type III prestressed concrete sleeper. The arrangement for the test follows EN13230-2 shown in Figure 7 [33]. The loading surface was placed at the bottom midspan of the sleeper with a resilient pad. Two rail pads were placed between the rail-seat area and each support. Sadeghi and Barati stated that the normal failure load of concrete sleepers is 140 kN [34], thus the applied load range was between 0 to 140 kN. The speed of loading was no more than 120 kN/min. The experimental load–deflection results in terms of the DIC will be compared with the numerical results for the sleeper model validation.

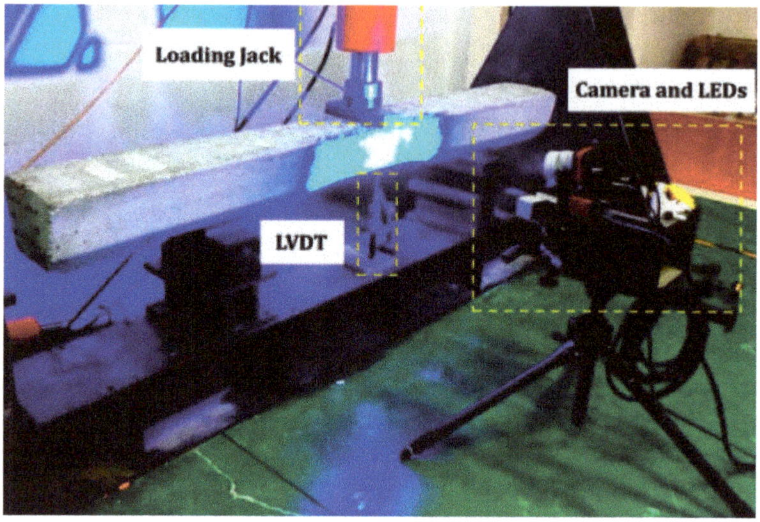

(**a**)

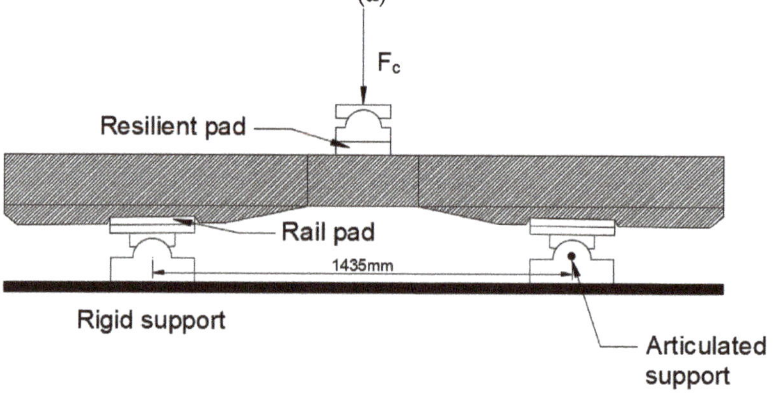

(**b**)

Figure 7. Arrangement for the bending moment test at midspan of the prestressed concrete sleeper [33,35]. (**a**) Experimental apparatus of the bending moment test. (**b**) Setup for the bending moment test.

4.4. FE Sleeper Model Validation

The material and structural performance of the FE sleeper model needs to be validated for further simulation. Previous research has presented the validation of the FE sleeper model in reference [23]. The mesh study of the FE sleeper model is also carried out for the models with 20 mm, 30 mm, and 40 mm mesh size. Figure 8 illustrates the load–deflection responses of the experimental and numerical results. The mesh study shows slight differences between the mesh size 20 mm, 30 mm, and 40 mm. Therefore, the closest results of mesh size 30 mm are selected (with 5.99% max error calculated from the deviation of the experimental and FEM results). The numerical performance indicates a good agreement with the experimental results.

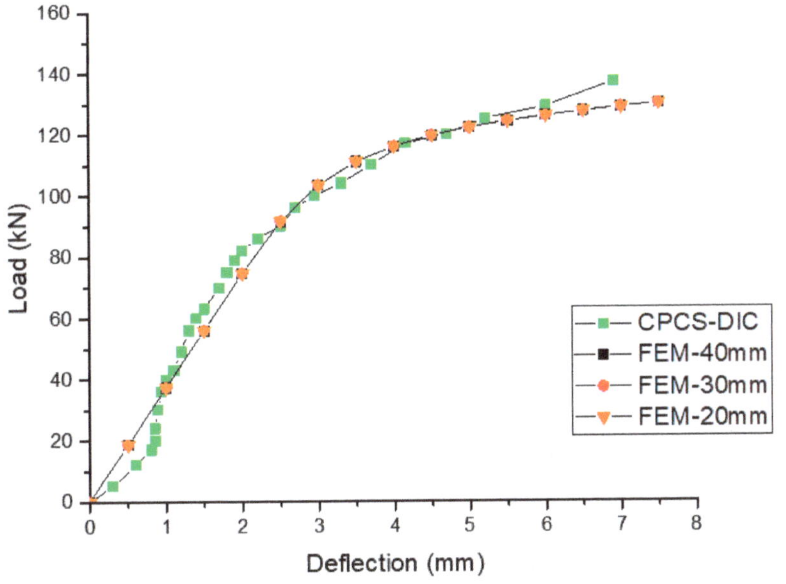

Figure 8. Load–deflection responses of experimental and numerical results.

4.5. Fatigue Model Validation

The life-cycle is evaluated by numerical and theoretical approaches with the cyclic loads between 55 kN and 365 kN. Each cyclic load (constant) can only calculate one fatigue life for both the theoretical and numerical methods. These results present the available life of the prestressed concrete sleepers for the fatigue analysis. The cyclic load is applied at the rail-seat area shown in Figure 9. Note that the static and fatigue tests are in a different test set up. The negative bending test was performed for the static tests (in accordance with EN 13230), whilst the positive rail-seat tests were considered for the fatigue tests. The theoretical results of the life-cycle were according to the damage accumulation method. The calculation details were presented in reference [29]. The results of the numerical and theoretical calculation are demonstrated in Figure 10. Life cycles are inversely proportional with the constant cyclic loads for both the numerical and theoretical results. In comparison between numerical and theoretical results, the average error is only 13.03%.

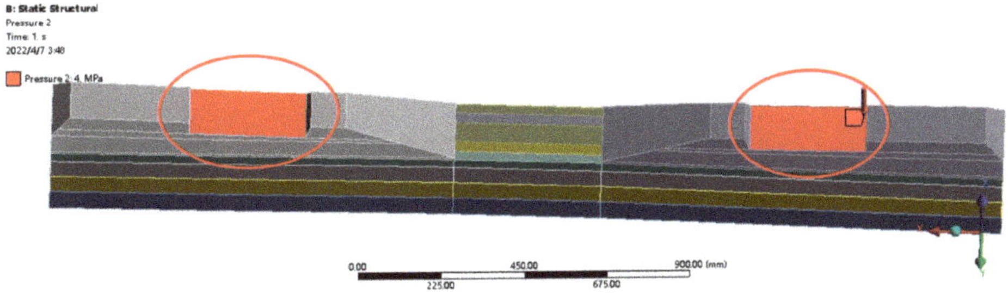

Figure 9. Setup for boundary conditions in the fatigue life simulation. The sleepers in the model are fully supported by ballast and substructure.

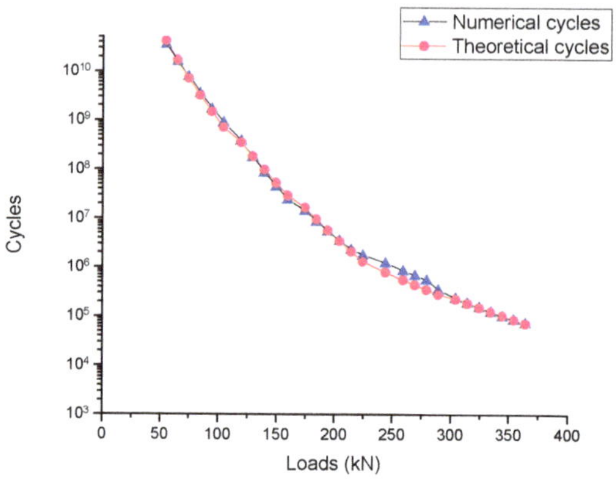

Figure 10. Comparison between numerical and theoretical fatigue-life results.

The experiments conducted by Parvez and Foster investigated failure cycles of prestressed concrete sleepers under cyclic loads [16,36,37]. Two specimens were selected from their experiments in order to validate the numerical fatigue model. The average of the failure cycles of the specimens are calculated and presented in Table 4. Table 5 shows the comparison between the experimental, theoretical, and numerical results.

Table 4. Experimental failure cycle of prestressed concrete sleepers [29].

Specimen ID	Failure Cycles	Average	Standard Deviation
SF2-a	773,793	896,290	173,236
SF3-a	1,018,787		

Table 5. Comparison between experimental, theoretical, and numerical results.

Failure Cycles				
Experimental Result	**Theoretical Result**	**Deviation Ratio%**	**Numerical Result**	**Deviation Ratio%**
896,290	889,577	0.75	849,000	5.28

From Figure 10 and Table 5, it is seen that the numerical results have a good correlation with the experimental and theoretical results, which provides a reliable method for the life-cycle in the surface abrasion study.

5. Influence of Surface Abrasions on Fatigue Life

In reality, railway sleepers experience aggressive dynamic loading, especially in sharp curves and high gradients, which causes rail-seat abrasion. The ballast angularity also results in differential abrasion on the soffit abrasion. The most common surface abrasions of railway sleepers are the rail-seat abrasion, the soffit abrasion at rail-seat, and the soffit abrasion at midspan [10]. Figure 11 shows the typical abrasions of a railway sleeper. This section investigates the life-cycle of the railway sleeper with each typical surface abrasion pattern using the numerical fatigue-life model. In the simulation, abrasions are applied in the FE sleeper model by changing the geometry of the rail-seat area and bottom surface at midspan. It is clear that the life-cycle can be affected by the surface abrasions.

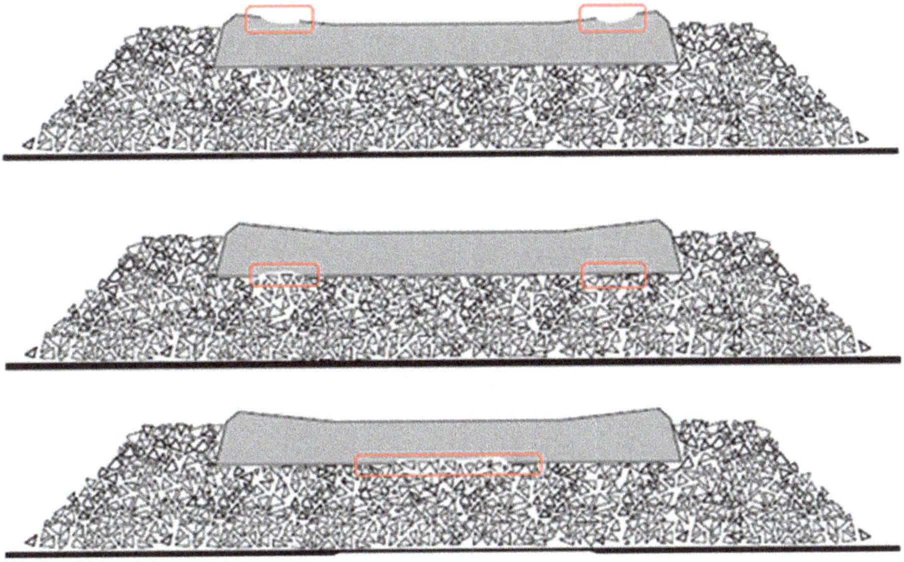

Figure 11. Typical surface abrasions of railway sleepers [10].

5.1. Rail-Seat Abrasion Results

Figure 12 shows the location of the rail-seat abrasion of the railway sleeper model. The depths of the rail-seat abrasions are chosen as 5 mm, 15 mm, and 30 mm to analyze the effect of the life-cycle on the railway sleeper. Loads of between 55 kN to 325 kN are applied in the simulation. Table 6 and Figure 13 indicate the results of the life-cycle with rail-seat abrasion.

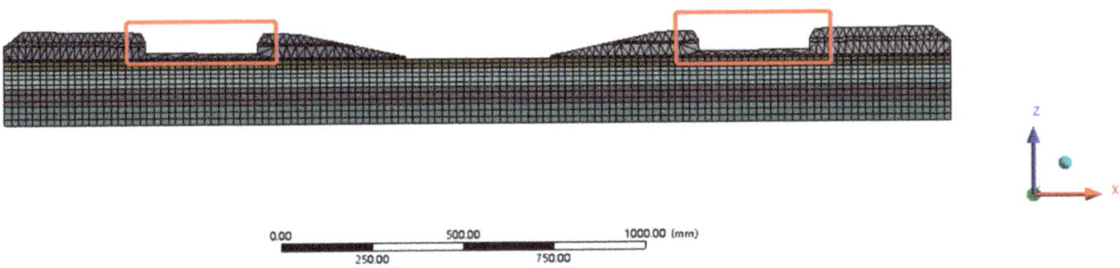

Figure 12. Rail-seat abrasion of the railway sleeper model.

Table 6. Life-cycle results of the railway sleeper with rail-seat abrasion.

Load (kN)	No Abrasion	5 mm Rail-Seat Abrasion	15 mm Rail-Seat Abrasion	30 mm Rail-Seat Abrasion
55	3.47×10^{10}	2.12×10^{10}	1.75×10^{10}	4.15×10^{9}
105	8.79×10^{8}	3.27×10^{8}	2.23×10^{8}	1.83×10^{7}
160	2.39×10^{7}	9.68×10^{6}	6.81×10^{6}	1.01×10^{6}
215	2.30×10^{6}	1.25×10^{6}	9.90×10^{5}	1.23×10^{5}
270	6.79×10^{5}	2.38×10^{5}	1.82×10^{5}	36,627
325	1.52×10^{5}	78,521	62,593	14,068

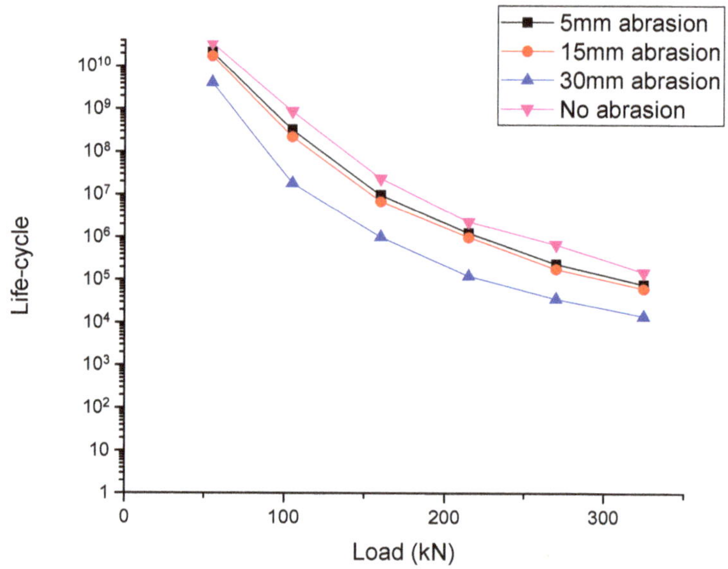

Figure 13. Comparison of life-cycle results with rail-seat abrasion.

5.2. *Soffit Abrasion at Rail-Seat Results*

Figure 14 shows the location of the soffit abrasion at the rail seat of the railway sleeper model. The depths of the soffit abrasions at the rail seat are chosen as 5 mm, 15 mm, and 30 mm to analyze the effect on the life-cycle of the railway sleeper. Loads of between 55 kN to 325 kN are applied in the simulation. Table 7 and Figure 15 indicate the results on the life-cycle with the soffit abrasion at the rail seat.

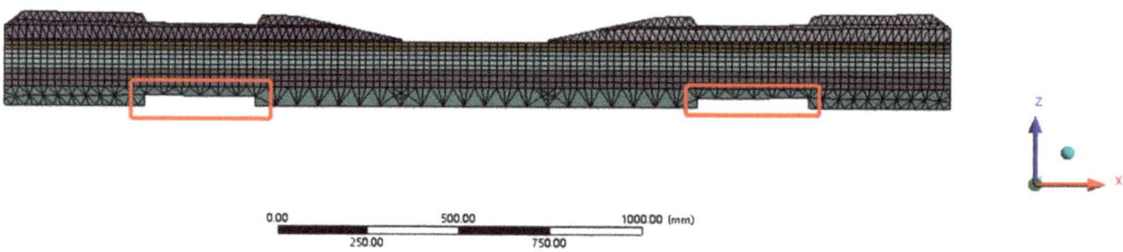

Figure 14. Soffit abrasion at rail seat of the railway sleeper model.

Table 7. Life-cycle results of the railway sleeper with soffit abrasion at rail seat.

Load (kN)	No Abrasion	5 mm Soffit Abrasion at Rail Seat	15 mm Soffit Abrasion at Rail Seat	30 mm Soffit Abrasion at Rail Seat
55	3.47×10^{10}	2.85×10^{9}	1.08×10^{9}	5.04×10^{8}
105	8.79×10^{8}	1.10×10^{7}	2.99×10^{6}	1.63×10^{6}
160	2.39×10^{7}	7.19×10^{5}	1.87×10^{5}	1.02×10^{5}
215	2.30×10^{6}	85,372	36,816	22,380
270	6.79×10^{5}	26,466	11,413	4527
325	1.52×10^{5}	10,165	1673	936

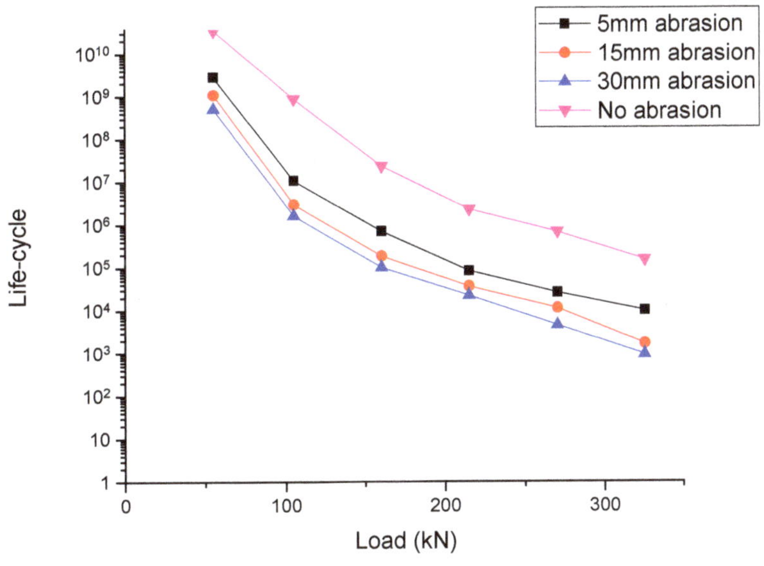

Figure 15. Comparison of life-cycle results with soffit abrasion at rail seat.

5.3. Soffit Abrasion at Midspan Results

Figure 16 shows the location of the soffit abrasion at the midspan of the railway sleeper model. The depths of the soffit abrasion at midspan are chosen as 5 mm, 15 mm, and 30 mm to analyze the effect on the life-cycle of the railway sleeper. Loads of between 55 kN to 325 kN are applied in the simulation. Table 8 and Figure 17 indicate the results on the life-cycle with the soffit abrasion at midspan.

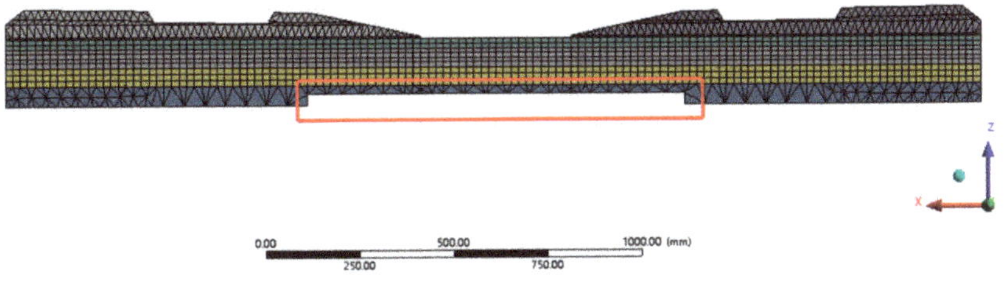

Figure 16. Soffit abrasion at midspan of the railway sleeper model.

Table 8. Life-cycle results of the railway sleeper with soffit abrasion at midspan.

Load (kN)	No Abrasion	5 mm Soffit Abrasion at Midspan	15 mm Soffit Abrasion at Midspan	30 mm Soffit Abrasion at Midspan
55	3.47×10^{10}	4.59×10^{9}	3.04×10^{9}	1.54×10^{9}
105	8.79×10^{8}	2.09×10^{7}	1.20×10^{7}	4.81×10^{6}
160	2.39×10^{7}	1.10×10^{6}	7.61×10^{5}	2.68×10^{5}
215	2.30×10^{6}	1.37×10^{5}	90,168	50,015
270	6.79×10^{5}	39,963	27,953	15,505
325	1.52×10^{5}	15,349	10,736	3250

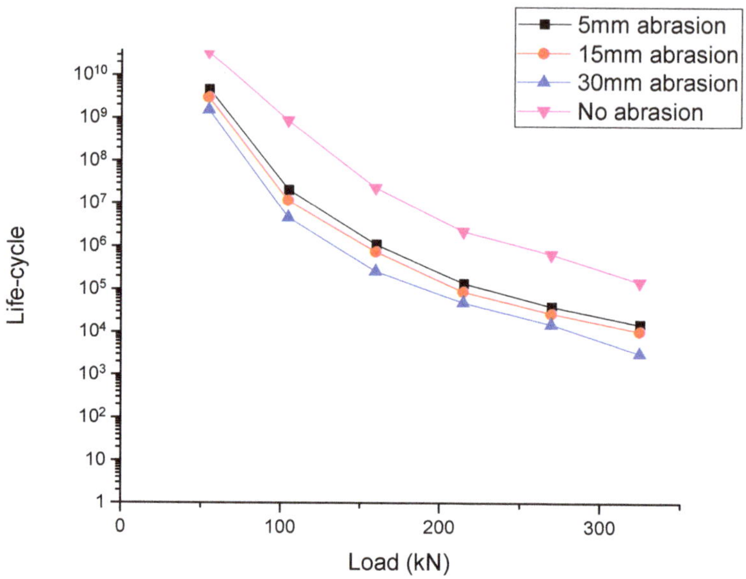

Figure 17. Comparison of life-cycle results with soffit abrasion at midspan.

5.4. Rail-Seat Abrasion and Soffit Abrasion at Rail Seat Results

Figure 18 shows the location of the rail-seat abrasion and the soffit abrasion at the rail seat of the railway sleeper model. The depths of the abrasions are chosen as 5 mm & 5 mm, 15 mm & 15 mm, and 30 mm & 30 mm to analyze the effect on the life-cycle of the railway sleeper. Loads of between 55 kN to 325 kN are applied in the simulation. Table 9 and Figure 19 indicate the results on the life-cycle with the rail-seat abrasion and soffit abrasion at the rail seat.

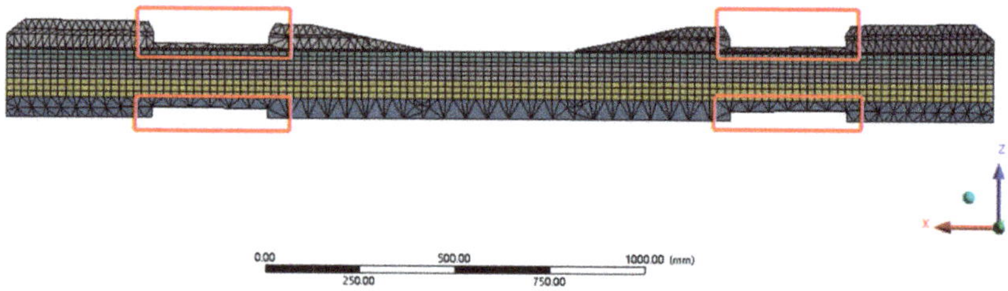

Figure 18. Rail-seat abrasion and soffit abrasion at rail seat of the railway sleeper model.

Table 9. Life-cycle results of the railway sleeper with rail-seat abrasion and soffit abrasion at rail seat.

Load (kN)	No Abrasion	5 mm & 5 mm	15 mm & 15 mm	30 mm & 30 mm
55	3.47×10^{10}	2.25×10^{9}	3.12×10^{8}	3.25×10^{7}
105	8.79×10^{8}	8.01×10^{6}	1.22×10^{6}	1.90×10^{5}
160	2.39×10^{7}	5.80×10^{5}	76,391	20,139
215	2.30×10^{6}	69,485	16,878	1728
270	6.79×10^{5}	21,541	2455	792
325	1.52×10^{5}	6630	871	621

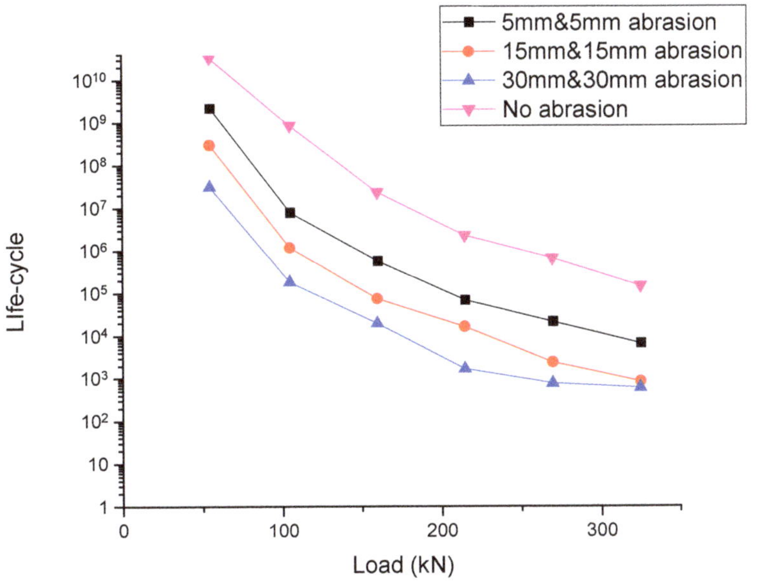

Figure 19. Comparison of life-cycle results with rail-seat abrasion and soffit abrasion at rail seat.

5.5. Rail-Seat Abrasion and Soffit Abrasion at Midspan Results

Figure 20 shows the location of the rail-seat abrasion and the soffit abrasion at midspan of the railway sleeper model. The depths of the abrasions are chosen as 5 mm & 5 mm, 15 mm & 15 mm, and 30 mm & 30 mm to analyze the effect on the life-cycle of the railway sleeper. Loads of between 55 kN to 325 kN are applied in the simulation. Table 10 and Figure 21 indicate the results on the life-cycle with the rail-seat abrasion and soffit abrasion at midspan.

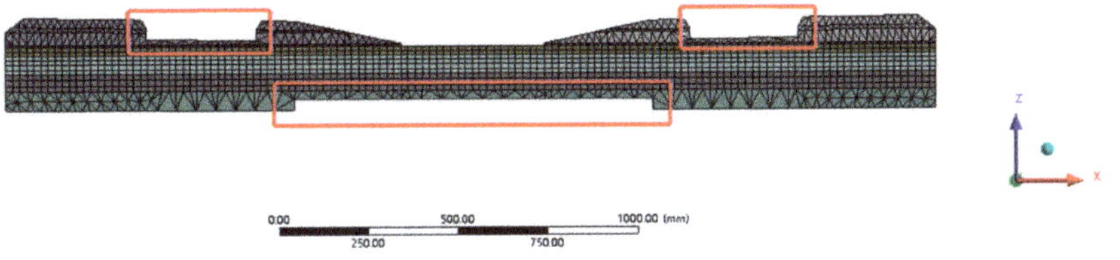

Figure 20. Rail-seat abrasion and soffit abrasion at midspan of the railway sleeper model.

Table 10. Life-cycle results of the railway sleeper with rail-seat abrasion and soffit abrasion at midspan.

Load (kN)	No Abrasion	5 mm & 5 mm	15 mm & 15 mm	30 mm & 30 mm
55	3.47×10^{10}	2.73×10^{9}	1.33×10^{9}	7.74×10^{8}
105	8.79×10^{8}	1.04×10^{7}	3.95×10^{6}	2.12×10^{6}
160	2.39×10^{7}	6.91×10^{5}	2.31×10^{5}	1.39×10^{5}
215	2.30×10^{6}	82,248	44,032	28,812
270	6.79×10^{5}	25,498	13,650	7828
325	1.52×10^{5}	9557	2466	998

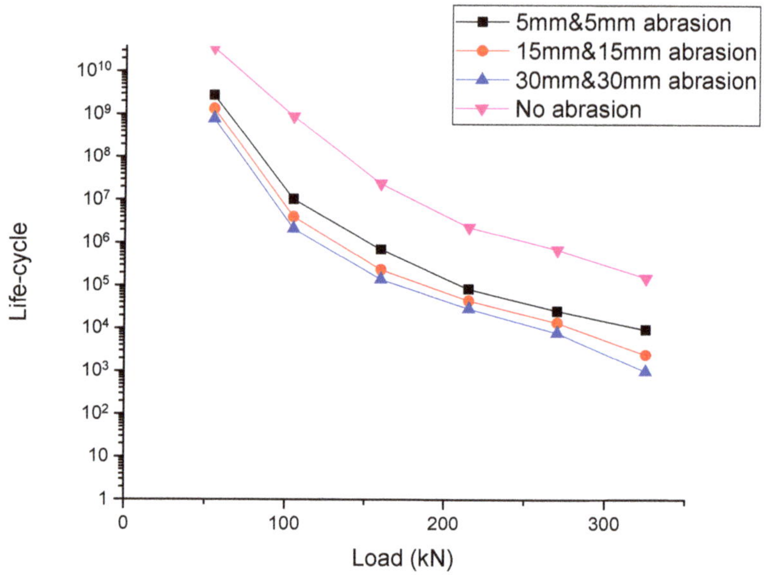

Figure 21. Comparison of life-cycle results with rail-seat abrasion and soffit abrasion at midspan.

5.6. Soffit Abrasion at Rail Seat and Soffit Abrasion at Midspan Results

Figure 22 shows the location of the soffit abrasion at the rail seat and the soffit abrasion at midspan of the railway sleeper model. The depths of the abrasions are chosen as 5 mm & 5 mm, 15 mm & 15 mm, and 30 mm & 30 mm to analyze the effect on the life-cycle of the railway sleeper. Loads of between 55 kN to 325 kN are applied in the simulation. Table 11 and Figure 23 indicate the results on the life-cycle with the soffit abrasion at the rail seat and the soffit abrasion at midspan.

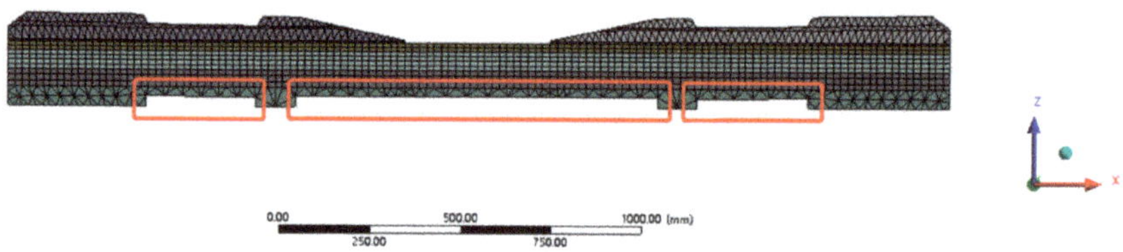

Figure 22. Soffit abrasion at rail seat and soffit abrasion at midspan of the railway sleeper model.

Table 11. Life-cycle results of the railway sleeper with soffit abrasion at rail seat and soffit abrasion at midspan.

Load (kN)	No Abrasion	5 mm & 5 mm	15 mm & 15 mm	30 mm & 30 mm
55	3.47×10^{10}	2.49×10^{9}	6.26×10^{8}	3.37×10^{8}
105	8.79×10^{8}	9.23×10^{6}	1.86×10^{6}	1.28×10^{6}
160	2.39×10^{7}	6.38×10^{5}	1.19×10^{5}	79,896
215	2.30×10^{6}	76,113	25,415	17,652
270	6.79×10^{5}	23,596	5963	2706
325	1.52×10^{5}	8078	966	881

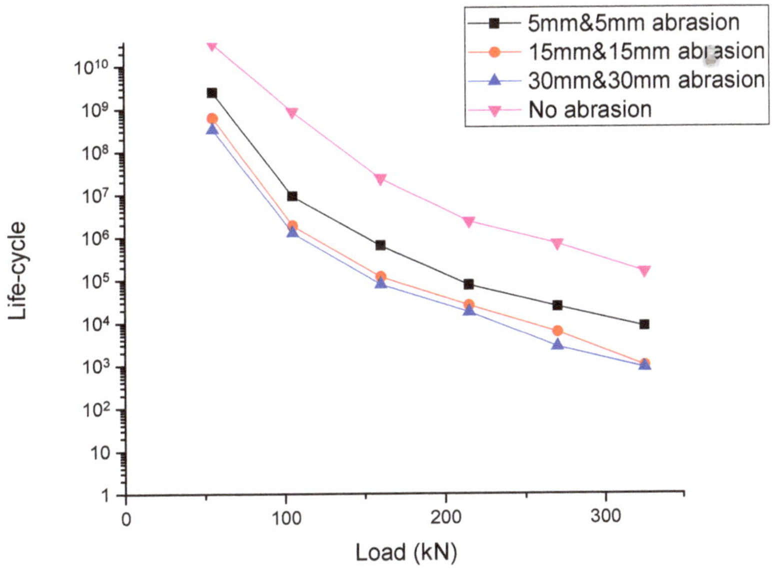

Figure 23. Comparison of life-cycle results with soffit abrasion at rail seat and soffit abrasion at midspan.

5.7. Rail-Seat Abrasion, Soffit Abrasion at Rail Seat, and Soffit Abrasion at Midspan Results

Figure 24 shows the location of the rail-seat abrasion, the soffit abrasion at rail seat, and the soffit abrasion at midspan of the railway sleeper model. The depths of the abrasions are chosen as 5 mm & 5 mm & 5 mm, 15 mm & 15 mm & 15 mm, and 30 mm & 30 mm & 30 mm to analyze the effect on the life-cycle of the railway sleeper. Loads of between 55 kN to 325 kN are applied in the simulation. Table 12 and Figure 25 indicate the results on the life-cycle with the rail-seat abrasion, the soffit abrasion at rail seat, and the soffit abrasion at midspan.

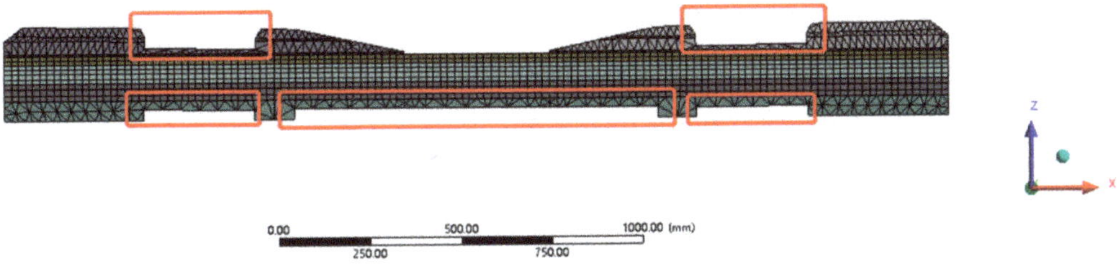

Figure 24. Rail-seat abrasion, soffit abrasion at rail seat and soffit abrasion at midspan of the railway sleeper model.

Table 12. Life-cycle results of the railway sleeper with rail-seat abrasion, soffit abrasion at rail seat and soffit abrasion at midspan.

Load (kN)	No Abrasion	5 mm & 5 mm & 5 mm	15 mm & 15 mm & 15 mm	30 mm & 30 mm & 30 mm
55	3.47×10^{10}	2.49×10^{9}	6.26×10^{8}	3.37×10^{8}
105	8.79×10^{8}	9.23×10^{6}	1.86×10^{6}	1.28×10^{6}
160	2.39×10^{7}	6.38×10^{5}	1.19×10^{5}	79,896
215	2.30×10^{6}	76,113	25,415	17,652
270	6.79×10^{5}	23,596	5963	2706
325	1.52×10^{5}	8078	966	881

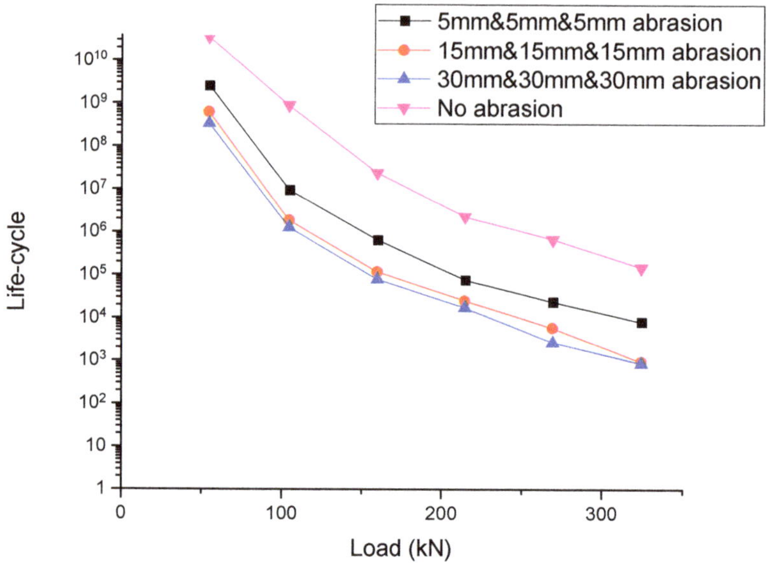

Figure 25. Comparison of life-cycle results with rail-seat abrasion, soffit abrasion at rail seat and soffit abrasion at midspan.

5.8. Discussions

Sections 5.1–5.3 demonstrate the life-cycle results with the single abrasion pattern of the railway sleeper. Sections 5.4–5.7 present the life-cycle results with the combination of the different abrasion patterns of the railway sleeper. The abrasion depth of the prestressed concrete sleeper is reduced by 5 mm, 15 mm, and 30 mm, respectively. In general, the life-cycle of the worn railway sleeper is less than an undamaged sleeper. The life-cycle decreases with an increase in abrasion, especially large depth abrasion (30 mm). Higher dynamic loads result in a reduction in the life-cycle. However, at the same abrasion depth, the performance of the sleeper varies with different abrasion patterns.

On comparison with each single abrasion patterns, the effect of the soffit abrasion at rail seat on the railway sleeper is the most significant. The soffit abrasion at midspan also has a very significant influence on the sleeper. For both of the soffit abrasion patterns, the life-cycle falls by more than 90%. The change of the life-cycle due to rail-seat abrasion is relatively low, with only a 53.32% change in comparison with an undamaged sleeper. However, when the rail-seat abrasion rises to 30 mm, the life-cycle also reduces by up to 93.63%.

Sections 5.4–5.7 show that the combination of the different abrasion patterns affects the life-cycle of the prestressed concrete sleeper. The results demonstrate more than one abrasion pattern occurring in the sleeper, and the life-cycle decreases sharply. There are four groups of multi-abrasion patterns analyzed. The triple abrasion (combination of rail-seat abrasion, soffit abrasion at rail seat, and soffit abrasion at midspan) is the most critically damaging. With a 5 mm abrasion for each position, the life-cycle falls by 97.16%. The worst case in this study is the triple abrasion with 30 mm depth, where the life-cycle undergoes a 99.88% change. For the double abrasion patterns (rail-seat abrasion/soffit abrasion at rail seat, rail-seat abrasion/soffit abrasion at midspan, and soffit abrasion at rail seat/soffit abrasion at midspan), the results are not very different. The combination of soffit abrasion at rail seat and soffit abrasion at midspan pattern is relatively worse than the other two double abrasion patterns, which reduces the life-cycle by up to 96.16%.

6. Conclusions

In this research, 3D finite element models were developed to investigate the fatigue behavior of the prestressed concrete sleeper considering surface abrasion. The theoretical life-cycle assessment method based on the damage accumulation concept was also demonstrated. The fatigue behavior of the prestressed concrete sleepers with surface abrasions were assessed by a numerical fatigue model. Seven cases of surface abrasion pattern were considered in this study. For each case analysis, a series of abrasion depths and dynamic loads were applied in numerical model in order to investigate the performance of the prestressed concrete sleeper. Based on the obtained results of this study, the key findings are revealed as follows:

- Surface abrasion significantly influences the structural performance of prestressed concrete sleepers. From the results, undamaged sleepers have much more service life than worn sleepers. Therefore, track maintenance should be carried out regularly to prevent loss of life-cycle from surface abrasion;
- Rail-seat abrasion has a relatively low influence on railway sleepers in comparison with soffit abrasion. Soffit abrasion at the rail seat can critically reduce the life-cycle;
- The risk of more than one abrasion pattern happening in the railway sleeper is far greater than a single abrasion pattern;
- In this study, the life-cycle was found to largely depend on the magnitude of the dynamic load and abrasion depth. Both large dynamic loads and abrasion depths can result in serious decreases in life-cycle.

This study is the first to present a finite element modelling for determining the life-cycle of the prestressed concrete sleepers with surface abrasion. The numerical fatigue model was validated using the theoretical life-cycle assessment method. The FE sleeper model was also validated by comprehensive experimental data. It should be noted that, in general, the lifespan of prestressed concrete sleepers can also be affected by factors such as bond slip, environmental conditions, chemical attack, etc. For example, the bond between the strand and the concrete is a factor which can directly or indirectly influence the lifespan of prestressed concrete sleepers. Therefore, more factors affecting the fatigue life of prestressed concrete sleepers are suggested for investigation in future research.

This paper confirms that inspection of railway sleepers is essential and even small abrasions that occur in critical positions can still affect the performance of railway sleepers. It is obvious that surface abrasions can result in a significant reduction in the life-cycle. Hence, it is important to reduce the impact loading to prevent railway sleepers from failing. From this study, the railway concrete sleepers should be inspected regularly, for example, every 10 to 15 years (or around 300 to 450 MGT or million gross tons). The outcomes of this study lead to a better insight into the effects of surface abrasion on prestressed concrete sleepers. This insight will enhance the inspection criteria of railway sleepers and track maintenance.

Author Contributions: Conceptualization, D.L. and S.K.; methodology, D.L.; software, D.L.; validation, D.L., R.Y. and S.K.; formal analysis, D.L.; investigation, D.L. and R.Y.; resources, R.Y.; data curation, D.L.; writing—original draft preparation, D.L.; writing—review and editing, S.K.; visualization, D.L.; supervision, S.K.; project administration, S.K.; funding acquisition, S.K. All authors have read and agreed to the published version of the manuscript.

Funding: This research was funded by the European Commission, grant number: H2020-MSCA-RISE No. 691135.

Institutional Review Board Statement: Not applicable.

Informed Consent Statement: Not applicable.

Data Availability Statement: The data that support the findings of this study are available from the corresponding author upon reasonable request.

Acknowledgments: The authors are grateful to the Track Engineering and Operations for Future Uncertainties (TOFU) Lab, University of Birmingham, for support throughout this study. The authors would like to thank the Commission for H2020-MSCA-RISE, Project No. 691135 "RISEN: Rail Infrastructure Systems Engineering Network" [38]. In addition, the authors wish to thank the China Railway (No K2021G015) and China Academy of Railway Sciences Corporation Limited (No 2020YJ031) for the experimental programs. The first author would like to thank the East China Architectural Design & Research Institute (ECADI) for technical support.

Conflicts of Interest: The authors declare no conflict of interest.

References

1. Steffens, D.M. Identification and Development of a Model of Railway Track Dynamic Behaviour. Ph.D. Thesis, Queensland University of Technology, Brisbane City, QLD, Australia, 2005.
2. Taherinezhad, J.; Sofi, M.; Mendis, P.; Ngo, T. A review of behaviour of prestressed concrete sleepers. *Electron. J. Struct. Eng.* **2013**, *13*, 1–16. [CrossRef]
3. Esveld, C.; Esveld, C. *Modern Railway Track*; MRT-Productions: Zaltbommel, The Netherland, 2001; Volume 385.
4. Kaewunruen, S.; Remennikov, A.M. Impact capacity of railway prestressed concrete sleepers. *Eng. Fail. Anal.* **2009**, *16*, 1520–1532. [CrossRef]
5. Remennikov, A.; Murray, M.H.; Kaewunruen, S. Reliability-based conversion of a structural design code for railway prestressed concrete sleepers. *Proc. Inst. Mech. Eng. Part F J. Rail Rapid Transit* **2012**, *226*, 155–173. [CrossRef]
6. Van Dyk, B.J.; Dersch, M.S.; Edwards, J. *International Concrete Crosstie and Fastening System Survey–Final Results*; University of Illinois at Urbana-Champaign: Champaign, IL, USA, 2012.
7. Ferdous, W.; Manalo, A. Failures of mainline railway sleepers and suggested remedies–review of current practice. *Eng. Fail. Anal.* **2014**, *44*, 17–35. [CrossRef]
8. Van Dyk, B.J.; Edwards, J.R.; Dersch, M.S.; Ruppert, C.J., Jr.; Barkan, C.P. Evaluation of dynamic and impact wheel load factors and their application in design processes. *Proc. Inst. Mech. Eng. Part F J. Rail Rapid Transit* **2017**, *231*, 33–43. [CrossRef]
9. Remennikov, A.M.; Kaewunruen, S. A review of loading conditions for railway track structures due to train and track vertical interaction. *Struct. Control Health Monit. Off. J. Int. Assoc. Struct. Control Monit. Eur. Assoc. Control Struct.* **2008**, *15*, 207–234. [CrossRef]
10. Li, D.; You, R.; Kaewunruen, S. Mechanisms and Evolution of Cracks in Prestressed Concrete Sleepers Exposed to Time-Dependent Actions. *Appl. Sci.* **2022**, *12*, 5511. [CrossRef]
11. Remennikov, A.; Kaewunruen, S. Determination of dynamic properties of rail pads using an instrumented hammer impact technique. *Acoust. Aust.* **2005**, *33*, 63–67.
12. Kaewunruen, S.; Remennikov, A. On the residual energy toughness of prestressed concrete sleepers in railway track structures subjected to repeated impact loads. *Electron. J. Struct. Eng.* **2013**, *13*, 41–61. [CrossRef]
13. Kaewunruen, S.; Remennikov, A.M.; Murray, M.H. Introducing a new limit states design concept to railway concrete sleepers: An Australian experience. *Front. Mater.* **2014**, *1*, 8. [CrossRef]
14. Chen, Z.; Shin, M.; Wei, S.; Andrawes, B.; Kuchma, D.A. Finite element modeling and validation of the fastening systems and concrete sleepers used in North America. *Proc. Inst. Mech. Eng. Part F J. Rail Rapid Transit* **2014**, *228*, 590–602. [CrossRef]
15. César Bastos, J.; Dersch, M.S.; Edwards, J.R. Degradation Mechanisms of Concrete Due to Water Flow in Cracks of Prestressed Railroad Sleepers under Cyclic Loading. *J. Mater. Civ. Eng.* **2022**, *34*, 04022025. [CrossRef]
16. Parvez, A.; Foster, S.J. Fatigue of steel-fibre-reinforced concrete prestressed railway sleepers. *Eng. Struct.* **2017**, *141*, 241–250. [CrossRef]
17. Riding, K.A.; Peterman, R.J.; Guthrie, W.S.; Brueseke, M.; Mosavi, H.; Daily, K. *A Study of Environmental and Track Factors That Contribute to Abrasion Damage of Concrete Ties*; Department of Transportation, Federal Railroad Administration: Washington, DC, USA, 2019.
18. Kernes, R.G.; Shurpali, A.A.; Edwards, J.R.; Dersch, M.S.; Lange, D.A.; Barkan, C.P. Investigation of the mechanics of rail seat deterioration and methods to improve the abrasion resistance of concrete sleeper rail seats. *Proc. Inst. Mech. Eng. Part F J. Rail Rapid Transit* **2014**, *228*, 581–589. [CrossRef]
19. Ngamkhanong, C.; Li, D.; Remennikov, A.M.; Kaewunruen, S. Dynamic capacity reduction of railway prestressed concrete sleepers due to surface abrasions considering the effects of strain rate and prestressing losses. *Int. J. Struct. Stab. Dyn.* **2019**, *19*, 1940001. [CrossRef]
20. Kaewunruen, S.; Ngamkhanong, C.; Lim, C.H. Damage and failure modes of railway prestressed concrete sleepers with holes/web openings subject to impact loading conditions. *Eng. Struct.* **2018**, *176*, 840–848. [CrossRef]
21. You, R.; Goto, K.; Ngamkhanong, C.; Kaewunruen, S. Nonlinear finite element analysis for structural capacity of railway prestressed concrete sleepers with rail seat abrasion. *Eng. Fail. Anal.* **2019**, *95*, 47–65. [CrossRef]
22. Li, D.; Ngamkhanong, C.; Kaewunruen, S. Influence of surface abrasion on creep and shrinkage of railway prestressed concrete sleepers. In *IOP Conference Series: Materials Science and Engineering*; IOP Publishing: Bristol, UK, 2017.

23. Li, D.; Kaewunruen, S.; You, R. Time-dependent behaviours of railway prestressed concrete sleepers in a track system. *Eng. Fail. Anal.* **2021**, *127*, 105500. [CrossRef]
24. *EN 1992-1-1*; Eurocode 2: Design of Concrete Structures—Part 1-1: General Rules and Rules for Buildings. European Union: Maastricht, The Netherlands, 2004.
25. Nussbaumer, A. EN1993-1-9: A code for fatigue design. In Proceedings of the Conference Fatigue Design, Denver, CO, USA, 17–21 July 2005.
26. Rao, V.; Talukdar, S. Prediction of fatigue life of a continuous bridge girder based on vehicle induced stress history. *Shock Vib.* **2003**, *10*, 325–338. [CrossRef]
27. Kaewunruen, S. Monitoring structural deterioration of railway turnout systems via dynamic wheel/rail interaction. *Case Stud. Nondestruct. Test. Eval.* **2014**, *1*, 19–24. [CrossRef]
28. Nielsen, J.C.; Li, X. Railway track geometry degradation due to differential settlement of ballast/subgrade–numerical prediction by an iterative procedure. *J. Sound Vib.* **2018**, *412*, 441–456. [CrossRef]
29. You, R.; Li, D.; Ngamkhanong, C.; Janeliukstis, R.; Kaewunruen, S. Fatigue life assessment method for prestressed concrete sleepers. *Front. Built Environ.* **2017**, *3*, 68. [CrossRef]
30. You, R.; Kaewunruen, S. Evaluation of remaining fatigue life of concrete sleeper based on field loading conditions. *Eng. Fail. Anal.* **2019**, *105*, 70–86. [CrossRef]
31. Béton, C.E.-I.d. *CEB-FIP Model Code 1990: Design Code*; Thomas Telford Publishing: London, UK, 1993.
32. Browell, R.; Hancq, A. *Calculating and Displaying Fatigue Results*; Ansys Inc.: Canonsburg, PA, USA, 2006; p. 2.
33. *EN 13230-2*; Railway Applications-Track-Concrete Sleepers and Bearers Part 2: Prestressed Monoblock Sleepers. Standardization, E.C.f., 2009. Available online: https://standards.iteh.ai/catalog/standards/cen/afc7f289-4b2c-4620-9849-1b5e6a7b865b/en-13230-2-2016 (accessed on 2 August 2022).
34. Sadeghi, J.; Barati, P. Comparisons of the mechanical properties of timber, steel and concrete sleepers. *Struct. Infrastruct. Eng.* **2012**, *8*, 1151–1159. [CrossRef]
35. Jing, G.; Yunchang, D.; You, R.; Siahkouhi, M. Comparison study of crack propagation in rubberized and conventional prestressed concrete sleepers using digital image correlation. *Proc. Inst. Mech. Eng. Part F J. Rail Rapid Transit* **2022**, *236*, 350–361. [CrossRef]
36. Parvez, A.; Foster, S.J. Fatigue behavior of steel-fiber-reinforced concrete beams. *J. Struct. Eng.* **2015**, *141*, 04014117. [CrossRef]
37. Li, D.; Kaewunruen, S.; You, R.; Liu, P. Fatigue life modelling of railway prestressed concrete sleepers. *Structures* **2022**, *41*, 643–656. [CrossRef]
38. Kaewunruen, S.; Sussman, J.M.; Matsumoto, A. Grand challenges in transportation and transit systems. *Front. Built Environ.* **2016**, *2*, 4. [CrossRef]

MDPI
St. Alban-Anlage 66
4052 Basel
Switzerland
www.mdpi.com

Sustainability Editorial Office
E-mail: sustainability@mdpi.com
www.mdpi.com/journal/sustainability

www.ingramcontent.com/pod-product-compliance
Lightning Source LLC
Chambersburg PA
CBHW061441210926
51850CB00008B/1574
9783036591056